BRAIN AND BEHAVIOUR

Los Angeles | London | New Delhi
Singapore | Washington DC | Melbourne

Sara Miller McCune founded SAGE Publishing in 1965 to support the dissemination of usable knowledge and educate a global community. SAGE publishes more than 1000 journals and over 800 new books each year, spanning a wide range of subject areas. Our growing selection of library products includes archives, data, case studies and video. SAGE remains majority owned by our founder and after her lifetime will become owned by a charitable trust that secures the company's continued independence.

Los Angeles | London | New Delhi | Singapore | Washington DC | Melbourne

BRAIN AND BEHAVIOUR

MOLECULAR MECHANISMS OF NEUROTRANSMISSION AND THEIR ROLE IN DISORDER

MARTIN CLARK

Los Angeles | London | New Delhi
Singapore | Washington DC | Melbourne

SAGE Publications Ltd
1 Oliver's Yard
55 City Road
London EC1Y 1SP

SAGE Publications Inc.
2455 Teller Road
Thousand Oaks, California 91320

SAGE Publications India Pvt Ltd
B 1/I 1 Mohan Cooperative Industrial Area
Mathura Road
New Delhi 110 044

SAGE Publications Asia-Pacific Pte Ltd
3 Church Street
#10-04 Samsung Hub
Singapore 049483

Editor: Donna Goddard
Editorial assistant: Emma Yuan
Production editor: Rachel Burrows
Copyeditor: Sarah Bury
Proofreader: Brian McDowell
Indexer: Silvia Benvenuto
Marketing manager: Fauzia Eastwood
Cover design: Sheila Tong
Typeset by: C&M Digitals (P) Ltd, Chennai, India

Library of Congress Control Number: 2022933644

British Library Cataloguing in Publication data

A catalogue record for this book is available from
the British Library.

ISBN 978-1-5297-6280-8
ISBN 978-1-5297-6279-2 (pbk)

CONTENTS

DETAILED CONTENTS

LIST OF FIGURES

ABOUT THE AUTHOR

Martin Clark is a lecturer of Neurobiology in the department of Psychology and Computing at the University of Central Lancashire. He is also an associate researcher at the University of Sheffield. Martin completed his PhD in the laboratory of Dr Enrico Bracci at the University of Sheffield. During his PhD he employed electrophysiological methods, including multi-electrode arrays, to investigate the circuit dynamics and molecular mechanisms that modulate activity within the ventral pallidum and ventral striatum.

His current research interests are focused on molecular mechanisms underpinning ASD (autism spectrum disorder) and bvFTD (behaviour variant frontal temporal dementia). He is particularly interested in GABAergic interneurons and the alterations in these populations that may underpin behavioural change or act as key pre-symptomatic biomarkers.

Martin is a Fellow of the Higher Education Academy and currently teaches on a range of Neuroscience and Psychology courses and modules.

ACKNOWLEDGEMENTS

Thank you to Dr Rachel Crosby, whose time and thoughts were invaluable.

1

AN INTRODUCTION TO NEUROTRANSMITTERS AND THEIR MACHINERY

Chapter outline

In this chapter we will cover:

- The neuron
- Outline of what constitutes a neurotransmitter
- Mechanisms of their:
  - synthesis
  - packaging
  - reuptake
- Outline of the neurotransmitter's receptors (ionotropic and metabotropic)

The neuron

Our brain is made up of a variety of different types of cells. The most abundant are glial cells such as astrocytes. However, much research has been focused on those highly specialised cells known as neurons.

The human brain contains hundreds of billions of these neurons that are all connected to many thousands of other neurons, forming a complex interconnected web of circuits. These complex interconnected circuits, and the 'neurons' they are made of, are believed to be the main component responsible for the brain's processing capabilities. Indeed, neurons' specialised anatomy is such that they are well equipped to deal with the transmission of signals and the integration of these signals. Many see neurons as the basic data processing unit in the brain due to this ability to integrate information from multiple inputs and produce output based upon this.

So, what is it about the neuron's anatomy that makes it good at its job? Well, first, emanating mainly from the cell body are highly branched processes known as dendrites. Because they are highly branched, they are able to communicate and receive multiple inputs from other neurons. They then have other projections, known as axons, which convey these messages away from the cell body and towards other neurons in the circuit. Axons, too, can be highly branched and therefore the conveyed message can be passed onto multiple other neurons and circuits. One of the key anatomical specialisations of the axon is that it can project short distances, but also can project relatively large distances, conveying the signals both within and between brain regions. All these features make neurons the perfect machinery to integrate, process and communicate throughout the brain and beyond.

━━ Test Yourself 1.1 ━━

Match up the terms in the table opposite with the numbers on the diagram of a neuron (Figure 1.1).

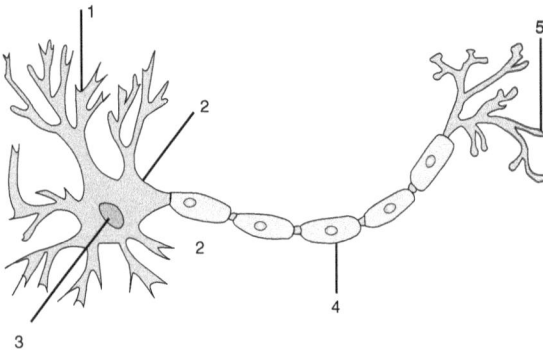

Figure 1.1　A schematic of a neuron

Table 1.1

Cell body	Synaptic terminal	Dendrites
Axon	Nucleus	Myelin sheath

Answers to all Test Yourself questions are at the end of the chapter.

Neurons convey signals both electrically and chemically. Electrical signals are propagated along their processes and are known as action potentials. Action potentials are facilitated by the flow of ions from outside to inside the neuron and vice versa. This flow of ions causes alterations in the neurons' membrane potential and results in the all-or-none response of an action potential, if a threshold is passed. This is a key process in the neuron as the propagation of this action potential is essential for the eventual release of neurotransmitters from the neuron's synaptic terminal, in most cases. I will not linger on this here as many other books have covered it in great detail. If you are unsure of the mechanisms involved in the formation of action potentials or the propagation of action potentials along the axon, then I strongly recommend reading the textbook *Neuroscience* by Purves et al. (2018) or one of the classics on the topic, such as *Ionic channels of excitable membranes* by Hille (2001).

Test Yourself 1.2

Have a go at labelling the following diagram (Figure 1.2) using the terms in the table overleaf. If you find this easy, well done and continue reading. If you struggle with this, I recommend going to the textbook referenced above and refamiliarising yourself with these ideas before going any further in this book.

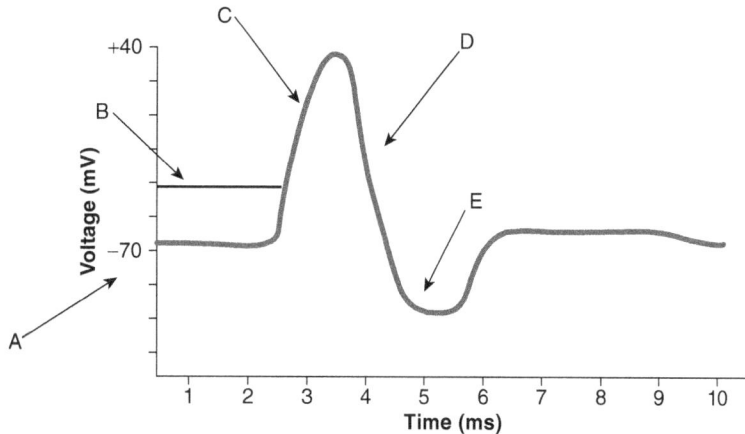

Figure 1.2   An indicative trace of an action potential

(Continued)

Table 1.2

the neuron is said to be depolarised	more permeable to potassium than sodium	action potential threshold	hyperpolarisation
this is roughly at the equilibrium potential for potassium	typically, -55 mV	Sodium channels open and the intracellular voltage moves towards the equilibrium potential for sodium	
refractory period	resting state/resting potential	voltage returns towards the equilibrium potential for potassium	Sodium channels close

Although neurons are undoubtably important, it is worth keeping in mind, as we go through this book, that neurons do not function in isolation, and that they are influenced by other neurons within the circuit as well as other brain cells, such as glia. This is especially relevant in the context of neurotransmission, as neurons are sometimes reliant on other types of cells in the brain, as well as other processes beyond the neuron and brain, for neurotransmitters synthesis and degradation/recycling.

Introduction to neurotransmitters

As mentioned above, neurons are highly connected, and form tangled circuits both within brain regions and between brain regions. In order for neurons to 'communicate' with each other in these circuits they must deliver a signal. The main method by which neurons interact and communicate is via the molecular transmission of a signal. This involves converting the electrical signal, the action potential, into a chemical signal, via the release of neurotransmitters. This was a highly contentious issue until the 1920s, when the German physiologist Loewi provided solid evidence that electrical signals were converted into chemical signals for communication. Loewi did this with hearts from frogs that had the vagal nerve (a large bunch of neurons that run from the brain to the heart) attached. Loewi electrically stimulated the vagal nerve and noted that this resulted in changes to the beat rate of the heart. Loewi then took another heart and placed it in the same solution as the first heart, but this time did not stimulate the vagal nerve. What Loewi noted was that even though the vagal nerve was not stimulated on this second heart, being in the solution from the first heart, see Figure 1.3, resulted in alterations in the hearts beat rate. This led Loewi and others to conclude that stimulation of the vagal nerve, on the first heart, must have resulted in the release of a chemical substance that altered the heart beat rate, and that this chemical substance was still suspended in the solution, resulting in changes to the second heart. This chemical was initially called vagal substance, although it became clear that this was the first neurotransmitter to be identified, namely acetylcholine.

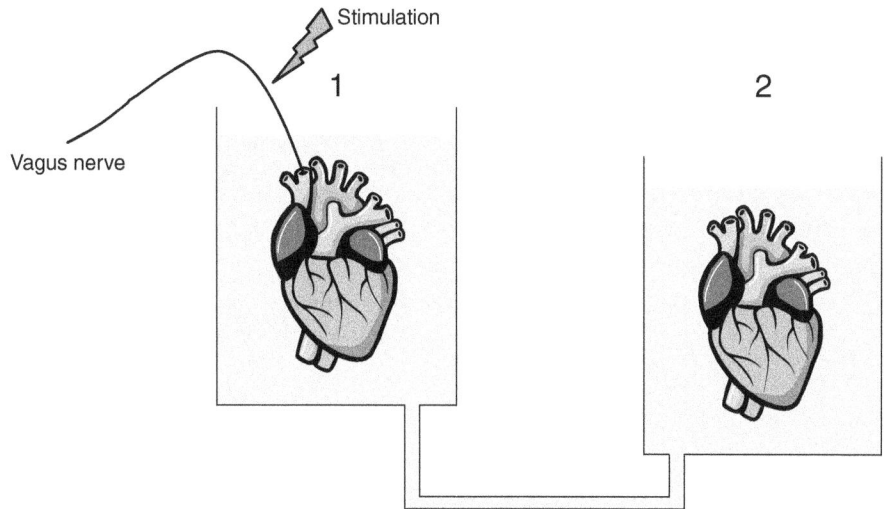

Figure 1.3 Schematic illustration of Loewi's experimental set up with connection between the two chambers to allow molecules to move from chamber 1 to chamber 2

Source: Heart image by Vectorportal.com

Neurotransmitters are commonly seen as being released from the neuron at specialised terminal ends. At these terminal ends the neurotransmitter is released into a small gap between neurons or the neuron and another cell, such as at neuron to heart muscle junction, as in Loewi's experiments. This gap is referred to as the synapse/synaptic cleft. Once released from the neuron's terminal end, the neurotransmitter diffuses across the synaptic cleft and binds to receptor molecules on the postsynaptic cell. We will discuss these receptors later in this chapter, in the context of the machinery behind neurotransmitters. First, we need to consider what defines a neurotransmitter.

Neurotransmitters are the core language for communication between neurons. Like languages there are many neurotransmitters, but what actually constitutes a neurotransmitter? Classically, it is considered that there are four key criteria that need to be met for a substance to be considered a neurotransmitter at the synapse. These are:

1. The substance must be found within the presynaptic terminal.
2. The substance is released by activation of the presynaptic neuron (arrival of an action potential at the synapse), which causes postsynaptic responses and its release is calcium dependent.
3. Its effects on the postsynaptic neuron are altered by naturally-occurring substances that are known to be agonists and antagonists of the substance.
4. There must be receptors for the substance on the postsynaptic cell and the substance must have the same effect on these receptors, and the neuron, if applied directly as it has when released by the presynaptic neuron.

Each criterion on its own is not enough to establish a substance as a neurotransmitter. For example, many molecules, such as the neurotransmitter glycine, are partially found in the presynaptic terminal (criterion 1) because it has an essential role to play in cell metabolism, beyond its role as a neurotransmitter. Therefore, if we only used criterion 1, there would be numerous other substances that were incorrectly classified as neurotransmitters.

Once we have identified a substance as a neurotransmitter, we can place it into various subtypes based upon a range of factors. This is important as there are currently more than 100 identified neurotransmitters, and this list is growing. This large number of different subtypes illustrates the modulatory role neurotransmitters play in signalling between neurons. Again, it is worth thinking of neurotransmitters as like a language. The more words you have the more accurately you can express your point; therefore, the more neurotransmitters that exit the more refined can be the communication between cells. So how do we place them into subtypes? In your previous reading on neurotransmitters, you may have heard them being referred to by their effect on the neuron, such as an excitatory neurotransmitter. Therefore, one, very simplistic, way of classifying them into different subtypes could be to consider their effect on the neuron. In this instance we could classify neurons into excitatory, inhibitory or modulatory subtypes. This is not a very successful classification system as the effect neurotransmitters have on a neuron is mediated by a number of factors, especially the type of receptor they target. A good example of this is the neurotransmitter glutamate, which is commonly said to be the main excitatory neurotransmitter in the brain. However, this is not always the case. As you will discover later in this book, even glutamate can sometimes have inhibitory effects on the neuron. We therefore need to consider other ways of subtyping neurotransmitters.

A more successful and generally accepted way to subtype neurotransmitters is based on their molecular structure. For example, we often separate them into two subtypes based simply on the size of the molecule, with neuropeptides being large and the second subtype, small molecule neurotransmitters, being smaller. It doesn't stop here, though, as there are further subdivisions (see Figure 1.4 for an example). It's good to think of this like a large company with many subdivisions, all working towards the common goal of communicating between neurons.

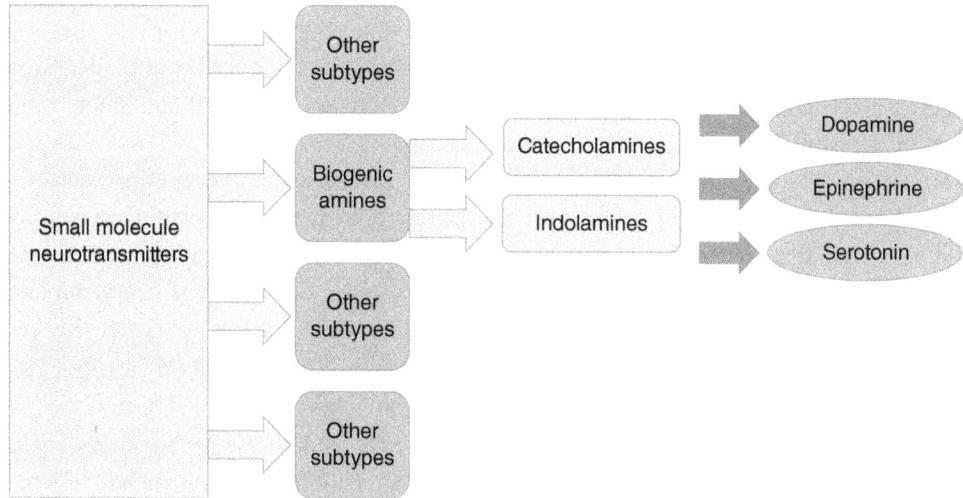

Figure 1.4 An example of the complex subtypes of neurotransmitters

Example flow chart illustrating the levels and large number of neurotransmitter subtypes synthesised in the brain as well as other mammalian tissue. As can be seen, amines are just one type, which can be further divided into at least two levels of subtypes.

The machinery behind neurotransmitters and neurotransmission

First, it is important to mention that not all neurotransmitters are created equal, or indeed function in the same manner. This means that the general principles and 'common' machinery we will cover in this introduction are true for most neurotransmitters. However, like the production of any product, there will be variance in the machinery that underpins them. We will address these differences as we move through the chapters.

Again, like any product, neurotransmitters require an often-complex machinery to be produced, packaged, delivered and returned once they have done their job. In this book we will focus specifically on this molecular machinery, considering the mechanistic aspects of its function before considering how alterations in these mechanisms can alter the behaviour of the cell and of the whole organism, resulting in our behaviour, both typical and atypical. Let's first look at the 'common' machinery that underpins neurotransmitters.

Mechanisms of their synthesis

Some neurotransmitters can be taken in through our diet, although the majority are too big to pass the blood–brain barrier. Therefore it is typical for precursor molecules (the building blocks) to be taken in through our diet and for the final neurotransmitter to be fully synthesised within the neuron. There are exceptions to this rule, however, with some neurotransmitters being synthesised in other brain cells and yet other neurotransmitters being translated and transcribed from cellular DNA, rather than being built up from 'imported' precursors.

So, where are they typically synthesised? This question can be answered in two ways. One way would be to focus on the brain regions where certain neurotransmitters are synthesised. Some neurotransmitters are synthesised in specific neurons that are localised to specific brain regions; for example, dopamine is predominantly synthesised in neurons that are localised to midbrain territories, such as the ventral tegmental area and the substantia nigra pars compacta. Synthesis of other neurotransmitters is much more diffuse, for example GABA-synthesising neurons are found in pretty much all brain regions. Another way we can answer this question is to consider where inside the neuron the neurotransmitters are synthesised. There is generally a consistent distinction that can be made here between the neuropeptides and the small molecule neurotransmitters. Because the neuropeptides are translated and transcribed from DNA, up in the cell body, in the subcompartments of the cell nucleus's, endoplasmic reticulum and Golgi body, these neuropeptide neurotransmitters are generally considered to be synthesised in the cell body and in the large vesicles that transport them down the axon and into the terminal end. The small molecule neurotransmitters, however, are commonly synthesised from precursors in the terminal end of the neuron.

So, the next question is how are they synthesised? Well, as already alluded to, there are two distinct processes, although these processes do have things in common, which we shall consider shortly. One process of synthesis is to produce the final neurotransmitter by altering precursor molecules. These precursor molecules are typically either a substance that can be taken in through our diet and can pass the blood–brain barrier, for example tryptophan in the case of the neurotransmitter serotonin's synthesis, or are a precursor molecule that is a product of cell metabolism, for example choline in the case of the neurotransmitter acetylcholine's synthesis. These precursors now need to be 'worked upon' to produce the final product. The work horses that produce the final product, namely neurotransmitters, are enzymes. Enzymes are essential to the synthesis of all neurotransmitters, with different types of neurotransmitters having specific enzymes that can catalyse their production.

These enzymes are often referred to as the rate limiting factor in neurotransmitter synthesis as the speed by which they can produce the final neurotransmitter determines the amount of the substance in the neuron. This means that the dysfunction of these enzymes can have dramatic effects on the ability of the neuron to 'communicate' with other neurons via synaptic transmission. They are well worth keeping in mind as we will focus on them at various points throughout the rest of the book, in the context of behavioural change. These enzymes work in a number of different ways to produce the final neurotransmitter and this differs between neurotransmitters, although they commonly either catalyse a reaction between the precursor and another molecule or they remove a side chain of the precursor to produce the final molecule. For example, the enzyme glutaminase catalyses a reaction between glutamine and H_2O, which results in the neurotransmitter glutamate and NH_3, while the enzyme aromatic L-amino acid decarboxylase is a lyase enzyme that catalyses the breaking of chemical bonds and results in serotonin from the precursor 5-hydroxytryptophan.

The other method of synthesis is from direct translation and transcription of cellular DNA. This is the method of synthesis for neuropeptides and, as mentioned previously, takes place up in the cell body, mainly in the endoplasmic reticulum and Golgi body. The results of this process are typically a large precursor molecule, which can be thought of as multiple copies of the neurotransmitter that needs 'breaking up'. This 'breaking up' of the large molecule typically happens in the packages that transport the molecule down the axon to the terminal end and is facilitated by those all-important enzymes. We will discuss these packages next.

Mechanisms of their packaging

It is essential for the health of the neuron that neurotransmitters are packaged and not free in the cytoplasm. This is more important for some neurotransmitters than others, but is true as a general rule. The reason for this is that the majority of neurotransmitters are actually neurotoxic and would ultimately result in cell death if not packaged up and controlled.

This process of packaging can either happen directly after the neurotransmitters are synthesised or can be one of the locations whereby the neurotransmitter is finally synthesised. The difference here is largely dependent on whether the neurotransmitter start its synthesis up in the cell body of the neuron or down in the terminal end. For those neurotransmitters, such as neuropeptides, that are produced from DNA translation and transcription, up in the cell body, the final enzymic reactions which produce the finished molecule are generally carried out in the vesicles.

These vesicles are typically referred to as large dense core vesicles (LDCVs) and bud of the Golgi body, up in the cell body. These LDCVs typically contain a large precursor molecule, which is essentially lots of copies of the final neurotransmitter bound together, and an enzyme that catalyses the cleavage of these molecules. This is typically called proteolysis. This process is done in the LDCV as it navigates its way down the axon, via microtubules, and finally reaches the terminal end, by which time the neuropeptide is in its finished state as an active neurotransmitter.

Many of the neurotransmitters we will talk about in this book are not synthesised by translation and transcription of DNA, like the neuropeptides, but are synthesised and packaged in the cytoplasm of the neuron's terminal ends. These neurotransmitters are packaged into what are commonly referred to as small clear vesicles. This packaging is done by channel proteins called vesicular transporters. The vesicular transporters are imbedded in the plasma membrane of the vesicle and work by exchanging hydrogen (H^+) ions, which are in the vesicle lumen (inside the vesicle) for the neurotransmitter molecule(s), which are in the neuron's cytoplasm, as in Figure 1.5. For this to happen, an electrochemical gradient must be set up across the membrane. For those of you who are familiar with the mechanistic aspects of action potential formation, the following will be very familiar as it is essentially the same mechanism as that which underpins the packaging of neurotransmitters into vesicles. The electrochemical gradient set up across the vesicle's membrane is called the membrane potential. This is primarily facilitated by another protein embedded in the membrane of the vesicle, which is referred to as a proton pump. The proton pump consists of the enzyme hydrogen ATPase, which splits ATP to provide the energy for active transport of hydrogen against a concentration gradient. This results in high levels of hydrogen packaged in the vesicle lumen for use by the vesicular transporter to move the neurotransmitter into the vesicle. As well as the transport of the neurotransmitter into the vesicle, some vesicular transporters also require the transport of other molecules, such as chloride. We will discuss these extra details in the relevant chapters later in the book.

The membrane potential is only one of two key mechanisms that underpin the transport of neurotransmitters into vesicles. The other main factor is PH. The packaging, into vesicles, of some neurotransmitters is more affected by membrane potential than PH. Others rely more on PH than membrane potential, while others are equally affected by both. The simple rule of thumb is that it depends upon the charge of the neurotransmitter. If the neurotransmitter has a negative charge, such as the neurotransmitter glutamate, then membrane potential is more important for effective packing into vesicles, whereas for neurotransmitters that are positively charged, such as monoamines or acetylcholine, then effective packaging is much more reliant on the relative PH levels inside the vesicle compared to outside.

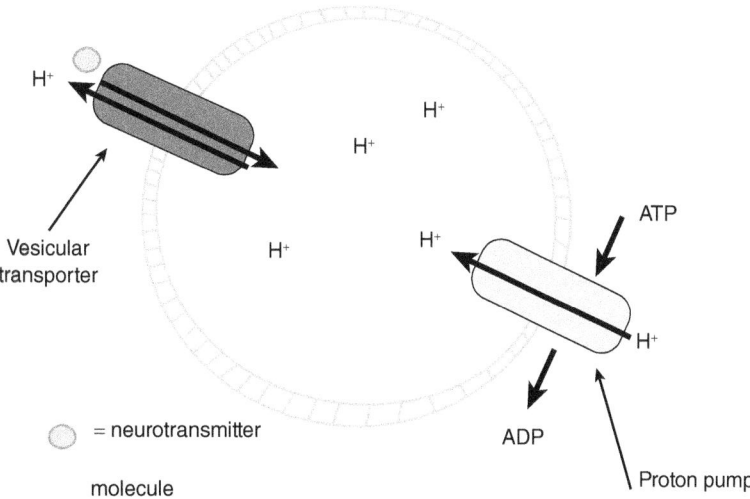

Figure 1.5 Illustration of a vesicle with the typical channels embedded in its membrane

Although there may be more things embedded in the membrane of vesicles, a proton pump, which uses ATP to transport H⁺ into the vesicle, and a vesicular transporter, which commonly exchanges H⁺ for the neurotransmitter that is to be packed, are found.

Source: Adapted from Chaudhry, F. A., Edwards, R. H., & Fonnum, F. (2008). Vesicular neurotransmitter transporters as targets for endogenous and exogenous toxic substances. *Annual Review of Pharmacology and Toxicology, 48*(1), 277–301. doi:10.1146/annurev.pharmtox.46.120604.141146

Further to this, neurotransmitters, such as GABA and glycine are considered neutral zwitterions. This results in a reliance on both membrane potential and PH differences set up by the concentration of H⁺ ions for their packaging into vesicles.

While these mechanisms for packaging neurons into vesicles are generally true, you must always keep an open mind for exceptions. As you will discover as you progress through this book, there are always exceptions and added levels of complexity.

Neurotransmitters at the synapse

Once neurotransmitters are packaged into vesicles, they are ready for release into the synapse. Again, this is a general rule and there are exceptions, for example nitric oxide, a neurotransmitter we discuss in Chapter 10, is not packaged into vesicles and can directly diffuse out of the neuron and into both the synapse and extra synaptic space. However, for the majority of neurotransmitters that are packaged into vesicles, an action potential must arrive at the terminal end, which activates calcium channels in the neuronal membrane, resulting in calcium influx into the terminal end. Finally, this promotes the movement of vesicle towards the plasma membrane where they fuse and release neurotransmitters into the synaptic cleft. This process of vesicle movement and fusion with the neuron's plasma membrane is referred to as exocytosis.

The next step in the process, once the neurotransmitters have been released into the synaptic cleft, is for them to diffuse across the cleft and bind with receptors. We will discuss these receptors next.

Mechanism of neurotransmission: Receptors

Receptors are generally unique for specific types of neurotransmitters, although there is some interaction between different types, as discussed in future chapters. The classic analogy here is that of the lock and the key, with the neurotransmitter being the key and the receptor being the lock. Once the neurotransmitter binds with its respective receptor, it starts a chain of events that alter the physical state of the target cell. In the case of neurons, the typical alteration is the electrical excitability of the target neuron. In other words, an altered likelihood that the voltage of the target neurons will pass threshold and an action potential will be produced. These sub-action potential changes in the neuron are commonly referred to as excitatory and/or inhibitory postsynaptic currents (EPSC and/or IPSC respectively), and you may come across this in your wider reading. It is important to remember that it is not all about electrical excitability. Activation of some receptors can have far more profound and far-reaching effects on the target neuron, which we will discuss shortly, and return to throughout the book.

It is not the case that there is one receptor for every neurotransmitter. In fact, the vast majority of neurotransmitters have multiple receptor types that they bind with, for example GABA is thought to have three principle different types, referred to as $GABA_a$, $GABA_b$ and $GABA_c$ receptors. Even this is not the end of the story as these different types are made up of subunits that can be arranged in different ways, resulting in types within types, that all have distinct effects on the targeted cell. Despite the large number of different receptors that this results in, receptors typically fall into one of two types. These types have very different mechanisms of action and, consequently, distinct effects on the target neuron once activated. For example, $GABA_a$ and $GABA_c$ are considered a type of receptor referred to as ionotropic, while $GABA_b$ is a type referred to as metabotropic. We will now discuss the mechanism of action for these ionotropic and metabotropic receptor types.

Ionotropic receptors

Ionotropic receptors, mechanistically, can be thought the simpler of the two types. This is because they are directly bound (receptor and channel are one) to a channel that is imbedded in the neuron's plasma membrane. The binding of the neurotransmitter to these receptors causes a direct, and rapid, conformational shape change in the channel protein's structure, and consequently the flow of ions between the extracellular space and the intracellular space (see Figure 1.6). Figure 1.6 suggests that this conformational shape change, caused by the binding of the neurotransmitter, results in the opening of the channel and the flow of ions into the cell. For most ionotropic receptors this is the case. However, it is not always the case. In fact, it is perfectly possible for the channel to be in the open state at rest and for binding of the receptor to cause the channel to close, consequently reducing the flow of ions. Either way, once activated by the binding of the neurotransmitter, the response of ionotropic receptors, and their effect on the neuron, is rapid. This is commonly referred to as a short latency; in other words, a short lag before they start having their effect. The responses they induce in the target neuron are similarly rapid. In fact, you might say everything about ionotropic receptors is rapid.

One other feature that distinguishes ionotropic receptors from the metabotropic receptors, which we will discuss in due course, is the fact that this conformational shape change and alteration in the openness of the pore only occurs when the neurotransmitter is bound to the receptor. Once unbound, the conformational shape change in the receptor is rapidly reversible. So, in the case of most ionotropic receptors, when the neurotransmitter unbinds the channel pore rapidly returns to the baseline state of being closed, restricting the future flow of ions into the intracellular space. This means that any 'effect' that the opening of the channel had on the neuron will be very short.

Obviously, the flow of ions, as we discussed at the start of this chapter, results in an alteration to the physical state of the targeted neuron, but what ions typically flow in through these channels? Well, this is dependent on what neurotransmitter the receptor is partnered with. As a good rule of thumb, you can think that, if the neurotransmitter has a general excitatory effect on the target neuron, in response to its binding to the receptor, then the ions flowing into the neuron will be positively charged: for example, ions such as sodium and/or calcium. However, if binding of the neurotransmitter to the receptor generally has a negative/inhibitory effect on the targeted neuron, then the ions flowing in are likely to be negatively charged: for example, ions such as chloride.

As well as the neurotransmitter binding to the ionotropic receptor there are also, typically, many other binding sites for other molecules. These extra binding sites typically allow molecules to bind that increase or decrease the sensitivity of the receptor to the neurotransmitter's binding. These binding sites, away from the main binding site (which is considered to be that for the neurotransmitter), typically allow the binding of molecules that are referred to as allosteric agonists and/or antagonist. Allosteric, roughly translated, means alongside, agonist means to work with, and antagonist means to work against. So this means that if an allosteric agonist binds to the receptor, it can increase the sensitivity and affinity of the receptor for the neurotransmitter, therefore increasing the responsiveness of the receptor and its channel. If an allosteric antagonist binds to the receptor, this will have the direct opposite effect, with the receptor being less sensitive and having a reduced affinity for the neurotransmitter molecule, ultimately reducing the responsiveness of the receptor and its channel. We will come back to these binding sites, but if you want to read forward in the book, then Chapter 3 is focused on the neurotransmitter GABA, especially GABA$_a$ receptors.

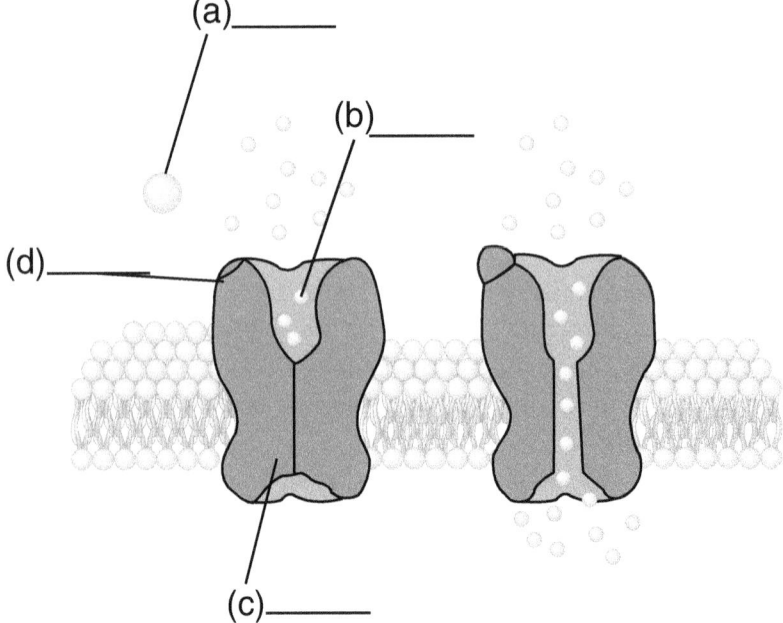

Figure 1.6   Schematic of a typical ionotropic receptor found bound in the plasma membrane of a neuron

Typically, a specific binding site is found on the extracellular domain of the receptor. Once the neurotransmitter binds, this causes a confirmation shape change in the receptor, which opens a central pore. This then allows the flow of ions from extracellular to intracellular, and in some cases in the opposite direction. (a) = neurotransmitter, (b) ion, (c) closed ion channel, (d) ionotropic receptor.

Source: Higgs, S., Cooper, A., & Lee, J. (2021). *The biopsychology colouring book*. London: Sage. Adapted from original creation by Hrejsa/Body Scientific Intl. for Sage Publishing.

Metabotropic receptors

While activation of ionotropic receptors, and the subsequent cellular effects on the target neuron, may be generally characterised as simplistic and rapid, the binding of neurotransmitters to metabotropic receptors and the subsequent cellular effects on the target neuron is a much slower and complex affair. So how are metabotropic receptors' effects more 'complex'? Well, for a start, metabotropic receptors are not directly bound to a channel protein (as illustrated in Figure 1.7), so any effect they have on the target neuron involves their interaction with other proteins embedded in the intracellular aspect of the plasma membrane. These are called guanyl nucleotide binding proteins, which are commonly referred to as G proteins. In fact, metabotropic receptors are commonly referred to as GPCRs (G protein coupled receptors). These G proteins are comprised of three subunits, referred to as the α (alpha), β (beta) and γ (gamma) subunits. The reason why activation of different metabotropic receptors results in a complex array of intracellular effects is because there are a number of different types of each of these α, β and γ subunits, which interact in different ways with various types of proteins in the neuron. This includes direct interaction with ion channels, but also with a host of different enzymes that catalyse the production of a variety of secondary messengers. These secondary messengers go on to have a myriad of different effects on the cell promoting process, such as transcription and translation of DNA.

So, mechanistically how does this all happen? Well, when the neurotransmitter molecule binds with the metabotropic receptor this causes a conformational shape change in the receptor, which in turn activates one of the three subunits (the α subunit) of its associated G protein. This activation results in the release of a bound molecule called GDP (guanosine diphosphate) in the alpha subunit, and its replacement with GTP (guanosine triphosphate), which is readily available in the intracellular cytoplasm. The binding of GTP results in the dissociation of the alpha subunit from the beta and gamma subunits of the G protein. Now the α subunit and the β/γ[1] complex are free to move around the intracellular domain of the plasma membrane and interact with other proteins embedded in the plasma membrane. This results in a whole host of different effects, including a range of intracellular secondary messengers that are free to move around the cell and effect numerous cellular processes. It is impossible to encapsulate all these processes and pathways in this book, but I will cover a few key pathways and proteins here that will be further alluded to as you go through the rest of the book.

G-protein subunits directly modulate membrane bound ion channels

Once the α subunit and the $\beta\gamma$ complex dissociate, there is significant evidence that they can both act as effectors directly onto ion channels that span the neurons

---

[1]The β and γ subunits typically stay united as a complex under normal biological conditions.

plasma membrane, as illustrated in Figure 1.7. This means that they can directly affect the ion flow across the plasma membrane, resulting in changes to the neuron's electrical activity. These protein–protein interactions between the G proteins and the plasma membrane protein (ion channel) are much slower and longer lasting than those mediated by activation of ionotropic receptors. One of the most well characterised ion channels directly modulated by G protein, in brain neurons, are GIRK (G protein-coupled inwardly rectifying potassium) channels. GIRKS have a strong hyperpolarising effect on the neuron when activated, and over the last 30 years have been heavily implicated in health and disease states. These GIRK channels are known to be directly modulated by multiple neurotransmitters, such as acetylcholine and serotonin, binding to their respective metabotropic receptors. It is also believed that it is the $\beta\gamma$ complex component of the G-protein complex which, when released by metabotropic receptor activation, directly modulates these GIRK channels.

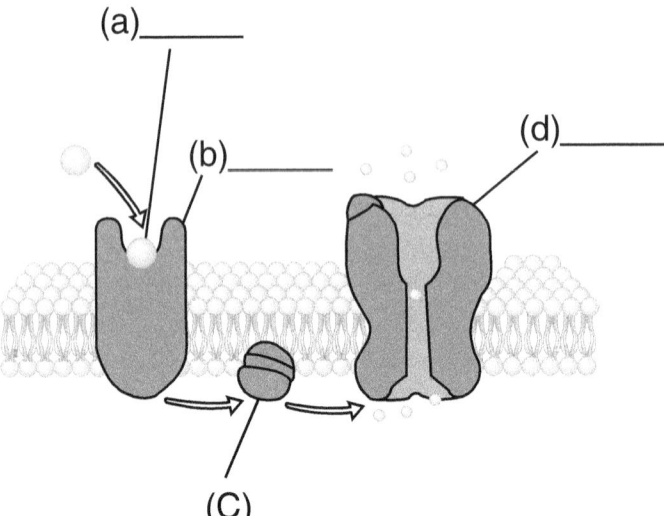

Figure 1.7   Schematic of a typical membrane bound metabotropic receptor

Notice that the binding site for the neurotransmitter is separate from the actual membrane-bound channel. Upon binding of the neurotransmitter, confirmation shape change in the receptor triggers the dissociation of intracellular-bound G proteins which, when released, interact with membrane-bound ion channels and a myriad of other proteins, both membrane-bound and free in the intracellular cytoplasm. (a) = neurotransmitter, (b) metabotropic receptor, (c) G protein, (d) G protein-gated ion channel.

Source: Higgs, S., Cooper, A., & Lee, J. (2021). *The biopsychology colouring book*. London: Sage. Adapted from original creation by Tomasikiewicz/Body Scientific Intl. for Sage Publishing.

As well as the G proteins directly acting on ion channels, the secondary messengers that are modulated by G proteins' interaction with enzymes can also have effects on membrane-bound ion channels. This mechanism of ion channel activation is much slower than any other and potentially far more long lasting, with it still having its effects on the ion channels, and the electrical activity of the neuron, minutes or even hours later. This is particularly striking when you consider that ionotropic receptor activation typically causes a response in the neuron that lasts in the millisecond range.

Secondary messengers: Adenylate cyclase and cAMP

Cyclic AMP is one of the key secondary messengers modulated as a result of metabotropic receptor activation by neurotransmitter binding.[2] Modulation of its production is typically caused by the alpha subunit of the G protein associating with an enzyme bound in the plasma membrane called adenylate cyclase (AC) which converts ATP into cAMP. Of course, this all depends on the G-protein subtype that is activated, with the production of cAMP being either increased or decreased depending on which specific G-protein subunit subtype binds with AC. For example, if the G protein α_s binds with AC, then it stimulates cAMP production, whereas if the G protein α_i subtype binds with AC, then it inhibits cAMP production. It is important to keep in mind that this is a simplification and sometimes, dependent on the type of neuron, the $\beta\gamma$ complex can be the G-protein subunit that regulates AC, and subsequent cAMP production.

So, what happens once cAMP is produced? Most downstream effects of cAMP are as a result of its effects on the enzyme PKA (protein kinase A). PKA catalyses many reactions in the neuron; for example it can phosphorylate several enzymes that are key to cellular metabolism. It can also, once activated, move to the cell nucleus and switch on the transcription process of many genes, such as genes coding for specific ion channels, fundamentally altering the function of the neuron.

Secondary messenger: PLC (phospholipase C), DAG and IP3

Certain α subunit of G proteins can also trigger an enzyme called phospholipase C (PLC) to hydrolyse the membrane-bound lipid phosphatidylinositol 4,5-bisphosphate, also known as PIP2. The consequence of this is the production of the secondary messengers DAG (diacylglycerol) and IP3 (inositol 1,4,5-trisphosphate). IP3 diffuses into the cytosol and triggers calcium release in the cell from the smooth ER (endoplasmic reticulum). This increase in calcium helps to activate protein kinase C (PKC). DAG is hydrophobic and remains bound to the plasma membrane, but it

---

[2]Neurotransmitter binding being considered the first messenger.

also goes on to activate PKC. PKC is then able to phosphorylate serine and threonine residues on target proteins, again resulting in a whole host of effects on the cell. One of the key things this does is to alter transcription factors, producing long-lasting effects on the cell, such as synaptogenesis and upregulation of channel proteins in the plasma membrane.

Secondary messenger: Calcium

As mentioned above, calcium is a key player in the activation of enzymes that are essential machinery in these intracellular signalling pathways. It is also considered a secondary messenger in its own right. Calcium levels can be increased, as mentioned above, by production of secondary messengers, such as IP3. However, it can also be brought into the cell from the extracellular space by activation of calcium channels, calcium pumps and calcium exchangers. Within the cell calcium has a few main targets. One, as mentioned above, is PKC. Others are the calcium-binding protein calmodulin and calpain, which is a protease dependent on calcium-binding. Calmodulin activation by calcium results in activation of a range of enzymes, such as AC, NOS (nitric oxide synthase) and calmodulin-dependent protein kinase (CaM kinase). This serves to illustrate how much influence calcium exerts on the cell, but also illustrates how it is essential to modulate other secondary messenger systems, such as AC and the production of cAMP. Further to this, calcium can also modulate the production of neurotransmitters, such as nitric oxide, which has its own chapter in this book. This means that as well as influencing intracellular processes and the 'health' of the cell, it can also facilitate the transmission of signals between cells. Indeed, it is worth remembering that calcium influx, at a basic level, is also essential for the movement of vesicles, containing neurotransmitters towards the cell membrane for release into the synapse.

As you will realise is a common theme throughout this book, these secondary messenger pathways are complex, and there are others, beyond those discussed above. Although I have presented them above as separate pathways, the reality is that they interact and affect each other. This is very much at the forefront of current molecular research, and clarifying these pathways may well be an important frontier for the development of novel therapeutics and rationale treatments for a myriad of atypical behavioural states associated with altered neuronal processes.

Location of receptors

Textbooks on this topic commonly simplify the explanation of neurotransmitters and receptors to discussing the effect once they bind to the postsynaptic receptor, after release from the terminal end of the presynaptic receptor and diffusion across the synapse. However, this can often lead to the false belief that receptors are only found on the postsynaptic terminal within the synapse. This is not true, and receptors can also be found in extra synaptic locations, on axons of the pre- and postsynaptic neuron, on the presynaptic terminal end and on other types of cells, such as glia and vascular cells. This is not an exhaustive list but aims to illustrate the fact that there are many more receptors than you might expect. Their location also leads to activation having quite different effects both on the targeted neuron and on a circuit level. On a circuit level, imagine you have a connection that releases an inhibitory neurotransmitter. When this neurotransmitter is released into the synapse, it diffuses and binds to the postsynaptic receptors where it inhibits the function of the postsynaptic neuron. If this presynaptic neuron continues to release this inhibitory neurotransmitter, we might expect that this causes sustained inhibition of the postsynaptic neuron. However, what might happen if there are also receptors for this inhibitory neurotransmitter on the presynaptic terminals? Well, in theory this would result, once the amount was high enough, in the presynaptic neuron also being inhibited, resulting in inhibition of inhibition, which is commonly referred to as disinhibition. This serves to illustrate how important the location of receptors is for the effects the neurotransmitter has on a circuit.

Once neurotransmitters have bound and had their effect on the receptor, they generally rapidly dissociate. Upon this dissociation, a cleaning up process needs to take place. This can be done in the synapse by enzymes free in the synapse and extra synaptic space. These enzymes typically denature the neurotransmitter or cleave it into its precursors, ready for reuptake by the cell. This process can also happen within the presynaptic neuron and within support cells, such as glia. We will discuss this process of 'cleaning up' and 'recycling' next.

Mechanisms of their reuptake/degradation

Once neurotransmitters have completed their job of binding to receptors, they dissociate from these receptors and the process of recycling begins. There are two common processes that happen for most neurotransmitters. One process is some form of degradation catalysed by an enzyme. This is commonly hydrolysis, which breaks the neurotransmitter up into the precursor molecule. This is done to 'inactivate' the molecule. As you may recall, many neurotransmitters are highly neurotoxic so it is important that they are rapidly denatured in this way. This can happen in the synaptic cleft and in the extra synaptic space outside the neuron, inside the presynaptic neuron and, in some cases, inside other brain cells, such as glia. Notice that this degradation sometimes occurs after the molecule has been transported back into the presynaptic terminal or another brain cell. This is the other part of the recycling process, namely the reuptake/transport of the neurotransmitter back into the cell. Most neurotransmitters have their own specific reuptake transporters. In fact, it is common for there to be several different transporters for specific neurotransmitters. These reuptake transporters are an active transport mechanism as they must transport the neurotransmitter molecule against a concentration gradient. This concentration gradient is generally very high, with estimates suggesting a typical neurotransmitter is up to 10,000 times more concentrated within the neuron compared to in the extracellular space. Many think that, like the enzymes that synthesis neurotransmitters, these reuptake transporters (sometimes referred to as plasma membrane transporters) are a key rate-limiting step in the synthesis of neurotransmitters. They are also known to have binding sites (much like the receptors we have discussed previously in this chapter) for a range of different molecules, which can affect the rate at which they reuptake the neurotransmitter. Therefore, they are an important focus of this book and will come up across multiple chapters, especially in the context of transporter alteration resulting in behavioural change for the neuron and the organism.

As well as transporting the neurotransmitter back into the cell (typically the presynaptic cell, but also sometimes support cells, such as glia), all reuptake transporters also co-transport chloride ions into the cell. This means that the transport of neurotransmitters by these reuptake transporters is electrogenic, which means it alters the membrane potential of the cell. Therefore, reuptake of neurotransmitters has a likely effect on the excitability of the neuron. In fact, it is not just chloride ions that are commonly co-transported; it is relatively common for both sodium and potassium also to be transported.

Conclusion

This first chapter is very much focused on your getting to grips with the biological machinery underpinning neurotransmitters and neurotransmission. In the following chapters we will be adding a little more biological detail onto these bones, but, fundamentally, if you have got to grips with this section, subsequent chapters will help you to see how this common biology underpins many of the different neurotransmitters expressed in our brains. In the following chapters we will also build on this and draw links between this machinery and behavioural states, both in terms of our typical and our atypical behaviour.

It is important to take your time and try to fully grasp everything we have covered in this introductory chapter as it will greatly help you in the forthcoming chapters to understand the real complexity of the biological processes that underpin our behaviour. It is therefore worth remembering that you can return to this chapter at any time, dipping in and out to help support your learning.

Test Yourself Answers

Test Yourself 1.1

Table 1.3

2. Cell body	5. Synaptic terminal	1. Dendrites
4. Axon	3. Nucleus	4. Myelin sheath

Test Yourself 1.2

A. A resting state/resting potential, more permeable to potassium than sodium, this is roughly at the equilibrium potential for potassium

B. Threshold, typically -55 mV

C. Sodium channels open and the intracellular voltage moves towards the equilibrium potential for sodium, the neuron is said to be depolarised

D. Sodium channels close, voltage returns towards the equilibrium potential for potassium

E. Hyperpolarisation, refractory period, during this period the neuron is insensitive to further input and action potential generation

References

Loewi, O. (1936). The chemical transmission of nerve action. *Nobel lecture.*

Loewi, O. (1949). On the antagonism between pressor and depressor in the frog's heart. *Journal of Pharmacology and Experimental Therapeutics, 96*(3), 295. Retrieved from http://jpet.aspetjournals.org/content/96/3/295.abstract

Purves, D., et al. (2018) *Neuroscience.* 6th Edition, Sinauer Associates, New York.

Hille, B. (2001). Ionic channels of excitable membranes, Sinauer Assoc. *Inc., Sunderland, MA.*

Fonnum, F. (1984). Glutamate: a neurotransmitter in mammalian brain. *J Neurochem, 42*(1), 1–11. doi:10.1111/j.1471-4159.1984.tb09689.x

Dikeakos, J. D., & Reudelhuber, T. L. (2007). Sending proteins to dense core secretory granules: still a lot to sort out. *The Journal of cell biology, 177*(2), 191–196. doi:10.1083/jcb.200701024

Boehning, D., & Snyder, S. H. (2003). Novel neural modulators. *Annu Rev Neurosci, 26*, 105–131. doi:10.1146/annurev.neuro.26.041002.131047

Lüscher, C., & Slesinger, P. A. (2010). Emerging roles for G protein-gated inwardly rectifying potassium (GIRK) channels in health and disease. *Nature reviews. Neuroscience, 11*(5), 301–315. doi:10.1038/nrn2834

2

GLUTAMATE

Introduction

Glutamate is found in almost every part of the brain; in fact, it is found in higher concentrations than any other amino acid. As such, alteration in its levels and function, which stem from alterations in its machinery, often result in global impairments in the normal function of the brain and result in profound behavioural effects.

Despite the fact that we can go all the way back to Krebs' pioneering research on metabolism and the tricarboxylic acid cycle in the 1930s, which identified the importance of glutamate in the brain, specifically for Krebs in the metabolic pathways. It was not until the late 1970s that research evidence started to build for its role as a neurotransmitter in the brain. At this point it was generally agreed that glutamate met the four key criteria for a molecule to be considered a neurotransmitter, as outlined in Chapter 1.

It is generally considered that glutamate is the main excitatory neurotransmitter in the brain, although, as we will discover later in the chapter, this is dependent on the type of glutamate receptor targeted by the molecule. Probably one of the main reasons glutamate has gained such levels of significance within the biomedical research community is because it, and its receptors, have been shown to be key players in neuroplasticity. Neuroplasticity involves changes in the 'strength of connections' due to neuronal activity, specifically related to activation of glutamate receptors. It is thought to be one of the key cellular processes that underpin learning and memory formation. Therefore, alterations in these mechanisms are key targets for our understanding of 'normal' learning and memory as well as pathology in these things, such as in Alzheimer's disease.

At the same time as glutamate playing a major part in neuronal adaptation and learning, it is also known to be highly neurotoxic. Like so many mechanisms in our body, for glutamate to function 'normally', it is very much about keeping it within a tight band, with both high and low levels having negative effects on neurons and, ultimately, our brain.

Outline of glutamate's synthesis, packaging and reuptake

Glutamate's synthesis

Glutamate can be taken in through our diet. In fact, to most of us, glutamate is better known by the name monosodium glutamate (MSG). MSG is typically used as a flavour enhancer and with other 'e' numbers has gained a rather bad reputation over the last 20 years. However, glutamate is a non-essential amino acid, and, as such, cannot cross the blood–brain barrier. This means that synthesis of glutamate must take place in the brain from precursors.

There is generally considered to be more than one pathway for glutamate to be synthesised within cells. Within neurons, most glutamate is synthesised from the precursor glutamine by the enzyme glutaminase. Glutamine must be actively transported into the neuron for this to happen and, interestingly, part of the chain of reactions that results in glutamate takes place outside the neuron in glial cells. This whole process is commonly referred to as the glutamate–glutamine cycle and we shall return to it when we look at the degradation and reuptake of glutamate.

As alluded to, there are also other mechanisms of glutamate synthesis. One is from an intermediate chemical in the Krebs cycle,[1] namely α-ketoglutarate. This is catalysed by the enzyme GABA α-oxoglutarate transaminase (GABA-T).

One interesting side point is that the synthesis of glutamate (see Figure 2.1) is not necessarily the end point for the molecule. Instead of glutamate being the final neurotransmitter, further enzyme-mediated reactions, specifically by the enzyme glutamic acid decarboxylase (GAD), can convert glutamate into GABA (gamma aminobutyric acid). It is important to keep in mind that these two neurotransmitters are heavily interlinked in both their synthesis pathways and their interaction on a higher 'circuit level' within the brain. We will come to GABA in the next chapter, but for now it is worth remembering that, in some respects, GABA is the antithesis of glutamate, with glutamate being the main excitatory neurotransmitter in the brain, while GABA is the main inhibitory.

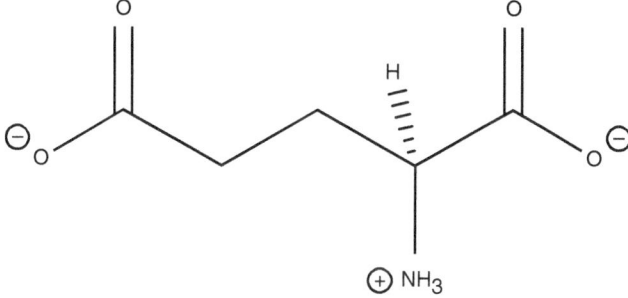

Figure 2.1   Molecular structure of glutamate

Molecular structure of glutamate, which, as well as being a neurotransmitter in its own right, can go through further enzyme-mediated alterations to produce other neurotransmitters, such as GABA.

Glutamate packaging

Once glutamate has been synthesised in the neuron it needs to be packaged into vesicles for short-term storage and eventual release. This packaging is done by channel proteins called VGLUTs, which stands for vesicular glutamate transporters. These channel

---

[1]The Krebs cycle is one of the key parts of metabolism within a cell and mainly takes place within the cell's mitochondria.

proteins reside in the membrane of the vesicles and work by exchanging hydrogen (H^+) ions for glutamate molecules. In order for this to happen, an electrochemical gradient has to be set up across the membrane. This is called the membrane potential. It is set up by the actions of the proton pump, which fundamentally brings hydrogen ions into the vesicle. This is especially important for the transport of glutamate into the vesicle as glutamate is negatively charged.

There are generally considered to be three main types of VGLUTs, aptly named VGLUT1, 2 and 3. VGLUT 1 and 2 are complimentary, expressed in the vast majority of glutamatergic neurons of the central nervous system. It is not just neurons that express VLGUTS in the brain. In fact, they are also consistently found to be expressed in astrocytes, which are a form of glial cell. Within astrocytes, VGLUT 1 and 2 are considered by many to have a heterogeneous distribution pattern across different brain regions. VGLUT1 is especially found to be distributed far more diffusely than VLGUT2, with evidence of its expression within the cortex and striatum, as well as various regions within the hippocampus. VGLUT2 is considered to be far more restricted in its expression pattern, with the majority found within the hippocampus. It is important here to think back to the section on glutamate synthesis. Remember that glia are essential to the synthesis of glutamate, so it makes complete sense that they would be heavily expressed in astrocytes.

So, what about VGLUT3? Well, these are a bit unusual as they are often found in neurons not known to be glutamatergic; for example VGLUT3 are known to be expressed in cholinergic neurons of the striatum and serotonergic neurons. I will return to this point many times throughout the book, but this is a good example of the complexity and interactivity of both neurotransmitters and their machinery in the brain. For example, Amilhon et al. (2010) used a VGLUT3 knockout animal[2] and showed that there were signifcant alterations in serotonergic tone within specific brain regions. We will discuss serotonin in a later chapter, but it is worth keeping this in mind as a point of critical commentary, especially when reading research focused on the role of serotonin in affective and anxiety disorders.

One really interesting thing about VGLUTs is that their ability to transport glutamate into vesicles seems to be heavily affected by extravesicular (outside the vesicle) free chloride levels (Cl⁻). It has been shown that when there is no Cl⁻, then vesicular uptake of glutamate is very low. However, with low levels of Cl⁻, then transport of glutamate into the vesicle is very high. When Cl⁻ ions increase further to moderate and high levels, then glutamate uptake into vesicles begins to gradually fall again. This biphasic dependence on Cl⁻ levels for the active transport of glutamate by

---

[2]the gene that codes for specific proteins has been removed

VLGUTs into vesicles is thought be due to VGLUT transporting Cl⁻ into the vesicle with glutamate in a process called allosteric modulation. This would make sense as the loading of Cl⁻ into the vesicle with glutamate would result in an alteration in the membrane potential as the negative ion builds up within the vesicle (see Figure 2.2).

As previously mentioned, the proton pump found imbedded into the vesicle membrane is considered key to setting up the membrane potential, which is especially important for the transport of negatively charged molecules, such as glutamate, into the vesicular lumen (space inside the vesicle). However, it is not just the proton pump bringing H⁺ ions into the vesicular lumen that is thought to modulate this membrane potential for glutamate transport. There is also strong evidence to suggest that, at least in the initial stages of packing the vesicles, efflux (movement out) of Cl⁻ ions from the vesicular lumen contribute to setting up this membrane potential and that this is, perhaps, mediated by the VGLUTs themselves.

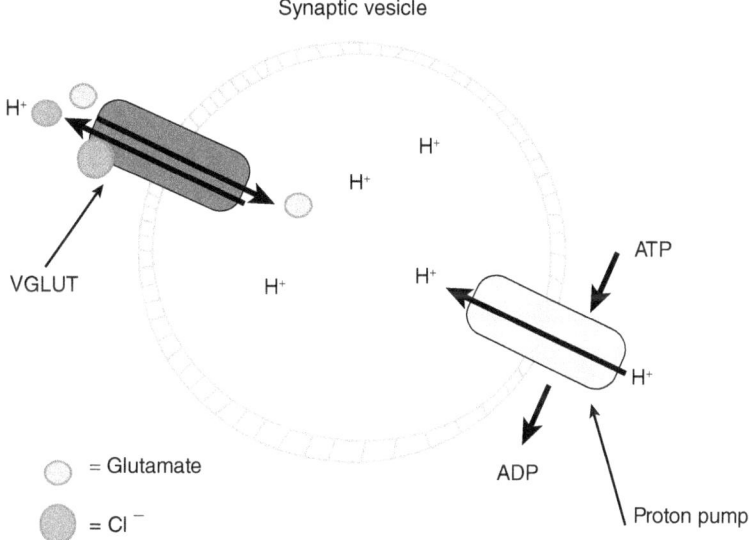

Figure 2.2 Illustration of a vesicle packaging the neurotransmitter glutamate

Typical channels are embedded in its membrane, as discussed in Chapter 1 of this book. Notice the co-transport of Cl⁻ irons with glutamate from the extracellular space to the vesicles lumen.

Source: Adapted from: Chaudhry, F. A., Edwards, R. H., & Fonnum, F. (2008). Vesicular neurotransmitter transporters as targets for endogenous and exogenous toxic substances. *Annual Review of Pharmacology and Toxicology, 48*(1), 277–301. doi:10.1146/annurev.pharmtox.46.120604.141146

Glutamate reuptake

As mentioned in the synthesis section, glial cells, including astrocytes, play an important role in the synthesis of glutamate and they are key players in the glutamate–glutamine cycle. They play this role primarily as most of the glutamate, released by neurons, rather than being taken back up by the neuron, or degraded by enzymes in the synapse, is taken up by glial cells. These glial cells then convert this glutamate to glutamine, ready to be exported back into the presynaptic neuron. We'll come back to this in a moment, but first, let's look at the machinery bringing glutamate into the glia.

The channels that mediate this uptake by glia are known as EAATs (excitatory amino acid transporters) and are generally considered to consist of five types, cunningly known as EAAT 1, 2, 3, 4 and 5. The first of these was identified in the early 1990s. This is now, confusingly, referred to as EAAT 2 in humans and as GLT-1 (glutamate transporter 1) in rodents. Each type of EAAT is known to have a different distribution pattern, both in terms of the type of cell/neuron they are expressed in and the brain region. Because of this distribution pattern in tissue, we will focus on EAAT 1, 2, 3 and 4, ignoring EAAT 5 as it is best characterised as being expressed in the retina in the photoreceptors and bipolar cells. It is therefore less relevant to our focus here, which is, predominantly, neurotransmitters of the brain. EAAT 1–4 are all expressed in the brain, although there is a distinct difference in the types of cells that express them, with EAAT 1 and 2 predominantly being expressed in glia, while EAAT 3 and 4 are predominantly expressed in neurons. It is important to remember here that this is not a hard-and-fast rule and there are exceptions, for example EAAT 2 are known to be expressed in some presynaptic neuronal terminals and EAAT 4, while expressed in neurons, are only thought to be expressed in specific Purkinje cells of the cerebellum.

Although there are clearly differences in the expression pattern of these glutamate transporters, they are believed to share a very similar mechanism of action. Indeed, like so many of these reuptake transporters that we will come across throughout this book, this is an active process as it involves moving glutamate against a concertation gradient and into the cell, be that cell a glial cell or a neuron. This active transport is reliant on the availability of Na^+ (sodium) ions, which are co-transported into the cell with the glutamate molecule. This co-transport of three Na^+ ions with the efflux of one K^+ (potassium) ion provides the main driving force for the influx (movement inwards) of glutamate (see Figure 2.3). This is, however, not the end of the story as there is a further passenger transported into the cell along with Na^+ and glutamate, namely one H^+ ion. This might seem like a bit of a side point, but it is very important as the transport into the cell of the H^+ ion results in a change in charge and ultimately a fluctuation in the membrane potential, which further helps to drive the movement of glutamate into the cell.

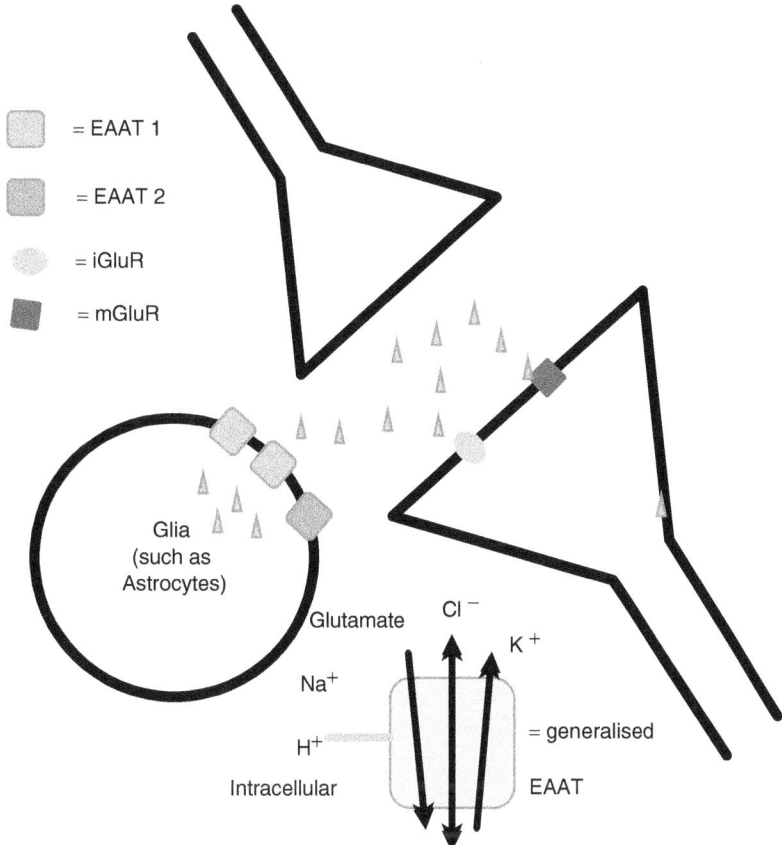

Figure 2.3 A minimal model illustrating the expression of EAATs on glia and the ions involved in the transport of glutamate by EAATs

After glutamate has dissociated from its receptors, the vast majority is taken into glia, such as astrocytes via the excitatory amino acid transporters (EAATs). There are a number of different EAATs that do this job, with differences in their mechanistic, but the common mechanism of transport involved the co-transport of Na^+ with glutamate into the glia and transport of Cl^- and K^+ ions into the extracellular space.

EAATs are highly effective at mopping up extracellular glutamate after release into the synapse. This is important largely because free glutamate in the extracellular space is highly neurotoxic. Research suggests that the fact they are highly efficient may also serve other functions. These are beyond the scope of the current book; however, an excellent introduction can be found in Magi, Piccirillo, Amoroso, and Lariccia, (2019) and Magi, Piccirillo, and Amoroso (2019).

So, what happens after glutamate is transported into the intracellular space by EAATs? Now we come back to the glutamate–glutamine cycle first mentioned in the section focused on glutamate synthesis at the start of this chapter. Once inside the glial cell, glutamate is readily converted back to glutamine by the enzyme glutamine synthase. Glutamine is then transported out of the glia and returned to the neuron via several different receptor families that include SNAT, LAT, y+LAT and ASC families of transporters. It is important to remember at this point that glutamate, once returned to the intracellular space, is not necessarily converted back to glutamine. In fact, this cannot happen for the glutamate transported back into the neuron directly as neurons do not express glutamine synthase. This enzyme is exclusively expressed by glia, such as astrocytes, as can be seen in Figure 2.4, and is the main biological mechanism, which means glia are essential for glutamine and eventually glutamate synthesis. So, what happens to some of the glutamate? Well, put simply, it's used in the cells' metabolic processes within those power plants of the cell known as the mitochondria. This is true both for the glutamate transported into glia and that transported directly into neurons.

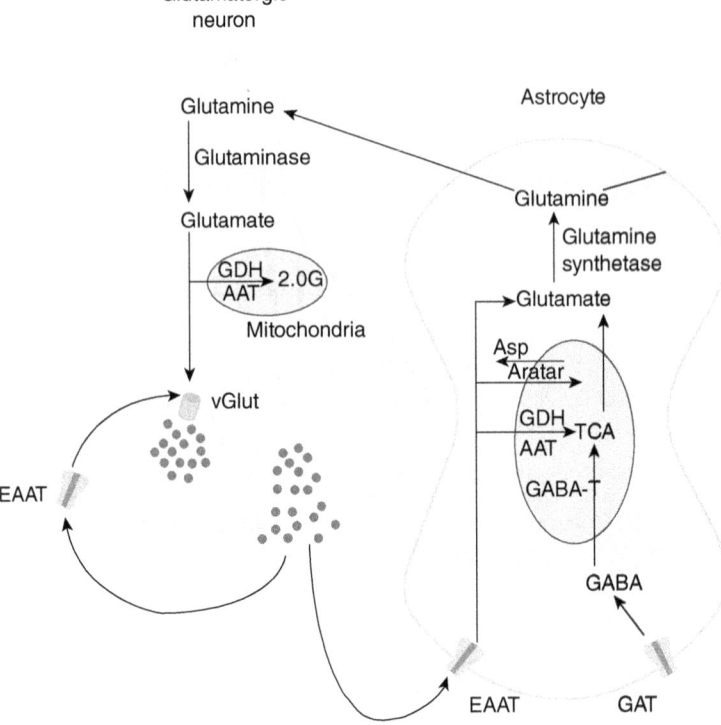

Figure 2.4   The cycle of glutamate uptake and synthesis in astrocytes

Once taken up into the glia/astrocytes by EAATs, glutamate can be fed back into the tricarboxylic acid cycle (TCA, also known as the Krebs cycle) or it can be directly used for the synthesis of glutamine by glutamine synthase. At this point, the cycle starts again and glutamine is actively transported out of the glia to the glutamatergic neuron for the resynthesis of glutamate within the neuron.

Source: Reprinted from Rowley, N. M., Madsen, K. K., Schousboe, A., & Steve White, H. (2012). Glutamate and GABA synthesis, release, transport and metabolism as targets for seizure control. *Neurochemistry International, 61*(4), 546–558. Copyright (2012), with permission from Elsevier.

Test Yourself 2.1

One of the key differences (compared to other neurotransmitters) between glutamate synthesis and reuptake is that part of these processes take part out of the neuron. Try answering the following questions to make sure you are clear about this:

1. If you wanted to label glutamatergic neurons, which enzymes would you target?
2. If you wanted to label the glia that are partly responsible for glutamate synthesis, which enzyme would you target?
3. Which molecule is transported from glia into neurons for the final stage of glutamate synthesis?
4. Which is the main transporter group responsible for glutamate uptake into glia?

Answers to all Test Yourself questions are at the end of the chapter.

Outline of glutamate receptors

Glutamate, like many neurotransmitters has both ionotropic and metabotropic receptors that are its target. Also, like many other neurotransmitters, it is not as simple as there being one ionotropic and one metabotropic receptor. In the case of ionotropic receptors, there are three types for glutamate, which are known as NMDA, AMPA and kainate receptors. It is worth noting that even this is a simplification as each type is made up of subunits that can be arranged in various different ways, resulting in lots of different subtypes. As discussed in Chapter 1, ionotropic and metabotropic receptors have different dynamics in their action and the effects on the neuron, once glutamate is bound. It is worth at this point, perhaps, refreshing your memory about the differences between these types of receptors.

Test Yourself 2.2

Sketch out a two-column table with the two headings below and sort the terms under the correct headings:

Table 2.1

Ionotropic	Metabotropic
Slowly reversible	
GPCRs	
Short latency	
Long latency	

(Continued)

Table 2.1 (Continued)

Secondary messengers involved
cAMP
They are ion channels
They are not directly connected to ion channels
Rapidly reversible

So, now to look at the ionotropic receptors in more depth. First, let's talk about NMDA receptors. Like all ionotropic receptors, these are essentially ion channels. Once activated, these channels open and allow Na^+ (sodium) and Ca^{2+} (calcium) to flow into the neuron (influx) and K^+ (potassium) to flow out of the neuron (efflux). This produces an excitatory postsynaptic current (EPSC) in the receptor-expressing neuron. This means that there is a small (measured in picoamps[3]) positive current that can be measured in the neuron, which, on mass, will produce a large increase in current, which eventually reaches a threshold to become an action potential, at which point we can say the neuron fires. The NMDA channel is, however, also blocked by Mg^+ (magnesium) and this is not easily displaced. For it to be displaced, and the above flow of ions to happen, there must be a change in membrane potential. This shunt requires the binding of both glutamate and glycine. Luckily, because the vesicular transporters, packaging glutamate for release, also package glycine, it is generally the case that where there is glutamate there is also glycine. One final interesting thing about the NMDA receptors relates to the flow of Ca^{2+} ions into the cell when they are open. Ca^{2+} is a very common secondary messenger within neurons. Therefore, although NMDA receptors don't directly activate the secondary messenger system within cells, like metabotropic receptors, this influx of calcium has lots of potential effects as a secondary messenger, resulting in a host of intracellular changes beyond the excitability of the neuron.

Next up are AMPA receptors. These are generally co-expressed at the synapse with NMDA receptors, as we will come back to shortly, and their interplay has huge implications for the neuron. While both NMDA and AMPA receptors produce excitatory synaptic currents when activated, AMPA receptors have quite distinct electrophysical effects on the neuron compared to NMDA receptors. Whereas the excitatory current produced by NMDA receptor activation lasts approximately 100 ms and reaches a peak ampage of approximately 25 pA, AMPA activation produces a much shorter duration excitatory current of approximately 25 ms, which reaches a peak amperage

---

[3]One amp is the equivalent to 1000000000000 picoamps.

of above 50 pA. To sum it up, the excitatory current produced by activation of AMPA receptors is faster, larger, but briefer than activation of NMDA receptors. This excitatory current is produced by the binding of glutamate alone to AMPA receptors and inflow of Na^+ and outflow of K^+.

Finally, to kainate receptors. Very much like AMPA receptors, kainate receptors are activated by glutamate binding and allow the inflow of Na^+ and outflow of K^+. Also, like AMPA receptors, and unlike NMDA receptors, they are not permeable to Ca^{2+}. In fact, in terms of structure and function, kainate receptors are very similar to AMPA receptors. The difference mainly seems to lie in their expression patterns. While AMPA and NMDA receptors are commonly co-expressed, kainate receptors are commonly found in regions of the brain, and on neurons, that do not express NMDA receptors. One of the main regions where extensive research has been done is the dorsal root ganglia. Despite the similarities between AMPA and kainate receptors, research suggests quite distinct differences in the kinetics of excitatory current produced by both types. For more detail on this please consult Lee, Labrakakis, Joseph, and Macdermott (2004).

Focus on research 🔍

Christian Lüscher and colleagues

Long-term plasticity in neurons has been a hot topic of research over the last 30 years. It essentially refers to changes in the neuron that are a result of continued activation. These changes seem to result in alterations to the effect further activation has on the neuron, in terms of its potential to 'fire'. If you consider that neurons synapse with other neurons, this long-term plasticity refers to the 'strength of the connection' between the two. It has been such a hot topic because these changes have been commonly considered a biological substrate for 'memory' and 'learning'.

Some of the best research done on this topic has been that by the group of Christian Lüscher. His work, as well as that by others, has established that alteration in the expression of NMDA and AMPA receptors is key to this long-term plasticity, both in terms of long-term potentiation (the increased likelihood of neurons firing) and long-term depression (the reduced likelihood of neuron firing).

For further reading on this topic, I highly recommend:

Lüscher, C., & Malenka, R. C. (2012). NMDA receptor-dependent long-term potentiation and long-term depression (LTP/LTD). *Cold Spring Harbor Perspectives in Biology, 4*(6), a005710. doi:10.1101/cshperspect.a005710

Lüscher, C., Xia, H., Beattie, E. C., Carroll, R. C., von Zastrow, M., Malenka, R. C., & Nicoll, R. A. (1999). Role of AMPA receptor cycling in synaptic transmission and plasticity. *Neuron, 24*(3), 649-658. doi:10.1016/S0896-6273(00)81119-8

Now is the time to move on to the metabotropic glutamate receptors. This is a very extensive group and consists of three distinct families, each with their own collection of family members. The families are logically named as groups 1, 2 and 3 and each member is referred to as mGlu followed by a number. For group 1, the family members are mGlu1 and mGlu5. For group 2, the family members are mGlu2 and mGlu3. For group 3, the family members are mGlu4, mGlu6, mGlu7 and mGlu8.

As with all metabotropic receptors, they are not directly coupled to ion channels and many of their intracellular effects are a result of activating a complex secondary messenger system within the neuron. This gets even more complex as the secondary messenger system involved is different for each of the groups (1, 2 and 3).

For group 1, activation by glutamate causes a chain of event that results in increased concentrations of secondary messengers that are essential for cell survival and 'normal' function. The first step in this chain is activation of the enzyme phospholipase Cβ by the uncoupling of the G proteins bound to mGlu1 and 5. Activation of this enzyme results in the formation of two key secondary messengers, namely IP3 and DAG. As outlined in Chapter 1, DAG's main job is to activate PKC, which results in activation of the MAPK pathway. IP3 formation results in intracellular release of Ca^{2+}, which, among other things, promotes vesicular movement towards the neuronal plasma membrane for release. It is important to remember that this is just a small number of processes and molecules that are resultant of group 1 activation (see the excellent paper by Nicoletti et al. (2011) for further information). As one final point, to emphasise the complexity of these receptors, it is also worth noting that group 1 receptors, specifically mGlu5, are also intracellularly linked to ionotropic NMDA receptors. There is therefore a complex pattern of interaction between these different classes of glutamate receptors as well as everything else.

Now for groups 2 and 3. Here is where we get to something very interesting. You will typically hear glutamate referred to as the main excitatory neurotransmitter in the brain, but its effects are not always excitatory. For example, activation of the group 2 and group 3 glutamate receptors actually results in downregulation of secondary messengers and depression of the neurons activity as well as suppression of key pathways that help the neuron survive and proliferate.

For group 2, activation results in downregulation of the enzyme adenylyl cyclase, which results in reduced levels of one of the key secondary messengers, namely cAMP. As well as this downregulation AC, group 2 activation also inhibits Ca^{2+} channels in the plasma membrane and increases the activity of K^+. Beyond this, group 2 activation has been shown to alter various other intracellular mechanisms that help the neuron survive and function 'normally', such as altering the function of calcium-binding proteins and supressing the MAPKsignalling cascade, which has fundamental roles in gene transcription and neuronal survival, as discussed in the introductory chapter of this book.

Finally, for group 3 we have similar effects on adenylyl cyclase as group 2 activation, ultimately reducing the intracellular levels of cAMP and having a down-regulation effect on the neuron. The majority of group 3 receptors are thought to be expressed on the presynaptic neuron in the active zone, so their suppressive role is thought to be a key negative feedback mechanism, reducing the chances of any further activity of the presynaptic neuron and, ultimately, suppressing further glutamate release.

Test Yourself 2.3

Look at the following figures and decide what the effect would be on each neuron if glutamate was released into this area. Try to express your answers in terms of presynaptic effects, postsynaptic effects, excitation, inhibition, suppression of glutamate release, slow excitation/inhibition, fast excitation/inhibition.

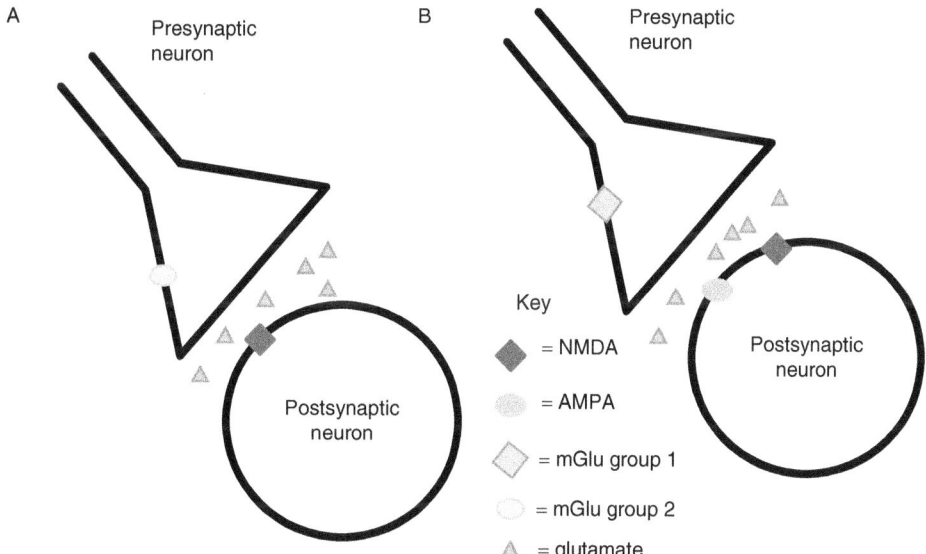

Figure 2.5 A minimal model illustrating glutamate receptor expression on pre- and postsynaptic terminals

Classical topics in glutamate research

It is always important to remember, when encountering any research on glutamate, that it is expressed widely across the brain and central nervous system. This fact has implications for behaviour. If you consider that it is expressed widely, then this means broad alterations in its levels should have very broad implications for behaviour. For example, one of the topics we are about to cover is glutamate's role in epilepsy and the resultant seizures. It makes sense that alterations in glutamate levels broadly would be involved, as epilepsy involves large parts of the brain and alteration of activity across all the brain. However, things such as addiction, which we are also going to cover, do not involve all brain regions, and 'drug addiction' is thought to be a result of drugs of abuse hijacking specific reward-related circuitry. So why would we cover research on glutamate and addiction? Well, the simple answer is, although global changes in glutamate levels might not be involved in addiction to drugs of abuse, we certainly see heterogeneous expression of glutamate receptors in the brain. Therefore, specific behaviour changes might well be a result of alterations in glutamate receptors.

Glutamate and epilepsy

Epilepsy is one of the most common neurological disorders, affecting approximately 1% of the world's population. It has a huge negative effect on a person's quality of life. Epilepsy is generally characterised by alterations in the excitability of neuronal populations. In fact, the prevailing view of epilepsy is that it is characterised by alterations in what is known as E/I balance. This stands for the balance of excitatory and inhibitory transmission in the brain. Fundamentally, the suggestion is that there is an increase in excitatory neurotransmission and/or a decrease in inhibitory neurotransmission. Either way the result is an alteration in E/I balance, increased excitability and, typically, seizures. It is therefore a logical conclusion that the machinery associated with the main 'excitatory' neurotransmitter in the brain, namely glutamate, may well underpin epilepsy and cause/contribute to the associated seizures.

There are several different antiepileptic drugs (AEDs) in use. Some function by targeting GABA receptors and reducing the brain's excitability by increasing inhibitory tone, while others' mode of action is to disrupt the function of glutamate receptors and/or disrupt the release of glutamate. We will return to the role the neurotransmitter GABA might play in epilepsy in the next chapter of this book. Now we will focus on glutamate and its machinery.

The fact that some anti-seizure medications disrupt glutamate release and glutamate receptor function is our first line of evidence that alterations in glutamate levels, or the function of its machinery, are involved in the aetiology (cause) of epilepsy or, as it is commonly referred to, epileptogenesis. However, it is important to keep in mind that medication that reduces symptomology doesn't necessarily tell us what the aetiology for a disorder is. For example, one drug used to treat generalised seizures is Perampanel, which is known to be an ionotropic glutamate receptor antagonist, specifically an AMPA receptor antagonist. It is conceivable that alterations in AMPA function are a consequence of the heightened excitability, which is the hallmark of epilepsy, rather than a cause. Certainly, we know from the section earlier in this chapter that focused on glutamate receptors that AMPA numbers are increased as a product of increased excitability in a network, rather than the cause of increased excitability in the network. During seizures, increased excitatory signals propagate across the brain. This means altered AMPA levels could easily be a result of seizures. Therefore, the medication works to reduce this propagation and, consequently, reduces the intensity of the seizures by inhibiting these receptors. Thus, any conclusions drawn about the aetiology of epilepsy, indeed any disease state, based on the action of medication should be taken with at least a degree of suspicion.

Despite this, there is much evidence to suggest that both ionotropic and metabotropic glutamate receptors are involved in seizures, which are an important part of epilepsy and have a dramatic impact on an individual's quality of life. For example, research has shown that many different glutamate receptor agonists induce seizures, while application of glutamate receptor antagonists, especially ionotropic receptor antagonists, are known to be potent anti-convulsants, when administered to animal models of epilepsy. For further information specifically on the role of ionotropic receptors in epilepsy, I recommend the recent review by Hanada (2020). There is also reasonable evidence from the use of genetically-manipulated animal models that suggests the involvement of altered ionotropic glutamate receptors. One particularly interesting piece of research by Chapman, Woodburn, Woodruff, and Meldrum (1996) involved an animal model that had a reduced function of the NR1 subunit of NMDA receptors. This was done by introducing what is called an antisense probe into the animal's DNA. This resulted in a complete inhibition of seizures induced by sound in an animal model that displayed sound-induced seizures without the alteration. Based on the known pharmacological effect of metabotropic receptors agonists and antagonist in modulating seizures, it would be fair to expect that genetic manipulation of these receptors would also alter seizure susceptibility, but this has not been found. Several studies have developed so-called knock out animals[4] that do not express metabotropic glutamate receptors, such as mGlu2 and mGlu5. These studies find alteration in things such as learning and alterations in synaptic plasticity (generally reductions in LTP), but they do not find any alteration in seizure susceptibility.

So, based on the evidence, it seems reasonable to assume that alterations in glutamate receptors are involved in seizures, although there is less definitive evidence to suggest they are the cause of epilepsy. Rather, it seems alterations in their number and function may be a consequence, which results in increased severity and/or propagation of seizures. So, because the involvement of receptors might be best viewed as a consequence of epilepsy rather than being key to epileptogenesis, what causal role could alterations in glutamate play in epileptogenesis? We will now cover two lines of evidence, one suggesting the involvement of alterations in vesicular transporters and another implicating glutamate reuptake transporters, specifically those glutamate reuptake transporters expressed on glia cells, such as astrocytes.

VGLUTs and epilepsy

As we have previously discussed, VGLUTs are responsible for the packaging of glutamate into vesicles for release into the synapse. It makes logical sense that if these are altered in their ability to package glutamate, then you will have alterations in the ability for glutamate to be released upon the arrival of an action potential at the synapse. This means that alterations in VGLUTs are a logical cause of epilepsy and/or seizures. The first line of evidence that supports VGLUTs' involvement in epileptogenesis is the fact that there has been shown to be a dramatic increase in glutamate release in patients diagnosed with epilepsy during both evoked and spontaneous seizures. This seems to imply that there is a higher level or a rapidly-ready pool of vesicles that are released during seizures. Further to this, there is a significant amount of evidence that suggests that VGLUT levels are altered in both animal models of epilepsy and humans with the diagnosis. This research is far less consistent than the latter, with some studies finding an increase in VGLUT expression, and mRNA for VGLUT, while other studies find the opposite.

Many of these studies, in humans, are heavily compromised as the samples are taken from medicated individuals. Animal research shows that these AEDs alter the expression levels of VGLUT. Thus, it is very difficult to say conclusively if the altered

---

[4]This is where the gene that programmes for the specific protein, in this case a metabotropic glutamate receptor, is edited out of the animal's DNA, resulting in either no expression of the protein or a dramatic reduction in expression.

levels of VGLUT are a result of this medication, rather than the disease state. It is also important to remember that most studies done with humans are with those patients known to have epilepsy. Therefore, the issue of causation rises again as we cannot be confident that the changes are not a result of the epilepsy/seizures rather than the cause (epileptogenesis). One way to generally try to resolve this issue is the use of genetically-altered animal models. Several of these have been developed over the last 10 years, with various alterations, including partial loss of VGLUT2 and total loss of VGLUT3. Much of this research seems to suggest an alteration in the threshold for seizure induction, with seizures generally being easier to induce in these animals. One example is the research of Schallier et al. (2009), whose study used a known convulsant and found that there was a reduced threshold to induce seizures in those animals with a partial loss of VGLUT2. Interestingly, research using VGLUT3 knockout animals has found increases in the occurance of specific absence seizures. This is particulary interesting as these VGLUT3 transporters are not typically expressed in glutamatergic neurons, as outlined in the earlier section on packaging in this chapter. This might suggest a complex interaction between glutamatergic neurons and others, such as cholinergic neurons, known to express VGLUT3 in the aetiology of epilepsy.

Novel roles for glutamate in epilepsy – glia (astrocytes), glutamate and epilepsy

As we discussed in the opening sections of this chapter, glia are essential for the synthesis of glutamate. Therefore, it should come as no surprise that alteration in glia, specifically the expression of glutamate receptors and transporters on glia, has been implicated in epileptogenesis.

Glutamate transporters, expressed on glia, are essential for the timely termination of glutamatergic transmission. Therefore, it makes sense that alteration in the function of these transporters would result in higher levels of free glutamate in the synapse and increased levels of excitability, as seen in epilepsy. If you think back to earlier in the chapter, there are several forms of these glutamate transporters, but as a group they are referred to as EAATs (excitatory amino acid transporters). The most commonly expressed in glia (including astrocytes) are EEAT1 and EAAT2. Evidence suggests an involvement for EAAT2 in epileptogenesis. One particularly strong piece of evidence comes from EAAT2 null mice. These are mice that have been genetically modified and do not express EAAT2. These mice are epileptic from early years, suggesting causal association in the aetiology of epilepsy. However, unfortunately, like much of the literature on epilepsy, the role EAATs play is inconsistent.

There is a strong line of evidence implicating alteration in glia rather than neurons in glutamate alteration and ultimately being a causal factor in epilepsy. So, what might it be about glia if it is not the expression of EAATs? Well, recent research suggests that it may be due to altered release of glutamate from glia. One excellent study that suggests this was done by Tian et al. (2005). I strongly recommend this paper as further reading. Other research, such as that by Woo et al. (2012), also suggests that altered glutamate release from glia might underpin epilepsy, although they suggest a different mechanism from Tian et al. (2005), specifically implicating two ion channels, one a calcium channel and one a potassium channel, known as best1 and TREK-1 respectively. This research fits with an old theory, namely that epilepsy is caused by alterations in potassium channels. This was first reported in the early 1960s. It therefore seems likely that epileptogenesis, like so many other disease states, is caused by a complex interaction between multiple molecules and multiple bits of cellular machinery.

Focus on research 🔍

E/I imbalance linking autism spectrum disorder and epilepsy

There is a significant amount of evidence building that suggests there is a comorbidity between epilepsy and autism spectrum disorder (ASD) with a higher expression of ASD in those diagnosed with epilepsy than its occurrence in the typically-developing population. So, what might underpin this relationship? Well, there is a line of thought that alteration in the levels of excitatory transmission in the brain, which we have discussed as underpinning epilepsy, may also underpin ASD. This ultimately comes back to the brain's E/I balance, which is kept in tight limits within the typically developing brain.

For further reading on this fascinating relationship, I suggest the following:

Besag, F. M. (2017). Epilepsy in patients with autism: links, risks and treatment challenges. *Neuropsychiatric Disease and Treatment*, *14*, 1-10. Doi:10.2147/NDT.S120509

Nelson, S. B., & Valakh, V. (2015). Excitatory/inhibitory balance and circuit homeostasis in autism spectrum disorders. *Neuron*, *87*(4), 684-698. Doi:10.1016/j.neuron.2015.07.033

So, much of this evidence suggests a role for glutamate in epilepsy. Yet, much of the evidence is inconclusive as to whether its alterations are a result of seizures or the cause of seizures. Many suggest that we are better not to think of glutamate as involved in epileptogenesis, but as involved in the initiation and spread of seizures.

However, there is a reasonably convincing line of evidence, from genetically-modified animal models, that seems to provide more solid support for the causal role that glutamate, and the machinery underpinning glutamate's release, may play in epileptogenesis. Finally, we will often focus on neurons in this book, but this topic brings home the fact that there are other types of cells in the brain, and these too can play a role in the function of the brain's neural networks by impacting neurotransmitters.

Metabotropic glutamate receptors and addiction

As pointed out at the start of this section, it is unlikely that alterations in glutamate levels broadly are a key factor in addiction to drugs of abuse, although there are several lines of evidence that suggest its receptors, especially the mGlu (metabotropic) receptors, may well be involved. Before we look at research focusing on drug addiction and glutamate, we first need to consider the brain circuits implicated in drug addiction.

Early research by Olds and Milner (1954) developed a technique known as intercranial self-stimulation (ICSS). This technique involved implanting an electrode into various brain regions of a rodent, and then giving the animal the ability to stimulate the brain regions by pressing a lever, which was attached to the implanted electrode. With this technique, Olds and Milner (1954) discovered that animals would work hard to stimulate some brain regions, but not others. The areas that animals would work hard to stimulate have now been identified as constituting the brain's rewards circuit, and include classic areas linked to responses to drugs of abuse, such as the nucleus accumbens and ventral tegmental area. An important issue to consider at this point is that most researchers believe that drugs of abuse have their addictive effects by essentially hijacking the normal function of the reward circuits. Therefore, ICSS has been employed widely to identify what neurotransmitters and receptors might be involved in this reward circuit, and ultimately reveal the machinery that might be hijacked by drugs of abuse to produce addiction.

Many of the studies in this area involve the application of pharmacological agonists and antagonists that target specific receptors. These are administered while allowing the animal to self-stimulate via the implanted electrode (ICSS). The application of various drugs of abuse, such as opioids, has been shown to consistently lower what is referred to as the ICSS threshold. The ICSS threshold is the small current which supports self-stimulation. When the ICSS threshold is increased (it takes more stimulation to be 'rewarding'), it is considered that the pharmaceutical manipulation must result in decreased levels of reward, whereas when the ICSS threshold is reduced (it takes less stimulation to be rewarding) it is considered that the pharmacological agent results in increased levels of reward. This gets more complicated as it also depends on whether the pharmacological agent is an agonist (works with) or an antagonist (works against). For example, if you look at Figure 2.6, you will see that if you apply an agonist for a specific receptor and get a decrease in ICSS threshold, then this implies the activation of the specific receptor is involved in increasing the reward value of self-stimulation. However, if you use an antagonist for a specific receptor, then an increase in ICSS threshold suggests activation of that receptor is more rewarding.

	Pharmaceutical for a specific receptor	
	Agonist	Antagonist
ICSS threshold increased	Decreased reward	Increased reward
ICSS threshold decreased	Increased reward	Decreased reward

Figure 2.6   The impact of specific receptor agonists and antagonists on intercranial self-stimulation (ICSS) threshold and the corresponding meaning for the value of rewards

So, what does the research using ICSS and agonist/antagonist pharmaceuticals, focused on glutamate receptors, suggest? Well, it largely seems to suggest a role for specific types of metabotropic glutamate receptors in reward, which suggests that they may be machinery 'hijacked' (in the reward circuits) by drugs of abuse. For example, ICSS research using the pharmacological agent MPEP (a group 1 mGlu5 receptor **antagonist**) produced significant increases in the ICSS threshold, suggesting that these specific receptors' 'normal' function is to increase reward value. However, modulation of other metabotropic glutamate receptors results in reduced reward value of ICSS. For example, ICSS research where LY314582 (an a**gonist** for the inhibitory group 2 metabotropic glutamate receptors) was administered significantly increased the ICSS threshold, suggesting that group 2 metabotropic glutamate receptors reduce the reward value of 'self-stimulation'.

However, there are obvious issues with these ICSS studies. First, we are assuming the ICSS is 'similar' to what we experience in response to naturally-occurring rewards, such as a drug of abuse, which is a jump at best. So, what can we do to solve this? Well, we can simply administer the drug of abuse to the animal and manipulate the levels/activity of glutamate receptors. This can be done in various ways. Early research did this by administering the drug of abuse while simultaneously administering various antagonists for the specific glutamate receptors. These kinds of studies typically found that blockade of mGlu5 receptors with antagonists reduced drug-related behaviour. In more recent years, this research has been achieved by employing genetically-altered animals, specifically animals where the gene that codes for specific types of receptors has been removed. These are typically referred to as genetic knockout animals. This knockout results in either no expression of the targeted receptors or dramatically reduced expression of the receptor. Research has shown that knockout of mGlu5 receptors significantly alters the addiction-like behaviour animals show to cocaine, reducing the amount they self-administer and reducing the increased locomotor activity typically produced in animals as a result of cocaine administration. This seems to fit well with the research outlined above using the chemical MPEP and provides a reasonably convincing argument for the involvement of mGlu5 receptors in drug addiction. However, this is a controversial area, and subsequent studies suggested that these same mGlu5 knockout animals show little change in their response to cocaine, still exhibiting a conditioned place preference[5] and displaying a reduced locomotor response. Crucially, this alteration in locomotor response to cocaine was reduced but still present. Therefore, it seems likely that mGlu5 receptors play some role in the rewarding aspects of drugs of abuse, but this might be better viewed as a modulatory role, rather than a causal role.

Indeed, the idea that mGlu receptors might play a modulatory role is borne out by the research evidence, with the suggestion being that they may regulate dopamine release in the 'reward circuit'. Early research by Taber and Fibiger (1995) suggests that this may well be the case. They administered a broad spectrum mGlu receptor agonist into the nucleus accumbens[6] and found that low doses of this agonist reduced the baseline levels of dopamine. By doing this, it is believed that mGlu receptors may modulate the 'reward-related' signal that is conveyed by dopamine. This is logical and many lines of evidence suggest that the dopamine signal is key to our experience of reward[7] and is indeed altered by drugs of abuse. Taber and Fibiger's (1995) research goes further and supports this assumption. When they stimulated other parts of the reward circuit (known to support self-stimulation, such as the ventral tegmental area) with an implanted electrode, the mGlu receptor agonists also managed to suppress the release of dopamine, which would normally be a consequence of ICSS of these regions. So, this seems like quite a convincing suggestion that drugs of abuse might 'hijack' the reward system and produce 'addiction' by altering the expression or sensitivity of these metabotropic glutamate receptors, most likely group 2 receptors, as

---

[5]Conditioned place preference (CPP) is a common experimental technique used to measure the effects of drugs on animal models. It involves injections of the drug being given to the animal in a specific environment and then the animal being allowed to choose where to spend their time, between this place and another environment where the drug wasn't injected. If the drug is 'rewarding', the animal will typically choose to spend time in the environment where they received the injections.

[6]The nucleus accumbens is one of the core regions of the 'reward circuit' and is heavily implicated in reward and pleasure responses and is known to be 'hijacked' by drugs of abuse.

[7]For further information, refer to Chapter 5 of this book, which is focused on dopamine.

outlined above: Specifically, those group 2 mGlu receptors that are expressed on dopaminergic terminals. Downregulation of these receptors would mean that dopamine release into the reward circuit was no longer inhibited by these mGlu receptors and therefore was theoretically 'more rewarding'. There is one major issue with this idea, namely that although group 2 mGlu receptor agonists increased the ICSS threshold (make self-stimulation less rewarding), they do not stop drugs of abuse from reducing the ICSS threshold (drugs of abuse are still rewarding). This suggests that mGlu receptors may play a modulatory role in addiction, but they are unlikely to be the main machinery 'hijacked' in the reward circuits by drugs of abuse.

It seems that group 1 and group 2 mGlu receptors might be involved to some degree in addiction to drugs of abuse. However, it is important to remember that addiction to drugs is often thought of as being made up of different stages. For example, George Koob (Wise & Koob, 2014) suggests that addiction consists of three stages: the initial stages of binge intoxication, followed by preoccupation (maintenance) and finally withdrawal and/or negative affect. It is therefore conceivable that different neurotransmitters and different receptor types might be involved in different aspects of addiction. For further reading on this, I suggest looking to Chapter 5 in this book, which has a focus on dopamine and reward (Wise & Koob, 2014; see also Kenny & Markou, 2004).

The reality is that addiction to drugs of abuse probably recruits multiple neurotransmitter systems. Therefore, as ever, the focus on a small family of receptors, such as metabotropic glutamate receptors, is unlikely to be the full story, and we must always remember that this research needs to be viewed as a part to the jigsaw rather than the whole jigsaw.

Glutamate and anxiety disorders

As a category of disorders, anxiety disorders are one of the most diagnosed in the western world and include various subtypes, such as generalised anxiety disorder (GAD) and post-traumatic stress disorder (PTSD). It is important to remember that a certain amount of anxiety is adaptive and can help you meet the challenges of a constantly changing world. However, in its disease states, it does the very opposite and reduces the ability to cope with the world, including reducing the ability to hold a job and sustain a 'normal' family life. Furthermore, those diagnosed with an anxiety disorder are far more likely to suffer from other biological illnesses, such as hypertension, and there is also a significant comorbidity with psychiatric disorders, such as the affective disorders.

Anxiety disorders are another set of disorders where alteration in E/I balance is considered a likely underlying biological pathology in their aetiology. Because of this, it makes sense that both glutamate and GABA alterations would be naturally implicated. Along these lines, multiple different drugs have been developed as anxiolytic agents, with most of the well-known ones, such as benzodiazepines (BZ – street name PAMS (positive allosteric modulators) and one of the most heavily prescribed) agonising GABA function in some way, either being specific receptor agonists or altering reuptake. In the case of PAMS, these act by increasing the action of the GABA$_a$ receptors. However, the focus of this section is glutamate and, more recently, numerous anxiolytic agents that target the glutamate system have been identified, with some being agonists and others being antagonists for specific types of glutamate receptors. The literature generally seems to suggest that specific antagonists at NMDA receptors seem to be effective anxiolytics. For example, Fendt, Koch, and Schnitzler (1996) showed that the NMDA-specific antagonists AP5 reduced the fear startle response in rodents. Further evidence to support these conclusions come from the generally consistent findings that NMDA agonists increase anxiety reponses. For example, Ho et al. (2005) found that a specific NMDA agonist reduced the amount of time animals spent in the open arms of the elevated plus maze, which is a well-used measure of 'anxiety', with less time suggesting more anxiety.

However, research focused on the role of metabotropic glutamate receptors seems to suggest that it is agonists at these sites that have anxiolytic effects. For example, Stachowicz et al.'s (2009) pharmacological experiments suggest that group 3 metabotropic glutamate receptor agonists reduce anxiety reponses measured in various ways, including the use of the elevated plus maze. The fact the animals show reduced anxiety in multiple experimental paradigms brings a degree of validity to the study, which is further increased by the fact affective responses (measured by the use of the tail suspension task) were not altered by the metabotropic glutamate receptor agonists. However, an important point here is that this research, and others, points towards metabotropic glutamate receptor agonists having anxiolytic effects by modulating the release of other neurotransmitters, specifically GABA and serotonin. This highlights one of the most important points we will return to throughout this book, namely that neurotransmitters and their machinery do not work in isolation.

An important point to consider, with research using animal models of anxiety, is the techniques that are employed to induce anxiety and/or measure anxiety in the animals. There are several commonly used methods to do this, including:

1. Chronic immobility paradigms, which use restraint to induce anxiety.
2. Conflict/punishment paradigms, which pair up a reward with a naturally aversive stimuli, such as a foot shock, which pits the animal's drive towards the reward with the pain of the shock, or sleep deprivation, which has been shown to increase anxiety levels.

3. The elevated plus maze, which is a simple maze with two open and two closed arms. Researchers measure the number of times the animal enters the closed and open arms as a metric of anxiety. This is based on the principle that small rodents are naturally anxious about open spaces and the increased risk of predation.

There are many studies using these methods, some of which are identified earlier in this chapter. The issue is whether they are really a measure of anxiety or something else? It is difficult to be definitive with an answer to this question, but many researchers consider they are a measure of anxiety, but also likely measures of stress responses and affective responses.

The glutamate precursor glutamine is an interesting target for much research, especially in the area of depression, which, as previously mentioned, is commonly comorbid with a diagnosis of anxiety disorder. The reason glutamine is such a target for research is that its increased levels, relative to the level of glutamate, have been shown to contribute to the excitotoxicity. It may therefore play a key role in the biological alterations seen in those with an anxiety disorder diagnosis. There is a significant weight of evidence that suggests glutamine levels are much higher in those diagnosed with depression, and evidence to suggest this is also the case in those diagnosed with specific anxiety disorders. For example, Pollack et al. (2008) using brain scanning techniques in humans diagnosed with social anxiety disorder and found an increased level of both glutamate and glutamine, with glutamine being specifically high in key brain regions associated with anxiety, such as the thalamus. What was particularly interesting about this study was the GABA levels were not found to be altered, suggesting that, at least for specific types of anxiety, the glutamate/glutamine system is a better target for anxiolytic drugs.

The study by Pollock et al. (2008) focuses on the specific nnominy disorder: social anxiety disorder. The type of anxiety disorder focused on is an important issue to consider in the research literature. Sometimes, for the sake of simplicity, anxiety disorders are treated as though they are one homogeneous group, but it is perfectly conceivable that different anxiety disorders have a different aetiology. If you think for a moment about the fact that altered E/I balance is believed to be the 'cause' of anxiety disorders. This may well be true, but E/I balance can be altered by numerous different biological mechanisms. For example, it could be reduction in GABAergic interneuron function, or it could be increases in glutamatergic transmission, or it could be a third factor, such as the decrease in glia function, which result in alteration in E/I balance. This means, while alterations in E/I balance may underpin all anxiety disorders, the biological mechanism may be different between the different subtypes. This is even more important when we consider animal research as it is seldom as nuanced as considering the different 'types' of anxiety, mainly reporting the effects on 'generalised' anxiety in the animals.

Conclusion

Glutamate is one of the big two neurotransmitters, expressed and released widely across the brain. As such, alterations in its expression, and particularly in the responsiveness of its receptors, of which there are many, play a key role in modulating complex behaviours. However, we cannot ignore the role that glutamate may play in these behaviours as a complex interaction with other neurotransmitters. Indeed, it seems likely that many of the alterations in glutamate may ultimately be as a result of alterations in other neurotransmitters, or result in alterations in other neurotransmitters, which ultimately combine to produce changes in our behavioural states. Indeed, as mentioned in this chapter, glutamate, and its machinery, may ultimately have its main effects by altering E/I balance. In the next chapter, we will move our attention to the second of the two 'big' neurotransmitters, which is the other key player in E/I balance, namely GABA.

Test Yourself Answers

Test Yourself 2.1

1. Glutaminase
2. Glutamine synthase
3. Glutamine
4. EAATs (excitatory amino acid transporters)

Test Yourself 2.2

Table 2.2

Ionotropic	Metabotropic
	Slowly reversible
	GPCRs
Short latency	
	Long latency
	Secondary messengers involved
	cAMP
They are ion channels	
	They are not directly connected to ion channels
Rapidly reversible	

(Continued)

Test Yourself 2.3

a.   Glutamate binding with the postsynaptic NMDA receptor will produce a slow excitatory current. As the glutamate level builds up, it will bind with the mGluR on the presynaptic neuron and produce a slow inhibition on the presynaptic neuron, resulting in a reduction in further release of glutamate from this neuron.

b.   Glutamate binding with the postsynaptic AMPA receptor will produce a fast excitatory current, while glutamate binding with the postsynaptic NMDA receptor will produce a slow excitatory current. In total, this will last approximately 125 ms. As the glutamate level builds up, it will bind with the mGluR on the presynaptic neuron and produce and excitatory response, facilitating further glutamate release from the presynaptic neuron.

References

Key information

Fonnum, F. (1984). Glutamate: a neurotransmitter in mammalian brain. *J Neurochem*, *42*(1), 1–11. doi:10.1111/j.1471-4159.1984.tb09689.x

Martineau, M., Guzman, R. E., Fahlke, C., & Klingauf, J. (2017). VGLUT1 functions as a glutamate/proton exchanger with chloride channel activity in hippocampal glutamatergic synapses. *Nature Communications*, *8*(1), 2279. doi:10.1038/s41467-017-02367-6

Amilhon, B., Lepicard, E., Renoir, T., Mongeau, R., Popa, D., Poirel, O., . . . El Mestikawy, S. (2010). VGLUT3 (vesicular glutamate transporter type 3) contribution to the regulation of serotonergic transmission and anxiety. *J Neurosci*, *30*(6), 2198–2210. doi:10.1523/JNEUROSCI.5196-09.2010

Magi, S., Piccirillo, S., Amoroso, S., & Lariccia, V. (2019). Excitatory amino acid transporters (EAATs): Glutamate transport and beyond. *International journal of molecular sciences*, *20*(22). doi:10.3390/ijms20225674

Magi, S., Piccirillo, S., & Amoroso, S. (2019). The dual face of glutamate: from a neurotoxin to a potential survival factor-metabolic implications in health and disease. *Cell Mol Life Sci*, *76*(8), 1473–1488. doi:10.1007/s00018-018-3002-x

Ryan, R., & Boudker, O. (2013). Glutamate transporter family. In G. C. K. Roberts (Ed.), *Encyclopedia of Biophysics* (pp. 893–900). Berlin, Heidelberg: Springer Berlin Heidelberg.

Rowley, N. M., Madsen, K. K., Schousboe, A., & Steve White, H. (2012). Glutamate and GABA synthesis, release, transport and metabolism as targets for seizure control. *Neurochem Int*, *61*(4), 546–558. doi:10.1016/j.neuint.2012.02.013

Leke, R., & Schousboe, A. (2016). The glutamine transporters and their role in the glutamate/GABA-glutamine cycle. *Adv Neurobiol, 13*, 223–257. doi:10.1007/978-3-319-45096-4_8

Lüscher, C., & Malenka, R. C. (2012). NMDA receptor-dependent long-term potentiation and long-term depression (LTP/LTD). *Cold Spring Harbor perspectives in biology, 4*(6), a005710. doi:10.1101/cshperspect.a005710

Lee, C. J., Labrakakis, C., Joseph, D. J., & Macdermott, A. B. (2004). Functional similarities and differences of AMPA and kainate receptors expressed by cultured rat sensory neurons. *Neuroscience, 129*(1), 35–48. doi:10.1016/j.neuroscience.2004.07.015

Nicoletti, F., Bockaert, J., Collingridge, G. L., Conn, P. J., Ferraguti, F., Schoepp, D. D., . . . Pin, J. P. (2011). Metabotropic glutamate receptors: From the workbench to the bedside. *Neuropharmacology, 60*(0), 1017–1041. doi:10.1016/j.neuropharm.2010.10.022

Glutamate and epilepsy

Van Liefferinge, J., Massie, A., Portelli, J., Di Giovanni, G., & Smolders, I. (2013). Are vesicular neurotransmitter transporters potential treatment targets for temporal lobe epilepsy? *Front Cell Neurosci, 7*(139). doi:10.3389/fncel.2013.00139

Barker-Haliski, M., & White, H. S. (2015). Glutamatergic mechanisms associated with seizures and epilepsy. *Cold Spring Harbor Perspectives in Medicine, 5*(8), a022863–a022863. doi:10.1101/cshperspect.a022863

Cho, C.-H. (2013). New mechanism for glutamate hypothesis in epilepsy. *Front Cell Neurosci, 7*, 127–127. doi:10.3389/fncel.2013.00127

Umpierre, A. D., West, P. J., White, J. A., & Wilcox, K. S. (2019). Conditional knock-out of mGluR5 from astrocytes during epilepsy development impairs high-frequency glutamate uptake. *J Neurosci, 39*(4), 727–742. doi:10.1523/jneurosci.1148-18.2018

Crino, P. B., Jin, H., Shumate, M. D., Robinson, M. B., Coulter, D. A., & Brooks-Kayal, A. R. (2002). Increased expression of the neuronal glutamate transporter (EAAT3/EAAC1) in hippocampal and neocortical epilepsy. *Epilepsia, 43*(3), 211–218. doi:10.1046/j.1528-1157.2002.35001.x

Chapman, A. G. (2000). Glutamate and epilepsy. *The Journal of Nutrition, 130*(4), 1043S–1045S. doi:10.1093/jn/130.4.1043S

Chapman, A. G., Woodburn, V. L., Woodruff, G. N., & Meldrum, B. S. (1996). Anticonvulsant effect of reduced NMDA receptor expression in audiogenic DBA/2 mice. *Epilepsy Res, 26*(1), 25–35. doi:10.1016/s0920-1211(96)00036-8

Hanada, T. (2020). Ionotropic glutamate receptors in epilepsy: A review focusing on AMPA and NMDA receptors. *Biomolecules, 10*(3), 464. doi:10.3390/biom10030464

Schallier, A., Massie, A., Loyens, E., Moechars, D., Drinkenburg, W., Michotte, Y., & Smolders, I. (2009). vGLUT2 heterozygous mice show more susceptibility to clonic seizures induced by pentylenetetrazol. *Neurochemistry International, 55*(1), 41–44. doi:https://doi.org/10.1016/j.neuint.2008.12.019

Tanaka, K., Watase, K., Manabe, T., Yamada, K., Watanabe, M., Takahashi, K., . . . Wada, K. (1997). Epilepsy and exacerbation of brain injury in mice lacking the glutamate transporter GLT-1. *Science, 276*(5319), 1699–1702. doi:10.1126/science.276.5319.1699

Tian, G.-F., Azmi, H., Takano, T., Xu, Q., Peng, W., Lin, J., . . . Nedergaard, M. (2005). An astrocytic basis of epilepsy. *Nat Med, 11*(9), 973–981. doi:10.1038/nm1277

Woo, D. H., Han, K.-S., Shim, J., Yoon, B.-E., Kim, C., Bae, J., . . . Lee, C. J. (2012). TREK-1 and best1 channels mediate fast and slow glutamate release in astrocytes upon GPCR activation. *Cell, 151*, 25–40. doi:10.1016/j.cell.2012.09.005

Glutamate and addiction

Kenny, P. J., Boutrel, B., Gasparini, F., Koob, G. F., & Markou, A. (2005). Metabotropic glutamate 5 receptor blockade may attenuate cocaine self-administration by decreasing brain reward function in rats. *Psychopharmacology (Berl), 179*(1), 247–254. doi:10.1007/s00213-004-2069-2

Kenny, P. J., Gasparini, F., & Markou, A. (2003). Group II metabotropic and alpha-amino-3-hydroxy-5-methyl-4-isoxazole propionate (AMPA)/kainate glutamate receptors regulate the deficit in brain reward function associated with nicotine withdrawal in rats. *J Pharmacol Exp Ther, 306*(3), 1068–1076. doi:10.1124/jpet.103.052027

Kenny, P. J., & Markou, A. (2004). The ups and downs of addiction: role of metabotropic glutamate receptors. *Trends Pharmacol Sci, 25*(5), 265–272. doi:10.1016/j.tips.2004.03.009

Olds, J., & Milner, P. (1954). Positive reinforcement produced by electrical stimulation of septal area and other regions of rat brain. *Journal of Comparative and Physiological Psychology, 47*(6), 419–427. doi:10.1037/h0058775

Bespalov, A., Dumpis, M., Piotrovsky, L., & Zvartau, E. (1994). Excitatory amino acid receptor antagonist kynurenic acid attenuates rewarding potential of morphine. *European Journal of Pharmacology, 264*(3), 233–239. doi:https://doi.org/10.1016/0014-2999(94)00462-5

Bespalov, A. Y., Dravolina, O. A., Sukhanov, I., Zakharova, E., Blokhina, E., Zvartau, E., . . . Markou, A. (2005). Metabotropic glutamate receptor (mGluR5) antagonist MPEP attenuated cue- and schedule-induced reinstatement of nicotine self-administration behavior in rats. *Neuropharmacology*, *49* Suppl 1, 167–178. doi:10.1016/j.neuropharm.2005.06.007

Chiamulera, C., Epping-Jordan, M. P., Zocchi, A., Marcon, C., Cottiny, C., Tacconi, S., . . . Conquet, F. (2001). Reinforcing and locomotor stimulant effects of cocaine are absent in mGluR5 null mutant mice. *Nat Neurosci*, *4*(9), 873–874. doi:10.1038/nn0901-873

Fowler, M., Varnell, A., & Cooper, D. (2011). mGluR5 knockout mice exhibit normal conditioned place-preference to cocaine. *Nature Precedings*. doi:10.1038/npre.2011.6180.1

Paterson, N. E., & Markou, A. (2005). The metabotropic glutamate receptor 5 antagonist MPEP decreased break points for nicotine, cocaine and food in rats. *Psychopharmacology*, *179*(1), 255–261. doi:10.1007/s00213-004-2070-9

Bäckström, P., Bachteler, D., Koch, S., Hyytiä, P., & Spanagel, R. (2004). mGluR5 antagonist MPEP reduces ethanol-seeking and relapse behavior. *Neuropsychopharmacology*, *29*(5), 921–928. doi:10.1038/sj.npp.1300381

Taber, M. T., & Fibiger, H. C. (1995). Electrical stimulation of the prefrontal cortex increases dopamine release in the nucleus accumbens of the rat: modulation by metabotropic glutamate receptors. *The Journal of Neuroscience*, *15*(5), 3896. doi:10.1523/JNEUROSCI.15-05-03896.1995

Harrison, A. A., Gasparini, F., & Markou, A. (2002). Nicotine potentiation of brain stimulation reward reversed by DH beta E and SCH 23390, but not by eticlopride, LY 314582 or MPEP in rats. *Psychopharmacology (Berl)*, *160*(1), 56–66. doi:10.1007/s00213-001-0953-6

Wise, R. A., & Koob, G. F. (2014a). The development and maintenance of drug addiction. *Neuropsychopharmacology*, *39*(2), 254–262. doi:10.1038/npp.2013.261

Glutamate and anxiety disorders

Pollack, M. H., Jensen, J. E., Simon, N. M., Kaufman, R. E., & Renshaw, P. F. (2008). High-field MRS study of GABA, glutamate and glutamine in social anxiety disorder: Response to treatment with levetiracetam. *Progress in Neuro-Psychopharmacology and Biological Psychiatry*, *32*(3), 739–743. doi:https://doi.org/10.1016/j.pnpbp.2007.11.023

Hasler, G., Buchmann, A., Haynes, M., Müller, S. T., Ghisleni, C., Brechbühl, S., & Tuura, R. (2019). Association between prefrontal glutamine levels and neuroticism determined using proton magnetic resonance spectroscopy. *Translational Psychiatry*, *9*(1), 170. doi:10.1038/s41398-019-0500-z

Stachowicz, K., Kłodzińska, A., Palucha-Poniewiera, A., Schann, S., Neuville, P., & Pilc, A. (2009). The group III mGlu receptor agonist ACPT-I exerts anxiolytic-like but not antidepressant-like effects, mediated by the serotonergic and GABA-ergic systems. *Neuropharmacology*, *57*(3), 227–234. doi:10.1016/j.neuropharm.2009.06.005

Fendt, M., Koch, M., & Schnitzler, H. U. (1996). NMDA receptors in the pontine brainstem are necessary for fear potentiation of the startle response. *Eur J Pharmacol*, *318*(1), 1–6. doi:10.1016/s0014-2999(96)00749-2

Ho, Y. J., Hsu, L. S., Wang, C. F., Hsu, W. Y., Lai, T. J., Hsu, C. C., & Tsai, Y. F. (2005). Behavioral effects of d-cycloserine in rats: the role of anxiety level. *Brain Res*, *1043*(1–2), 179–185. doi:10.1016/j.brainres.2005.02.057

Wierońska, J. M., Stachowicz, K., Pałucha-Poniewiera, A., Acher, F., Brański, P., & Pilc, A. (2010). Metabotropic glutamate receptor 4 novel agonist LSP1-2111 with anxiolytic, but not antidepressant-like activity, mediated by serotonergic and GABAergic systems. *Neuropharmacology*, *59*(7–8), 627–634. doi:10.1016/j.neuropharm.2010.08.008

Cortese, B. M., & Phan, K. L. (2005). The role of glutamate in anxiety and related disorders. *CNS Spectr*, *10*(10), 820–830. doi:10.1017/s1092852900010427

Linden, A. M., Shannon, H., Baez, M., Yu, J. L., Koester, A., & Schoepp, D. D. (2005). Anxiolytic-like activity of the mGLU2/3 receptor agonist LY354740 in the elevated plus maze test is disrupted in metabotropic glutamate receptor 2 and 3 knock-out mice. *Psychopharmacology (Berl)*, *179*(1), 284–291. doi:10.1007/s00213-004-2098-x

Ramos-Prats, A., Kölldorfer, J., Paolo, E., Zeidler, M., Schmid, G., & Ferraguti, F. (2019). An appraisal of the influence of the metabotropic glutamate 5 (mGlu5) receptor on sociability and anxiety. *Front Mol Neurosci*, *12*, 30. Retrieved from www.frontiersin.org/article/10.3389/fnmol.2019.00030 www.ncbi.nlm.nih.gov/pmc/articles/PMC6401637/pdf/fnmol-12-00030.pdf

Ferraguti, F. (2018). Metabotropic glutamate receptors as targets for novel anxiolytics. *Curr Opin Pharmacol*, *38*, 37–42. doi:10.1016/j.coph.2018.02.004

3

GABA (GAMMA AMINOBUTYRIC ACID)

Introduction

Along with glutamate, covered in Chapter 2, GABA (gamma aminobutyric acid) is one of the most heavily expressed neurotransmitters in the brain; in fact, it is more concentrated in the mammalian brain than it is in any other mammalian tissue. To give a comparative example of how heavily expressed it is, GABA concentrations are thought to be approximately 1,000 times higher in the human brain than monoaminergic neurotransmitters, such as serotonin and/or dopamine. GABA is both heavily expressed and diffusely expressed in the brain. It is also expressed in a variety of different neuron types, including projection neurons and interneurons. Because of its diffuse expression within the brain, like glutamate, it is best to think about it as we discussed glutamate, namely that any changes in its expression are likely to have global effects on brain function and the organism's behaviour. However, its receptor subtypes are expressed heterogeneously throughout the brain. These GABA receptors are also expressed at various locations on different types of neurons. What this means is that the effect alterations in GABA on brain function and the organism's behaviour are likely to be mediated by alterations in specific subtypes of these receptors.

GABA is generally considered to be the main inhibitory neurotransmitter in the brain. This is generally true apart from under specific intracellular conditions, which we will return to when we discuss GABA receptors, later in this chapter. This inhibitory effect of GABA has a profound effect on controlling the level of excitatory transmission in the brain. Many consider GABAergic neurons to act as local brakes at various points in the brain's many neuronal circuits, ultimately controlling the flow of information and restraining the excitability of neuronal networks.

One emerging theory of brain function/dysfunction is that of E/I balance. This relates to the general level of excitation and inhibition (E/I) within the brain, with brain activity maintained within a narrow parameter, which we might say is balanced. Common belief is that many atypical behavioural states are characterised by alteration in this E/I balance. The resulting imbalance is typically characterised by reduced inhibition and subsequent increase in excitatory neuronal discharge. Much of the evidence for this emerging theory of atypical behavioural states, therefore, implicates alterations in GABA and its underpinning machinery. It is therefore an exciting time for GABA research, as understanding the machinery that underpins its synthesis, packaging, reuptake and receptors can, perhaps, be viewed as more important than ever.

Outline of GABA's synthesis, packaging and reuptake

GABA has many interesting differences in its machinery compared to other neurotransmitters that we will cover in this book. One of the interesting starting points is that its synthesis is inherently intertwined with that of glutamate.

GABA synthesis

GABA is synthesised from glutamate in a process typically referred to as the GABA shunt. This process, like glutamate synthesis, partially involves astrocytes and the TCA (tricarboxcylic acid cycle). As part of reuptake and degradation, GABA is taken up by astrocytes and fed into the TCA cycle, where it is converted to succinic semialdehyde (SSA) by the enzyme GABA-T. This then goes through several other stages of the TCA before the formation of glutamine. The story should now be familiar as glutamine is transported out of the astrocyte and into the neuron for its final conversion, into glutamate, by the enzyme glutaminase.

The next step is probably best considered as the most important for GABA as it is the step that most limits the rate of synthesis. Once in the neuron terminal end, glutamate can now be converted into GABA by the enzyme glutamic acid decarboxylase (GAD). One important point surrounding GAD is the expression of two different types. One type is known as GAD65 and the other is GAD67. These two types are thought to have quite distinct modulatory roles. It is generally considered that GAD67 is involved in maintaining a basal (minimal) level of GABA and therefore contributes to GABAergic tone within the brain's neural networks. Indeed, genetically-modified animals where GAD67 has been 'knocked out' are not viable organisms. This provides quite stark evidence of how essential GAD67 is to the organism's 'normal' function.

GAD65, however, has a much more reactive role to play in GABA synthesis. Typically, GAD65 is found bound to the outside membrane of vesicles. Many think that this allows it to play a rapid role in upregulating and synthesising GABA when an increased amount of GABA is suddenly needed. This is likely done under circumstances when excitatory transmission is rising rapidly or is beginning to dominate, therefore GAD65 helps to rapidly upregulate GABA levels. This provides a key control mechanism for surges in excitatory transmission, ultimately bringing the neuronal circuits back into balance. In other words, it can be considered much like a rapid response force that can control excitatory transmission that is rapidly becoming 'out of control'. The altered function and expression of this enzyme (both GAD65 and GAD67) has been strongly linked to various atypical behaviour states. Later in this chapter we will explore this further in the context of epilepsy. For now, this is the end of the process of GABA synthesis and it is now ready for packaging into vesicles for transport out of the neuron and for the cycle to begin again.

—**Focus on research methods** —

Immunohistochemistry (IHC)

As an interesting side point, many researchers want to be able to visualise GABAergic neurons in order to investigate multiple factors, such as if their number are altered in various atypical behaviour states, among other things. In order to do this, a technique called
immunohistochemistry (IHC) is commonly used. This involves the use of antibodies that
specifically target certain proteins within the neuron. This is much like when your body's immune system produces antibodies that can recognise and attach to foreign bodies, such as
bacteria and viruses. Commonly, when researchers want to investigate GABA neurons,
they will expose a tissue sample to antibodies that are specific for GAD. The antibodies will
bind to GAD and you can then get a secondary antibody, which has a fluorescent marker
attached, which will bind to the antibodies specific to GAD. Once this has been done, you
will now be able to see the neurons, as in Figure 3.1, when it is placed under a specialist
fluorescent microscope.

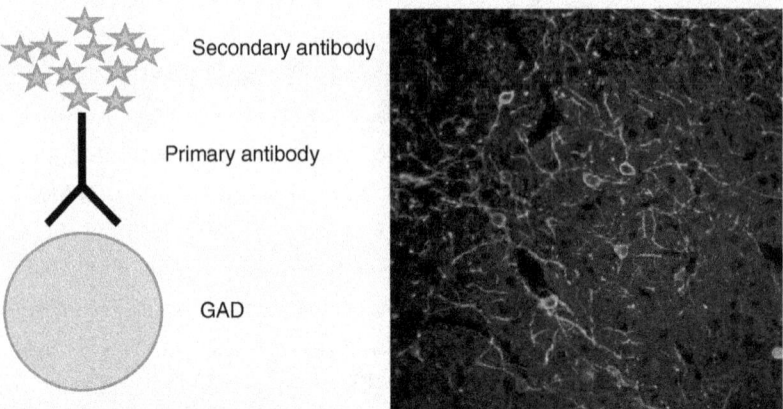

Secondary antibody

Primary antibody

GAD

Figure 3.1 A simplified schematic representing the binding of antibodies to a target
protein (GAD) using IHC techniques, and the resultant image visualised using a
fluorescent microscope

The figure illustrates how the antibodies bind to the targeted molecule. The primary antibody binds directly
to the targeted molecule, in this case the enzyme GAD, expressed within GABAergic neurons. The secondary
then binds to the primary antibody and builds a complex which can then be visualised down the microscope.
The Figureillustrates the results of such a protocol, with the secondary being fluorescent, requiring excitation
by a specific wavelength of light. In this case, the neurons that are fluorescing green are those that express
the GAD.

If you want to learn more about IHC, I recommend the following book: Polak and Van
Noorden (2003). This certainly is an important technique to learn if you aspire to move into
molecular neuroscience and laboratory-based work.

GABA packaging

Many of the neurotransmitters we will discuss in this book have more than one type of vesicular transporter that help to package the neurotransmitter into vesicles. GABA is unusual as it has only one type of vesicular transporter that is solely responsible for its packaging into vesicles. This vesicular transporter is called VGAT. VGAT is transcribed, in humans, from the SLC32 gene family; therefore, in your wider reading, it may be worth keeping an eye out for this gene as you can be sure any research focused on this gene will be related to VGAT levels/activity.

Another quality that is unusual about GABA is that it is what is known as a zwitterion. This means it carries no charge. As discussed in the introductory chapter, the charge of the molecule often determines whether the PH or the membrane potential is the main driving force allowing the vesicular transporter to bring the neurotransmitter into the vesicle. However, for zwitterions, both PH and membrane potential affect their transport. The membrane potential is set up in the typical manner you will come across throughout this book, and outlined in Chapter 1, namely via the action of the vesicle's proton pump bringing H^+ ions into the vesicular lumen.[1] PH differences between the vesicular lumen and the extra vesicular space is, however, modulated by a different process. The 'acidification' of the vesicle lumen is generally considered to be brought about by the transport of Cl^- ions into the lumen. This is done by a further voltage-gated vesicular membrane channel referred to as CLC-3. It is important to keep the role of Cl^- ions in mind when considering atypical states linked to altered GABA. Indeed, several genetically-modified animal models with reduced expression of CLC-3 channels are known to have an increased susceptibility to epilepsy. This may be linked, ultimately, to a chain of events that reduce GABA levels ready for release. It is also worth reflecting on how crowded the vesicle's membrane is getting, as there are VGATs, proton pumps, CLC-3 Cl^- channels and GAD65 expressed on and across the membrane. This serves to emphasise how complex all of these processes are and how much machinery is involved in the seemingly simple act of packaging neurotransmitters into vesicles.

The complexity does not stop there for GABA and its packaging. VGATs are also known to transport other neurotransmitters into the vesicle along with GABA. One example is the other commonly-occurring zwitterion, namely glycine. Recent research (Juge, Omote, & Moriyama, 2013) has provided convincing evidence to suggest these VGATs further contribute to the packaging of the neurotransmitter β-alanine. This is always worth remembering as it adds a layer of complexity to the

---

[1]The lumen is the space inside the vesicle.

findings and conclusions drawn from any research focused on VGATs and GABA. In your wider reading, it will serve you well to keep this in mind as any conclusions drawn regarding the impact of alterations in VGATs and GABA, and consequently the function of the neuron or whole organism, must also consider how alterations to VGATs' function may impact on the levels of glycine and β-alanine, and ultimately how alterations in their levels may contribute to the behavioural state.

Further complexity to this story comes when we note that the transport of these other neurotransmitters into vesicles is not as dependent on both PH and membrane potential as the transport of GABA. For example, only membrane potential seems to be

Test Yourself 3.1

Can you identify what is the VGAT, the proton pump and what is the CLC-3 channel in the Figure 3.2?

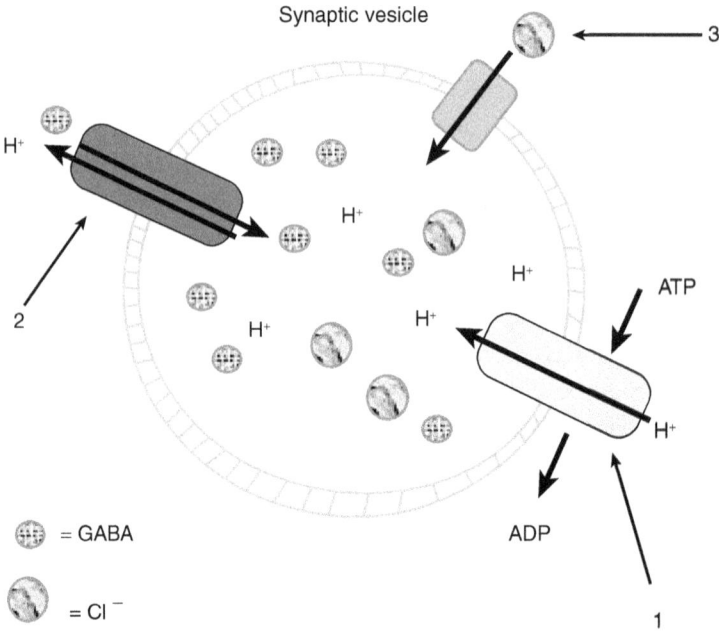

Figure 3.2   Illustration of a vesicle packaging the neurotransmitter GABA

This figure illustrates the many mechanisms that modulate the state of vesicles and set up the right conditions for GABA to be transported into the vesicle.

Source: Adapted from: Chaudhry, F. A., Edwards, R. H., & Fonnum, F. (2008). Vesicular neurotransmitter transporters as targets for endogenous and exogenous toxic substances. *Annual Review of Pharmacology and Toxicology, 48*(1), 277–301. Doi:10.1146/annurev.pharmtox.46.120604.141146

Answers to all Test Yourself questions are at the end of the chapter.

essential for the transport of β-alanine into the vesicle lumen. It also seems that this is at least partially aided by the transport of Cl⁻ ions. As for glycine, the fact that glycine is packaged into vesicles and is released alongside GABA serves to illustrate the complex intertwined relationship between GABA and glutamate. Glycine itself is known to have inhibitory effects when it is released and binds with the glycine receptor, although, as mentioned in Chapter 2 on glutamate, glycine is a key co-factor in the activation of glutamatergic NMDA receptors. Therefore, as well as GABA release having generally inhibitory effects, its release can also result in regulation of excitatory glutamate channels via glycine. This illustrates the complexity of the relationship between various neurotransmitters and the inherently intertwined bond between the machinery underpinning GABA and that of glutamate. Perhaps, in some respects, it is better to think of these two neurotransmitters (including their co-released neurotransmitters, such as glycine) and their respective machinery to be key players on the same team, ultimately both aiming to work in harmony to bring a balance to the brain's neuronal circuitry.

Outline of GABA receptors

After the complexity of GABA packaging into vesicles, it might be good to hear that GABA receptors are, relatively, more straightforward. However, there are some unusual complexities that we will discuss in this section.

First, GABA receptors are generally split into three types, appropriately named GABA$_a$, GABA$_b$ and GABA$_c$ receptors. Two of these are ionotropic receptors, namely GABA$_a$ and GABA$_c$, while GABA$_b$ receptors are metabotropic.

We will start by exploring the ionotropic receptor GABA$_a$. We will not explore the GABA$_c$ in this book as it is largely believed to be expressed in the visual system, and therefore, to some degree, is beyond the brain and subsequently beyond the focus of this book. GABA$_a$ is a 'typical' ionotropic receptor in the sense that it is a channel protein and binding of the neurotransmitter causes a rapid conformational shape change, which opens the channel pore and allows Cl⁻ ions to flow in. This produces an IPSC (inhibitory postsynaptic current) in the target neuron. Like most ionotropic receptors, this response has a short latency and the receptors rapidly become desensitised. Up to now this is straightforward. However, the complexity first comes when you consider how these ionotropic receptors are organised. Ionotropic GABA$_a$ receptors are typically made up of five subunits: two α subunits, two β subunits and a γ subunit (see Figure 3.3). These subunits have various isoforms, as suggested by the number after each, for the most commonly expressed type above. The α subunit has 1–6, while the β is 1–4 and the γ is 1–3. These can be arranged in various combinations. The most commonly expressed type of GABA$_a$ receptor is constructed of two $\alpha 1$ subunits, two $\beta 2$ subunits and a $\gamma 1$ subunit (because of this, $GABA_a$ receptors are referred to as a pentameric protein complex). This results in many different types of $GABA_a$ receptor. These different types, based on the combination of subunit isoforms, results in lots of different functional implications with distinct different opening and closing dynamics as well as different levels of affinity for GABA and other molecules that are known to bind to these receptors. To clarify, it is important to point out that there are several other subunits that $GABA_a$ receptor can be comprised of, but I will leave this to your wider reading.

Further complexity is added to $GABA_a$ receptors when you consider that GABA is not the only molecule that binds to these receptors; in fact, many molecules are known to bind to $GABA_a$ receptors and act as either allosteric (alongside) agonists or antagonists. Each of these agonists and antagonists have binding sites away from the main binding site.[2] These agonists and antagonists cause conformational shape changes in the tertiary structure of the channel protein, resulting in alterations to the affinity of the receptor for GABA at the main binding sites. They also make alterations to the channel itself, resulting in changes to the size of the channel and the subsequent ability of Cl ions to flow through this channel. Although the vast majority of receptors have allosteric binding sites, $GABA_a$ receptors have a high number. As a result, they have been extensively studied as targets for various pharmacological interventions aimed at altering the effects of GABA. One good example of an allosteric agonist[3] is the benzodiazepines. This group of compounds is used extensively for the treatment of anxiety-related disorders. They have a specific allosteric binding site on the $GABA_a$ receptor, located between the α and γ subunit (see Figure 3.3). As mentioned previously, there are many different isoforms of the $GABA_a$ receptor based on the expression of these subunits and the types of these subunits. This is important to remember because as benzodiazepines rely on the α subunit type to bind to $GABA_a$, not all $GABA_a$ receptors will have an allosteric binding site for benzodiazepines. When $GABA_a$ receptors do express an allosteric binding site for benzodiazepines, they are known to act as strong allosteric agonists. Two well-known examples of benzodiazepines are Librium and Valium. Of course, there are also numerous molecules that act as allosteric antagonists,[4] binding to the $GABA_a$ receptor and causing conformational shape changes in the receptor, resulting in a reduced ability for GABA to bind or reduced opening of the channel/the ability of Cl⁻ ions to flow through the channel and into the neuron. One example of an allosteric antagonist is Flumezanil. This is used to treat people who have

---

[2]The main binding site is that reserved for the neurotransmitter. In this case, GABA.

[3]These allosteric agonists are commonly referred to as PAMs (positive allosteric modulators).

[4]These allosteric antagonists are commonly referred to as NAMs (negative allosteric modulators).

overdosed on benzodiazepines or, in some instances, is used to inhibit general anaesthesia after surgery. It works by being a competitive antagonist at the benzodiazepine site, blocking the effects of benzodiazepine and, ultimately, not causing the same changes to the protein structure as benzodiazepine binding would. Unlike Flumezanil, other types of allosteric antagonists have their own specific binding sites. Again, many of these allosteric antagonists are only expressed on certain subtypes of GABA$_a$ receptors and it commonly depends on the type of α unit expressed by the GABA$_a$ isoform, of which there are 6, as we discussed above.

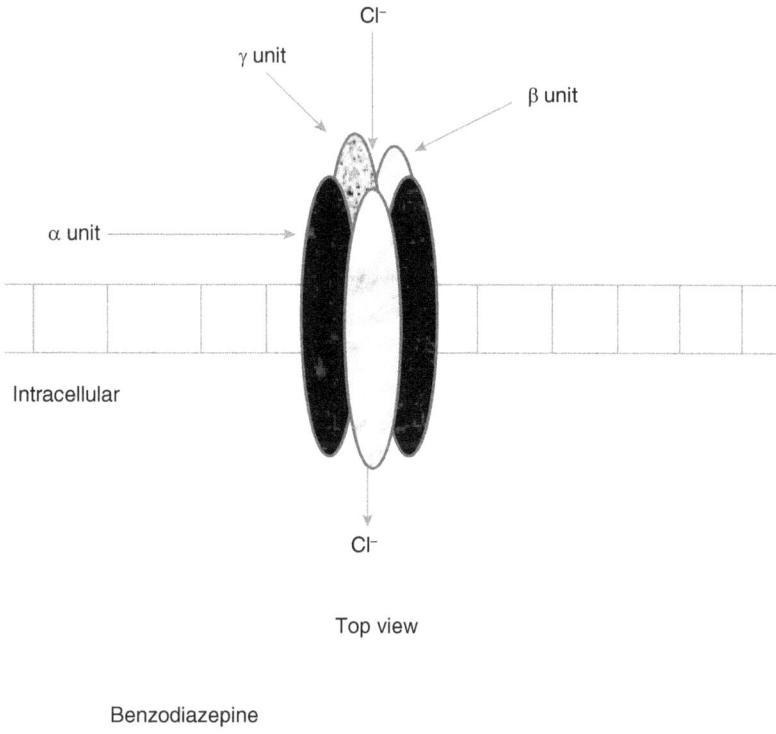

Top view

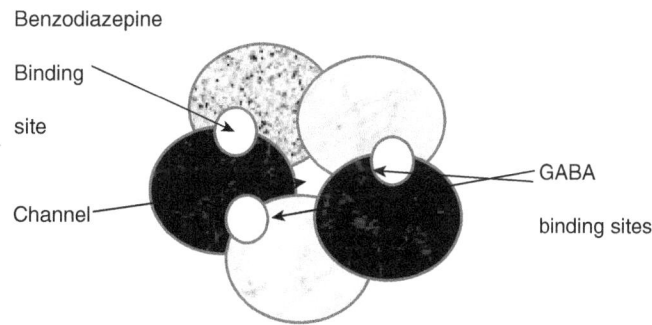

Figure 3.3 Side and top view of a GABA$_a$ receptor inserted in the neuronal plasma membrane

There are two distinct binding sites for GABA, but also a number of other binding sites for molecules, such as benzodiazepines.

As well as GABA$_a$ receptors having a number of allosteric binding sites, they also have two main binding sites for GABA (see Figure 3.3). This has functional implications, as when GABA concentrations are high, GABA binds to both sites, resulting in a prolonged opening of the pore and a subsequently increased flow of Cl⁻ ions into the neuron, resulting in a sustained IPSC. Although it would be logical to think that this would produce a sustained hyperpolarisation, under certain conditions when the Cl⁻ levels are already high within the neuron, this can actually result in a depolarising current and the opening of Ca^{2+} channels. This is the only circumstance when GABA will have an excitatory effect on the neuron and it is best to think of this as very much the exception, rather than the rule.

Now we have covered the apparently 'simple' ionotropic GABA$_a$ receptor, we will turn our attention to the metabotropic GABA$_b$ receptor. Unlike GABA$_a$ receptors, where under certain, exceptional, intracellular conditions the binding of GABA can cause depolarising currents (as mentioned above), GABA binding at GABA$_b$ receptors always has inhibitory effects on the neuron. GABA$_b$ receptors are what are commonly referred to as a functional heterodimer[5] as they are constructed of two key subunits, GABA$_{B1}$ and GABA$_{B2}$. The GABA$_{B1}$ subunit includes the binding site for GABA, while the GABA$_{B2}$ subunit is associated with the intracellular-bound G proteins. The GABA$_{B1}$ subunit has two isoforms GABA$_{B1a}$ and GABA$_{B1b}$, resulting in two structurally different types of GABA$_b$ receptor. Like the majority of GPCRs (G protein coupled receptors), the cellular responses they mediate are slow, when compared to those mediated by ionotropic receptors. It is generally considered that activation of GABA$_B$ receptors is key to maintaining an inhibitory tone in neuronal circuits, therefore supressing any possible spontaneous neuronal discharges. Like other metabotropic receptors, when the neurotransmitter binds there is aconformational shape change in the receptor. This results in the dissociation of the bound G proteins on the intracellular domain of the receptor. In the case of GABA$_b$ this results in the αi/o subunit dissociating from the $\beta\gamma$ complex (see Figure 3.4). Now the αi/o subunit moves through the intracellular domain of the plasma membrane where it associates with and deactivates the enzyme adenylate cyclase. The ultimate result of this is a reduction in the intracellular levels of the secondary messenger cyclic adenosine monophosphate (cAMP). This has numerous effects on the cell; one of the key effects is to downregulate protein kinase A (PKA), which is essential for the phosphorylation of a number of enzymes essential for cellular metabolism. Therefore, one

---

[5]A functional heterodimer is a protein composed of two distinct polypeptide chains that combine. Their combination results in functional alterations compared to the two polypeptide chains separately.

of the consequences of GABA binding to GABA$_b$ receptors is the reduction in cellular metabolism. This ultimately reduces the energy necessary to power many processes in the cell. Although the $\alpha i/o$ of the G protein is obviously a key player in enacting the effects of GABA binding to GABA$_b$ receptors, the $\beta\gamma$ complex also has an equally important role to play. Once the $\alpha i/o$ subunit has dissociated, the $\beta\gamma$ complex moves around the intracellular aspect of the plasma membrane and interacts with a number of different effector proteins, such as channel proteins. Two key channel proteins the $\beta\gamma$ complex modulates are G protein-modulated inwardly rectifying potassium channels (GIRKs) channels, which they activate, and Ca^{2+} channels, which they inhibit. Their activation of GIRK channels results in the extracellular flow of K$^+$, resulting in an inwardly rectifying current. This current, and these channels when activated, are known to suppress the backward propagation of action potentials and, ultimately, results in a hyperpolarised neuron, which is less likely to achieve the threshold required for action potential generation. The inhibition of Ca^{2+} channels results in the reduced likelihood of vesicular (if the GABA$_b$ receptor is presynaptic) fusion and the release of more GABA. However, if they are postsynaptic, then this reduces the ability of Ca^{2+} to act as a secondary messenger and has implications for the function of a number of enzymes, such as PKC and the calcium-binding proteins, such as calmodulin, which go on to modulate the Cam (calmodulin dependent) kinases.

Like the ionotropic GABA receptors, the metabotropic receptors also have sites for allosteric agonists and antagonists, the commonly known PAMs and NAMs. What is interesting about these is that the majority seem to have their effects by binding to the GABA$_{B2}$ subunit, specifically to its trans-membrane domain, as you can see in Figure 3.4. The majority of known allosteric agonists do not activate the receptor on their own but are generally considered to increase the affinity for the neurotransmitter at its binding site in the GABA$_{B1}$ subunit. One of the key allosteric modulators that seemingly has its effect at the GABA$_{B1}$ subunit is Ca^{2+}. Ca^{2+} has been shown to increase the affinity of GABA to the GABA$_b$ receptor. Therefore, high concentrations of extracellular Ca^{2+} further increases the inhibitory effects of GABA binding to GABA$_b$ receptors. If you can begin to join the dots here you will remember that one of the intracellular actions of GABA binding to GABA$_b$ receptors was the inhibition of Ca^{2+} channels by the $\beta\gamma$ complex of the dissociated G protein. As we discussed, this would reduce the intracellular levels of Ca^{2+} but subsequently increase the extracellular levels of Ca^2, which presumably would bind to the GABA$_{B1}$ subunit and increase its affinity for GABA, heightening the inhibitory effects of GABA binding further. This serves as an excellent illustration of how these processes are non-linear and the consequences of activating some pathways can often be modulation of the activation as well as downstream effects. It is perhaps best to think of all these effects like a rapidly running river. Most of the water flows towards the sea, but some at the edges gets caught up and causes whirlpools.

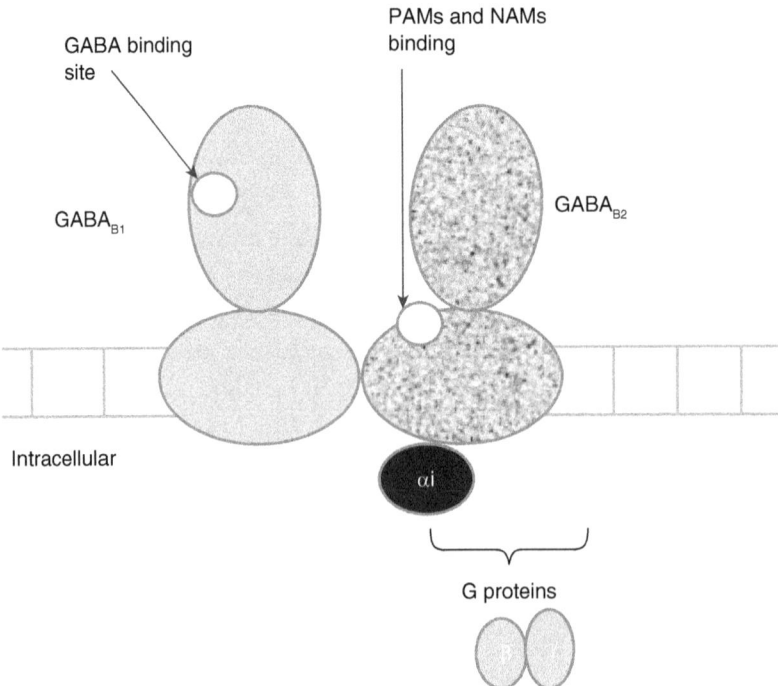

Figure 3.4   Side view of a GABA$_b$ receptor located in the neuron's plasma membrane

When GABA binds to the B1 binding site it results in confirmational shape changes which result in dissociation of the intracellular bound G proteins, located on the B2 subunit.

GABA reuptake

So, let's now consider the process of reuptake and the return to the GABA shunt, outlined earlier in the GABA synthesis section of this chapter. Like many other neurotransmitters, once GABA has dissociated from its target receptor it is taken back into the cell. As is the case with glutamate, GABA is taken up by astrocytes. However, unlike glutamate, a considerable amount is also taken back into the pre-synaptic neuron. GABA is typically taken into astrocytes when extracellular GABA is at very high levels. When this is the case, GABA is taken up into the astrocyte and fed into the TCA cycle to aid in cell metabolism. Ultimately the product of this is glutamic acid, which eventually is fed back into neurons. These neurons can be either GABAergic neurons, where glutamic acid is then converted into GABA by the enzyme GAD, or into glutamatergic neurons, where the glutamic acid is converted into glutamate by glutaminase. Again, this serves to illustrate a common theme within this chapter, namely the fate of GABA and glutamate is inherently intertwined, leading to their roles in excitatory/inhibitory balance, which we will discuss further, later in the chapter.

GABA transporters belong to the sodium symporter family. This means they share considerable functional similarity with other members of the symporter family, such as the monoamine reuptake transporters and the EAATs,[6] outlined in Chapter 2. GABA transporters function by transporting Na^+ and Cl^- in and out of the cell as well as GABA. For every molecule of GABA, two molecules of Na^+ and one molecule of Cl^- are transported. Because GABA is a zwitterion, this results in the influx of 1 net positive charge per cycle. This helps to set up a membrane potential, largely as Na^+ is transported down its concentration gradient and into the cell (see Figure 3.5). A key point to consider here is that GABA transporters can transport Na^+, Cl^- and GABA both into and out of the cell. This highlights their key role, which is to maintain the balance of GABA and its baseline concentration within the neuron.

In mice, there are generally six subtypes of these reuptake channels for GABA: GAT1, GAT2, GAT3, BGT1, CT1 and TauT. The nomenclature for GABA transporters can get rather confusing as it is different in humans, with each of these six being referred to as A followed by a number. For example, GAT1 is known as A1 in humans, but GAT2 is known as A13. The main two expressed in the brain and spinal cord are GAT1 (A1) and GAT3 (A13). These are both expressed on axon terminals and on astrocytes. GAT1 has a roughly 50/50 expression profile between astrocytes and axon terminals, whereas GAT3 is predominantly expressed on astrocytes. This is not to say that GAT2, BGT1 and the others are not expressed in the brain. They are, but in considerably lower concentrations than GAT1 or GAT3. It is generally believed that all these GABA transporters also transport other molecules. For further reading on this topic see the reference list.

------

[6]Excitatory amino acid transporter.

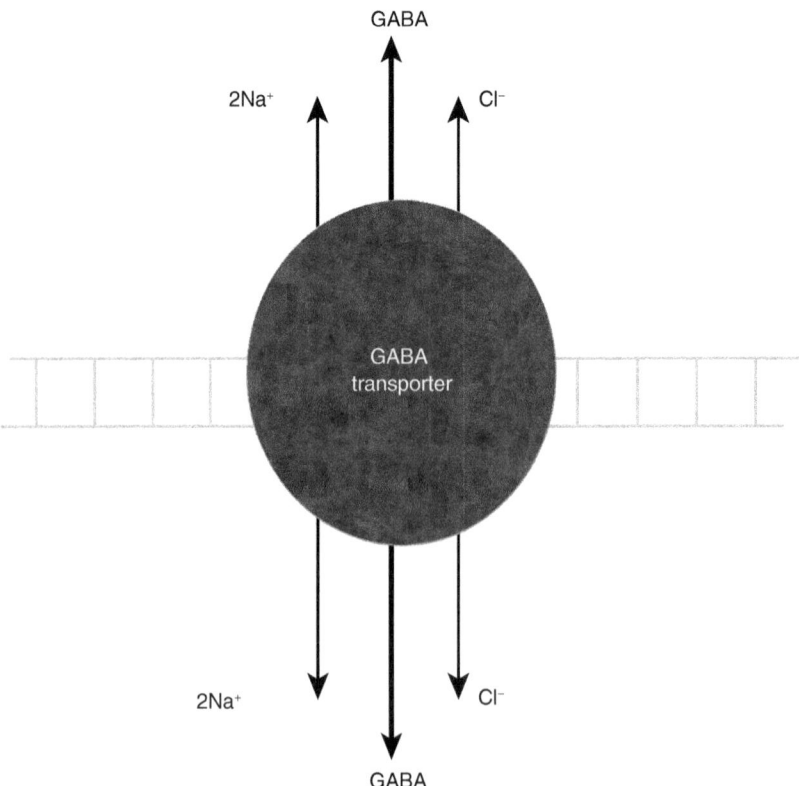

Figure 3.5   Side view of a GABA transporter located in the neuron's plasma membrane

A key point of note is that these can both reuptake GABA and transport it out of the neuron.

These GABA transporters are known to be expressed in different concentrations in different regions of the brain. This suggests a certain degree of functional heterogeneity and a potential mechanism by which alteration of GABA's machinery can have specific effects on the behaviour of the organism. This is because alterations in each type will influence certain, specific neuronal circuits and brain regions. A good example is GAT3, which is known to be expressed in high concentrations in the basal ganglia, especially the substantia nigra. Conversely, another GABA transporter, GAT1, is expressed in high concentrations in the hippocampus but not the substantia nigra. GAT1 is also found more in cortical territories than GAT3. It is worth keeping in mind when carrying out your further reading, that these transporters are heterogeneously distributed. This means that alteration in their function is more likely to have specific implications for the behaviours of both the targeted cell and the whole organism behaviour.

Classic topics in GABA research

As mentioned at the start of this chapter, it is likely that many alterations in the machinery that underpin GABA are likely to result in global impact and therefore produce global effects on behaviour. We will start by focusing on a condition characterised by global alterations in brain function, namely epilepsy. However, we will also look at alterations in more specific behaviours, such as depression and autism spectrum disorder (ASD). It is worth keeping in mind that these specific behavioural alterations are likely to be underpinned by changes in GABA machinery that are known to be heterogeneous, such as GABA receptors, or the previously mentioned GABA transporters.

GABA and epilepsy

Several times throughout this chapter we have discussed how GABAerigc neurons act as local brakes, controlling the flow of excitatory transmission in the brain. One condition characterised by uncontrollable spontaneous excitatory discharge of neurons is epilepsy. It therefore seems highly probable that epilepsy and the seizures that typify the condition will, at least to some degree, be characterised by alterations in the machinery that underpin GABA's function as a local brake.

Many of the more effective current antiepileptic drugs (AEDs) and seizure medication function by modulating the machinery of GABA and shifting the balance away from excitation towards inhibition. For example, AEDs such as sodium valproate act by reducing the degradation of GABA, therefore increasing the free GABA in the synapse. This, you could perhaps say, is the first circumstantial evidence that alterations in GABA's machinery is involved in epilepsy.

Early studies support this role for GABA alterations in epilepsy. The research of Tapia et al. (1975) and the work of Kash et al. (1997) clearly identified that inhibition of the enzyme GAD (which, as we discussed earlier, synthesises GABA) results in an increased susceptibility to seizures. However, most of these studies seemed to show that this effect was very much age-dependent, existing in young animals, typically in the first few weeks of life. When GAD is inhibited in adult rodents, it produced no clear effects on seizure susceptibility. This suggests that whatever effect inhibiting GAD has may produce seizures by altering another factor in the developing brain. You may also recall that there are two distinct isoforms of the GABA synthesising enzyme, namely GAD65 and GAD67. We previously discussed in this chapter that transgenic animals have been produced with functional knockout of both GAD isoforms, as well as one or the other. Unfortunately, complete knockout of GAD67 results in animals that are not viable, therefore a heterozygous[7] animal that expresses reduced levels of GAD67 was produced. This was found to not produce any increased susceptibility to seizures. A more likely candidate to be involved in seizure generation is GAD65. This is largely because of their expression in the vesicular membrane and the belief that they are involved in the rapid production of GABA when it is needed to 'put a brake on' spontaneous discharge in their target neurons. Viable animals with complete knockout of GAD65 is possible. These animals have been shown to express significant increases in spontaneous seizures as well as being more susceptible to induced seizures. An interesting study by Qi et al. (2018) investigated the relationship between stress and seizures in GAD65 knockout animals. This study's findings are interesting as it suggests that these animals displayed an increased level of anxiety before they started to display an increase in spontaneous seizures. This suggests that GAD65 might induce increased anxiety, which mediates the potential of seizure susceptibility. This is interesting as it suggests that controlling the environment and the exposure to anxiety may moderate the impact reduced expression of GAD65 has on seizures. Linked to these studies around alterations in GAD6, some researchers now think that epilepsy may be best considered as an autoimmune disorder. This was based on research that identified an increase in antibodies raised against GAD65 in those that had been diagnosed with epilepsy. This is particularly seen in those diagnosed with late onset epilepsy and temporal lobe epilepsy (TLE). Of course, as usual there is an issue here with direction of effect. The majority of the autoimmune studies involve patients who have been diagnosed with epilepsy and therefore have had seizures. It is therefore conceivable that the increased number of antibodies raised against GAD65 is a consequence of seizures rather than the cause of seizures.

We have focused a lot on the enzymes that synthesise GABA, but there is a large body of research that also heavily implicates alterations in GABA receptors, specifically the ionotropic receptor $GABA_a$. These studies typically investigate alterations in $GABA_a$ related to TLE, so it is worth keeping in mind that these studies and alterations in $GABA_a$ may not be taken as an explanation for the aetiology (cause) of all forms of epilepsy. We will discuss this more shortly in the context of $GABA_b$ and its involvement in absence seizures. There is a significant number of human receptor binding studies which convincingly suggests that $GABA_a$ levels are altered in those with a diagnosis of TLE. For example, multiple studies using molecules with a high binding affinity to $GABA_a$ receptors that have been labelled with a radioactive compound,

---

[7]Generally referring to a gene that has two different alleles.

such as ^{3}H-muscimol[8], have found a significant decrease in the binding and therefore either lower levels of the receptor or a reduced affinity of the receptor for GABA. Much of this research has been focused on the brain region known to be a focal point for seizure generation, namely the hippocampus. It is worth keeping in mind, with your critical hat on, that this is not the only brain region that is a focal point for seizure generation. Therefore, alterations here may be more related to certain 'types' of epilepsy than others. Indeed, there are other studies that have found alterations in GABA$_a$ receptor levels, in brain regions as diverse as the cerebellum after induction of seizures. Again, there is also another critical point to consider, namely that of cause and effect. We have mentioned above that when studying those who have been diagnosed with epilepsy, it is difficult to conclusively say if the alterations in GABA's machinery are a cause or consequence of the seizures. With receptors for neurotransmitters generally, we know that a consequence of reduced neurotransmitter release is downregulation by endocytosis[9] of the receptors. Therefore, whatever alterations in GABA receptors we may see in those diagnosed with epilepsy may actually be a consequence of alteration in the machinery that synthesises GABA, and so a reduced number of GABA receptors can simply be considered a 'knock on' effect of the root cause. This highlights how inherently intertwined the machinery involved in neurotransmitter synthesis is and how difficult it can often be to trace the root causes of behavioural change.

The thorny issue of GABAa receptor subunits

As mentioned earlier in the chapter, there are a lot of different types of GABA$_a$ receptors based on their expression of various subunits, generally different α (1–6) and β (1–3) subunits in the mammalian brain. A great deal of evidence from the research literature suggests that the relative number of several of these subunits is altered in both animal models of epilepsy and in human patients with TLE (temporal lobe epilepsy). What is interesting about this is that it is not just reductions of these subunits, as you might expect based on the previously outlined findings from binding studies. For example, in both humans with TLE and animal models there has been found to be an increase in the $\alpha 3$ subunit as well as the $\beta 3$. There has also been found to be a relative reduction in the $\alpha 1$ subunit. Although there is some consistency between studies using animals and human patients, there is also a considerable amount of

---

[8]This is a high affinity GABA$_a$ receptor agonist 'tagged' with a radioactive hydrogen molecule.

[9]This is the process by which the cell takes proteins embedded in the plasma membranes back into the cell and recycles them if they are not being used.

variation in the subunits that are seen to increase and decrease, both between studies and between brain regions focused on in the research. Be mindful of this in your wider reading. What we can say for a fact is that there is a definite alteration in the expression of certain GABA receptor subunits. This suggests that it may be the subunit 'makeup' of these $GABA_a$ receptors that contribute to epilepsy and seizures rather than the general loss of $GABA_a$ receptors. This also suggests that it may be more to do with the $GABA_a$ receptors' affinity for GABA than it is the number of receptors *per se*, as we know $GABA_a$ receptors with certain subunit compositions have a greater affinity than others for GABA. However, many researchers think that this change in expression levels of the various subunits may be a compensatory measure as a response to seizures, rather than the 'cause' of epilepsy and the subsequent seizures.

So how can we see if it is the $GABA_a$ receptors that are altered and not machinery earlier in the chain? Well, the best method we currently have is the use of either genetically-modified animal studies or whole genome studies with patients displaying epilepsy symptomology. Interesting studies using knockout animals suggest a more causal role for $GABA_a$ receptors and the composition of their subunits in epilepsy aetiology. For example, one recent study by Liao et al. (2019) employed a knockout of the $\gamma 2$ subunit in zebrafish larvae and found an increase propensity towards light-induced seizures. Other research employing rodents has also found that 'knockout' of other subunits, such as the $\beta 3$ subunit results in animals that display an epilepsy phenotype, including alterations to neuronal activity patterns and seizures. It is not just genetically-modified animal studies that suggest the connection between $GABA_a$ receptor subunits. Recently, a rare mutation in the gene coding the α-3 subunit has been identified that is expressed in several human families. Members of these families show an increased propensity towards epilepsy. Other whole genome sequencing studies have also identified mutations in the $\alpha 5$ subunit gene in patients that have extremely severe early-onset epilepsy.

GABAb receptors and absence epilepsy

What we often overlook when thinking about epilepsy is that it is not just one disorder but a group of disorders, all of which are characterised by alterations in the neuronal activity and discharge. We tend to think of epilepsy as being a disorder characterised by seizures and convulsions, but absence epilepsy also exists. Absence epilepsy is not typified by convulsions, but there are distinct changes in the electrical activity of the brain and the individual's behaviour. In fact, it is best characterised by momentary lapses in behaviour. In general, the effects of what is known as 'typical' absence epilepsy are not considered to be that severe. However, atypical absence epilepsy (ATAE) also exists and this has profound negative effects on the individual, including slow wave discharge in the brain, severe absence seizures, and significant cognitive impairment. Unusually, research findings are very consistent regarding $GABA_b$ involvement in both forms of absence epilepsy, with the vast majority of pharmacological studies finding that $GABA_b$ receptor agonists exacerbate seizures while $GABA_b$ receptor antagonists reduce the severity of seizures. So, what we see here is a potential dichotomy with alterations in $GABA_a$ receptors contributing to TLE epilepsy while $GABA_b$ receptors contribute to absence epilepsy. Unfortunately, as ever, this is far too simplistic and the reality is that there is a significant body of research that also implicates alterations in $GABA_b$ receptors in TLE and early-onset epilepsy.

All we can definitively conclude about GABA machinery and epilepsy is that it is likely this machinery is involved in either the aetiology of epilepsy and/or the propagation of seizures. What a lot of this research does do, though, is give us avenues for potentially novel and more 'efficacious' medication for the suppression of seizures, as well as other developmental disorders. For an excellent perspective on this, I recommend including Braat and Kooy's (2015) article in your wider reading. This perspective article is also highly useful when considering GABA's role in autism spectrum disorder (ASD), which we will discuss next.

Novel topics in GABA research

GABA and autism spectrum disorder

Autism spectrum disorder (ASD) is considered to be a developmental disorder typified by deficits in two core domains, according to the American Psychiatric Association's (APA) *Diagnostic and Statistical Manual of Mental Disorders*, namely the DSM-5 (American Psychiatric Association, 2013).[10] The two core behavioural domains that typify ASD are as follows:

1. Social, communication and interaction deficits.
2. Restrictive, repetitive behaviours and interests.

Social, communication and interaction deficits include, but are not limited to, altered eye contact, reduced understanding of reciprocity[11] and a lack of awareness of social boundaries, such as personal space. Restrictive repetitive behaviours and interests

---

[10]Diagnostic Skill Manual Version 5. This is the main document used by clinicians to identify and diagnose developmental, degenerative and psychiatric conditions in individuals.

[11]This is essential for the understanding of turn taking in conversation or social communication.

include, but are not limited to, intense interest in specific objects or activities, an insistence on routine/'sameness' and repetitive movement. It is important to remember that ASD exists on a continuum, with individually diagnosed symptoms varying greatly in severity. Some individuals manage the negative aspects of the diagnosis effectively and have a good quality of life, while at the other end of the spectrum individuals with an ASD diagnosis can have profound impairments in their ability to communicate and manage daily tasks.

As well as the behavioural phenotype[12] outlined above, people with ASD also have a biological phenotype, in that there is seen to be a consistent early overgrowth in the brain, especially within cortical regions. Interestingly, this overgrowth has also been observed by a number of researchers in subcortical structures, such as the striatum. Excellent research on this has been carried out by Langen et al. (2014). This overgrowth persists through life, compared to typically-developing individuals. The interesting thing about this overgrowth, which suggests that it is connected perhaps in a causal way to the behavioural phenotype of ASD, is that it emerges at a similar time to the identification of the behavioural phenotype. The question you might have in mind now is what is the overgrowth? We may naturally assume it is an increase in neuron number, but it is worth considering that it may not be number, but physiological changes in the neurons that increase their size. We will discuss this later. Further still, it could be that it is not neurons at all; indeed, it is worth remembering that a significant number of the cells in our brains are not neurons, but cells such as glia. This is beyond the scope of the current book, but a growing literature is relating alterations in brain network functions and subsequent alterations in behaviour to these glial cells.

One interesting avenue that may help us untangle the biological alterations that underpin ASD is the apparent connection between ASD and epilepsy. Recently, several studies have seemingly revealed that there is a relatively high degree of comorbidity, with estimates between 55 and 37% of those diagnosed with ASD also being diagnosed with epilepsy (Jeste & Tuchman, 2015). This seems to suggest that there may be a common underlying pathology underpinning both. Therefore, in the context of this chapter, and what we have previously discussed about the role of GABA in epilepsy (GAD65 and GABA receptors), it seems reasonable to assume that there may be some involvement of GABA alterations in ASD. One further interesting point that connects epilepsy with ASD is a commonly used AED, namely valproic acid (aka sodium valproate). What is so interesting about this is that animals exposed to

---

[12]These are the behaviours, including cognitions and emotions, that are associated with a specific clinical condition/atypical behavioural state.

valproic acid in utero have been shown to display behavioural phenotypes that are consistent with humans diagnosed with ASD. For further reading on this I would highly recommend Nicolini and Fahnestock's (2018) review. Indeed, in Jeste and Tuchman's (2015) review they hypothesise that ASD and epilepsy may be 'two sides of the same coin'. Therefore, what we see in epilepsy in terms of reduced GABA levels may be flipped by the administration of valproic acid, resulting, hypothetically, in an increase of GABA levels and the emergence of an ASD phenotype. In fact, most of the research literature on ASD does not suggest this. Instead, the majority of the literature implies that it is reduced GABA and specifically GABAergic innervation of projection neurons that underpins ASD. We will now discuss this further.

If we discount the evidence relating to valproic acid for the moment and continue the line of thought that the same machinery underpinning GABA's involvement in epilepsy may also underpin ASD, one of the prevailing current theories related to the aetiology of ASD centres around alterations in the brain's E/I balance. As previously discussed in this chapter, this theory posits that imbalance in the excitatory and inhibitory communication in the brain underpins atypical behaviour states. There are several good lines of evidence that suggest this is the case in ASD, with a particular focus on the idea that GABAergic interneurons and/or the machinery of GABA is affected, resulting in a shift towards reduced inhibition. We will start by looking at the research centred around electrophysiological recordings in humans.

In this book, when we have looked at alterations in human biology, rather than animals, we have typically focused on post-mortem research. However, it is perfectly possible to conduct electrical recordings of the brain activity in vivo[13] in humans. There are several techniques that you can employ to do this. One commonly used technique for this research is called EEG (electroencephalography) and involves placing numerous electrodes on the scalps of individuals. This allows you to record from neurons 'close to the surface' of the brain and measure population activity. This population activity is typically referred to as electrical activity that is separated into different bands. The band that is associated with GABAergic interneuron activity is called the GAMMA (γ) band. This is electrical activity at 30–80 Hz,[14] although some argue that it also includes higher electrical activity at 80Hz+. Alterations in the GAMMA band reveal the function of GABA and inhibitory transmission in the brain and their synchronisation with glutamatergic projection neurons. The literature on this is complex and there is quite a degree of contradiction, with many studies finding increases in the γ band in those

---

[13]This means 'within the living'. The opposite is invitro, which means 'outside the living' or 'in the dish'.

[14]Hz (Hertz) refers to the firing frequency of an electrical wave form. 1 Hz means one 'spike' per second, so 80 Hz means 80 spikes per second.

with an ASD diagnosis and yet many others finding decreases in the γ band range in those with an ASD diagnosis. What we can perhaps conclude is that there is an alteration in the γ band, which implies altered function in GABAergic interneurons. One of the interesting facts about the γ band is that it is generally considered to reveal the function of a specific type of GABAergic interneuron, known as a PV+ type. We will return to this type later in this chapter.

There are many other lines of research that implicate GABAergic interneurons and GABAs machinery in ASD. A significant amount of research finds that there is an altered expression and function of GABAergic interneurons in the brains of animal models of ASD. This reduced expression and/or function ultimately reduces the ability of these interneurons to carry out their job, which many people think is to act as 'local brakes' at various points throughout brain circuits, slowing down and/or controlling the flow of information through these circuits. There are numerous studies employing genetically-modified animal models of ASD (typically knockout animals) that show distinct alterations in the function of these GABAergic interneurons and their response to input from projection neurons. However, is this reduced function related to the GABAergic aspects of these interneurons' machinery? Well, many studies suggest yes, finding reductions in proteins like GAD65 expression. There are also studies that suggest a reduction in the cerebral cortex of VGATs in animal models of ASD. However, there is other research evidence, such as the recent study by Brandenburg et al. (2020), which finds that, at least in regions of the striatum, GAD65 and GAD67 expression levels are not altered in those with a diagnosis of ASD. Thus, this suggests the alteration in interneurons and the synthesis of GABA in those with ASD may be very region-specific at best.

An interesting issue with studies suggesting GABAergic interneurons are altered in ASD revolves around the molecules typically targeted to investigate these interneurons. You would be forgiven for thinking that the molecule targeted to visualise them would be related to GABA, but this is not typically the case. The method used to identify and investigate these GABAergic interneurons typically involves visualising them by immunohistochemical techniques, which we discussed earlier in this chapter. Much of the research literature has done this by targeting the calcium-binding proteins uniquely expressed by specific subtypes of these GABAergic interneurons. These calcium-binding proteins include calbindin, calretinin, calmodulin and parvalbumin (PV+). Recently, researchers have started to question whether it is really the GABA machinery in these interneurons that is altered, or whether it is actually loss/alteration of the calcium-binding proteins in these neurons that results in altered function and subsequently the behavioural phenotype of ASD. There are two recent, excellent studies that suggest this may be the case. For example, the work of Filice et al. (2016) shows that it is not the number of these neurons that are reduced but the calcium-binding protein levels that are lowered. Further evidence for the role of these calcium-binding proteins, rather than the GABA machinery, comes from electrophysiological recording studies using genetically-modified animals. One good example is the research of Wöhr et al. (2015). They employed a calcium-binding protein (PV) knockout mouse and confirmed that this animal showed many of the animal analogues of the human ASD behavioural phenotype. They also found that there were significantly reduced responses from these GABAergic interneurons to innervation from cortical projection neurons.

It seems that it may not be the GABA machinery within these Interneurons *per se* that is altered in ASD. So, what else could be altered in the GABA machinery? Let's now consider GABA heteroreceptors. These are receptors expressed on non-GABA neurons. There is a weight of evidence, using different methods, that suggest they are altered in epilepsy. So, what do we find with GABA receptor expression and affinity in those diagnosed with ASD and animal models of ASD? Well, we find, pretty consistently, that there is a reduction in the expression of $GABA_a$ receptors. This has been found frequently in humans with a diagnosis of ASD both invitro and in vivo. There is also a suggestion that it might actually be specifically related to the subunits expressed by these GABA receptors. For example, some research has shown specific reductions in the $α5$ $GABA_a$ subunit invitro in humans with a diagnosis of ASD.

But is it all about the machinery of GABA? As discussed above, it seems unlikely that it is solely down to alterations in GABA's machinery as there also seems to be a strong case for the involvement of calcium-binding proteins. But what else might contribute to a biological phenotype of ASD? Well, if you think about the nature of E/I balance, then it would seem reasonable that it is perhaps not an alteration in GABA's machinery *per se* that alters E/I balance and contributes to ASD, but perhaps alteration in the neurons targeted by the GABAergic 'local brakes'. This links to postsynaptic located GABA receptors, which are found on the spines of projection neurons. These spines are small projections that increase the surface area of the target neuron and therefore increase the amount of space for GABA receptors. A body of research has found that these spines are altered in number, but also in size and shape, in those diagnosed with ASD. For example, Hutsler and Zhang (2010) found that there was a significantly increased density of spines on cortical projection neurons. There is also research evidence which brings us back to the aforementioned valproic acid, with animals that had been exposed to VPA, in utero, displaying changes in behaviour akin to human ASD but also displaying dramatic increases in spine density.

What this may ultimately mean is that it is an interaction between GABA's molecular machinery, other molecular machinery in GABAergic interneurons, and the machinery of other neurons that may provide a 'fuller' explanation of the biological underpinnings of ASD. Indeed, if you make the connection now to the earlier known biological phenotype of ASD, namely 'brain overgrowth', it makes a lot of sense that this brain overgrowth is actually a representation of the increased number of spines on projection neurons, resulting in an alteration in the number of GABA receptors on these projection neurons. However, the issue here is that this logic does not fully add up, based on the research evidence. As we discussed above, and is reported in an excellent review of the human research literature carried out by Cellot and Cherubini (2014), the weight of literature points, therefore, towards a reduced expression of GABA$_a$ and GABA$_b$ receptors in the brains of those with an ASD diagnosis. All we can perhaps say is that there are several biological alterations that seem to underpin ASD and that some of these alterations relate to the machinery of GABA.

Conclusion

In conclusion, it is essential that we always keep in mind the inherently intertwined nature of GABA and glutamate transmission, not only in terms of the connection between synthesis, but also on a functional level in considering the necessity of a 'balance' being maintained within the brain, keeping the signalling of both within tightly controlled bounds to enable 'typical' function. Of course, as we have discussed, a loss of this balance and alterations in the machinery underpinning GABA can have profound global effects on the behaviour of the organism as a whole, ultimately underpinning disorders, such as epilepsy, but also contributing to more specific behaviour alterations, such as those seen in the case of autism spectrum disorder.

──Test Yourself Answers──

Test Yourself 3.1

1. Proton pump
2. VGAT
3. CLC-3

References

Key information

Juge, N., Omote, H., & Moriyama, Y. (2013). Vesicular GABA transporter (VGAT) transports β-alanine. *Journal of Neurochemistry, 127*(4), 482–486. doi:10.1111/jnc.12393

Frangaj, A., & Fan, Q. R. (2018). Structural biology of GABAB receptor. *Neuropharmacology, 136*, 68–79. doi:https://doi.org/10.1016/j.neuropharm.2017.10.011

Rowley, N. M., Madsen, K. K., Schousboe, A., & Steve White, H. (2012). Glutamate and GABA synthesis, release, transport and metabolism as targets for seizure control. *Neurochem Int, 61*(4), 546–558. doi:10.1016/j.neuint.2012.02.013

Polak, J. M., & Van Noorden, S. (2003). *Introduction to immunocytochemistry* (3rd ed. ed.). Oxford: BIOS Scientific Publishers.

Ahnert-Hilger, G., & Jahn, R. (2011). CLC-3 spices up GABAergic synaptic vesicles. *Nature Neuroscience, 14*(4), 405–407. doi:10.1038/nn.2786

Martin, L. J., Bonin, R. P., & Orser, B. A. (2009). The physiological properties and therapeutic potential of alpha5-GABAA receptors. *Biochem Soc Trans, 37*(Pt 6), 1334–1337. doi:10.1042/bst0371334

Wongsamitkul, N., Maldifassi, M. C., Simeone, X., Baur, R., Ernst, M., & Sigel, E. (2017). α subunits in GABA A receptors are dispensable for GABA and diazepam action. *Scientific Reports, 7*(1), 15498. doi:10.1038/s41598-017-15628-7

Scimemi, A. (2014). Structure, function, and plasticity of GABA transporters. *Front Cell Neurosci, 8*, 161–161. doi:10.3389/fncel.2014.00161

GABA and epilepsy

Asada, H., Kawamura, Y., Maruyama, K., Kume, H., Ding, R., Ji, F. Y., . . . Obata, K. (1996). Mice lacking the 65 kDa isoform of glutamic acid decarboxylase (GAD65) maintain normal levels of GAD67 and GABA in their brains but are susceptible to seizures. *Biochem Biophys Res Commun, 229*(3), 891–895. doi:10.1006/bbrc.1996.1898

Kash, S. F., Johnson, R. S., Tecott, L. H., Noebels, J. L., Mayfield, R. D., Hanahan, D., & Baekkeskov, S. (1997). Epilepsy in mice deficient in the 65-kDa isoform of glutamic acid decarboxylase. *Proceedings of the National Academy of Sciences of the United States of America, 94*(25), 14060–14065. doi:10.1073/pnas.94.25.14060

Daif, A., Lukas, R. V., Issa, N. P., Javed, A., VanHaerents, S., Reder, A. T., . . . Wu, S. (2018). Antiglutamic acid decarboxylase 65 (GAD65) antibody-associated epilepsy. *Epilepsy & Behavior, 80*, 331–336. doi:https://doi.org/10.1016/j.yebeh.2018.01.021

Dugladze, T., Maziashvili, N., Börgers, C., Gurgenidze, S., Häussler, U., Winkelmann, A., . . . Gloveli, T. (2013). GABAB autoreceptor-mediated cell type-specific reduction of inhibition in epileptic mice. *Proceedings of the National Academy of Sciences, 110*(37), 15073. doi:10.1073/pnas.1313505110

Fritschy, J.-M. (2008). Epilepsy, E/I Balance and GABA(A) Receptor Plasticity. *Front Mol Neurosci, 1,* 5–5. doi:10.3389/neuro.02.005.2008

Gaspard, N. (2020). How Much GAD65 Do You Have? High Levels of GAD65 Antibodies in Autoimmune Encephalitis. *Epilepsy currents, 20*(5), 267–270. doi:10.1177/1535759720949238

Hao, F., Jia, L.-H., Li, X.-W., Zhang, Y.-R., & Liu, X.-W. (2016). Garcinol Upregulates GABAA and GAD65 Expression, Modulates BDNF-TrkB Pathway to Reduce Seizures in Pentylenetetrazole (PTZ)-Induced Epilepsy. *Medical science monitor : international medical journal of experimental and clinical research, 22,* 4415–4425. doi:10.12659/msm.897579

Hunt, R. F., Girskis, K. M., Rubenstein, J. L., Alvarez-Buylla, A., & Baraban, S. C. (2013). GABA progenitors grafted into the adult epileptic brain control seizures and abnormal behavior. *Nature Neuroscience, 16*(6), 692–697. doi:10.1038/nn.3392

Magloire, V., Mercier, M. S., Kullmann, D. M., & Pavlov, I. (2019). GABAergic Interneurons in Seizures: Investigating Causality With Optogenetics. *Neuroscientist, 25*(4), 344–358. doi:10.1177/1073858418805002

Qi, J., Kim, M., Sanchez, R., Ziaee, S. M., Kohtz, J. D., & Koh, S. (2018). Enhanced susceptibility to stress and seizures in GAD65 deficient mice. *PloS one, 13*(1), e0191794–e0191794. doi:10.1371/journal.pone.0191794

Tapia, R., Pasantes-Morales, H., Taborda, E., & Pérez de la Mora, M. (1975). Seizure susceptibility in the developing mouse and its relationship to glutamate decarboxylase and pyridoxal phosphate in brain. *J Neurobiol, 6*(2), 159–170. doi:10.1002/neu.480060204

Massieu, L., Rivera, A., & Tapia, R. (1994). Convulsions and inhibition of glutamate decarboxylase by pyridoxal phosphate-gamma-glutamyl hydrazone in the developing rat. *Neurochem Res, 19*(2), 183–187. doi:10.1007/bf00966814

Sperk, G., Furtinger, S., Schwarzer, C., & Pirker, S. (2004). GABA and its receptors in epilepsy. *Adv Exp Med Biol, 548,* 92–103. doi:10.1007/978-1-4757-6376-8_7

Titulaer, M. N., Kamphuis, W., Pool, C. W., van Heerikhuize, J. J., & Lopes da Silva, F. H. (1994). Kindling induces time-dependent and regional specific changes in the [3H]muscimol binding in the rat hippocampus: a quantitative autoradiographic study. *Neuroscience, 59*(4), 817–826. doi:10.1016/0306-4522(94)90286-0

Bazyan, A. S., Zhulin, V. V., Karpova, M. N., Klishina, N. Y., & Glebov, R. N. (2001). Long-term reduction of benzodiazepine receptor density in the rat cerebellum by acute seizures and kindling and its recovery 6 months later by a pentylenetetrazole challenge. *Brain Res, 888*(2), 212–220. doi:10.1016/s0006-8993(00)03045-6

Loup, F., Wieser, H. G., Yonekawa, Y., Aguzzi, A., & Fritschy, J. M. (2000). Selective alterations in GABAA receptor subtypes in human temporal lobe epilepsy. *J Neurosci, 20*(14), 5401–5419. doi:10.1523/jneurosci.20-14-05401.2000

Liao, M., Kundap, U., Rosch, R. E., Burrows, D. R. W., Meyer, M. P., Ouled Amar Bencheikh, B., . . . Samarut, É. (2019). Targeted knockout of GABA-A receptor gamma 2 subunit provokes transient light-induced reflex seizures in zebrafish larvae. *Dis Model Mech, 12*(11). doi:10.1242/dmm.040782

Niturad, C. E., Lev, D., Kalscheuer, V. M., Charzewska, A., Schubert, J., Lerman-Sagie, T., . . . Leshinsky-Silver, E. (2017). Rare GABRA3 variants are associated with epileptic seizures, encephalopathy and dysmorphic features. *Brain, 140*(11), 2879–2894. doi:10.1093/brain/awx236

DeLorey, T. M., Handforth, A., Anagnostaras, S. G., Homanics, G. E., Minassian, B. A., Asatourian, A., . . . Olsen, R. W. (1998). Mice lacking the beta3 subunit of the GABAA receptor have the epilepsy phenotype and many of the behavioral characteristics of Angelman syndrome. *J Neurosci, 18*(20), 8505–8514. doi:10.1523/jneurosci.18-20-08505.1998

Braat, S., & Kooy, R. F. (2015). The GABAA Receptor as a Therapeutic Target for Neurodevelopmental Disorders. *Neuron, 86*(5), 1119–1130. doi:10.1016/j.neuron.2015.03.042

Han, H., Cortez, M., & Snead, O. (2012). GABAB receptor and absence epilepsy. In Cortez, M. A., McKerlie, C., & Snead, O. C., 3rd. (2001). A model of atypical absence seizures: EEG, pharmacology, and developmental characterization. *Neurology, 56*(3), 341–349. doi:10.1212/wnl.56.3.341

Abou-Khalil, B. W. (2016). Antiepileptic Drugs. *Continuum (Minneap Minn), 22*(1 Epilepsy), 132–156. doi:10.1212/con.0000000000000289

GABA and ASD (Autistic spectrum disorder)

American Psychiatric Association. (2013). Diagnostic and statistical manual of mental disorders (5th ed.) (DSM-5). Washington, DC: APA.

Mahdavi, M., Kheirollahi, M., Riahi, R., Khorvash, F., Khorrami, M., & Mirsafaie, M. (2018). Meta-analysis of the association between GABA receptor polymorphisms and autism spectrum disorder (ASD). *J Mol Neurosci, 65*(1), 1–9. doi:10.1007/s12031-018-1073-7

Marotta, R., Risoleo, M. C., Messina, G., Parisi, L., Carotenuto, M., Vetri, L., & Roccella, M. (2020). The neurochemistry of autism. *Brain Sciences, 10*(3). doi:10.3390/brainsci10030163

Besag, F. M. (2017). Epilepsy in patients with autism: links, risks and treatment challenges. *Neuropsychiatric Disease and Treatment, 14*, 1–10. doi:10.2147/NDT. S120509

Ewen, J. B., Marvin, A. R., Law, K., & Lipkin, P. H. (2019). Epilepsy and Autism Severity: A Study of 6,975 Children. *Autism Res, 12*(8), 1251–1259. doi:10.1002/aur.2132

Jeste, S. S., & Tuchman, R. (2015). Autism spectrum disorder and epilepsy: two sides of the same coin? *Journal of Child Neurology, 30*(14), 1963–1971. doi:10.1177/0883073815601501

Nicolini, C., & Fahnestock, M. (2018). The valproic acid-induced rodent model of autism. *Experimental Neurology, 299*, 217–227. doi:https://doi.org/10.1016/j. expneurol.2017.04.017

Gao, R., & Penzes, P. (2015). Common mechanisms of excitatory and inhibitory imbalance in schizophrenia and autism spectrum disorders. *Current Molecular Medicine, 15*(2), 146–167. doi:10.2174/1566524015666150303003028

Adorjan, I., Ahmed, B., Feher, V., Torso, M., Krug, K., Esiri, M., . . . Szele, F. G. (2017). Calretinin interneuron density in the caudate nucleus is lower in autism spectrum disorder. *Brain, 140*(7), 2028–2040. doi:10.1093/brain/awx131

Hutsler, J. J., & Zhang, H. (2010). Increased dendritic spine densities on cortical projection neurons in autism spectrum disorders. *Brain Res, 1309*, 83–94. doi:10.1016/j.brainres.2009.09.120

Selten, M., van Bokhoven, H., & Nadif Kasri, N. (2018). Inhibitory control of the excitatory/inhibitory balance in psychiatric disorders. *F1000Research, 7*, 23–23. doi:10.12688/f1000research.12155.1

Filice, F., Vörckel, K. J., Sungur, A., Wöhr, M., & Schwaller, B. (2016). Reduction in parvalbumin expression not loss of the parvalbumin-expressing GABA interneuron subpopulation in genetic parvalbumin and shank mouse models of autism. *Mol Brain, 9*, 10. doi:10.1186/s13041-016-0192-8

Wöhr, M., Orduz, D., Gregory, P., Moreno, H., Khan, U., Vörckel, K. J., . . . Schwaller, B. (2015). Lack of parvalbumin in mice leads to behavioral deficits relevant to all human autism core symptoms and related neural morphofunctional abnormalities. *Translational Psychiatry, 5*, e525. doi:10.1038/tp.2015.19

Cellot, G., & Cherubini, E. (2014). GABAergic signaling as therapeutic target for autism spectrum disorders. *Frontiers in Pediatrics, 2*, 70. Retrieved from www. frontiersin.org/article/10.3389/fped.2014.00070

Courchesne, E., Mouton, P. R., Calhoun, M. E., Semendeferi, K., Ahrens-Barbeau, C., Hallet, M. J., . . . Pierce, K. (2011). Neuron number and size in prefrontal cortex of children with autism. *Jama, 306*(18), 2001–2010. doi:10.1001/ jama.2011.1638

Langen, M., Bos, D., Noordermeer, S. D., Nederveen, H., van Engeland, H., & Durston, S. (2014). Changes in the development of striatum are involved in repetitive behavior in autism. *Biol Psychiatry, 76*(5), 405–411. doi:10.1016/j.biopsych.2013.08.013

Rojas, D. C., & Wilson, L. B. (2014). γ-band abnormalities as markers of autism spectrum disorders. *Biomarkers in Medicine, 8*(3), 353–368. doi:10.2217/bmm.14.15

Brandenburg, C., Soghomonian, J. J., Zhang, K., Sulkaj, I., Randolph, B., Kachadoorian, M., & Blatt, G. J. (2020). Increased dopamine type 2 gene expression in the dorsal striatum in individuals with autism spectrum disorder suggests alterations in indirect pathway signaling and circuitry. *Front Cell Neurosci, 14*, 577858. doi:10.3389/fncel.2020.577858

Orenbuch, A., Fortis, K., Taesuwan, S., Yaffe, R., Caudill, M. A., & Golan, H. M. (2019). Prenatal nutritional intervention reduces autistic-like behavior rates among Mthfr-deficient mice. *Front Neurosci, 13*, 383. doi:10.3389/fnins.2019.00383

Braat, S., & Kooy, R. F. (2015). The GABAA Receptor as a therapeutic target for neurodevelopmental disorders. *Neuron, 86*(5), 1119–1130. doi:10.1016/j.neuron.2015.03.042

Yang, E.-J., Ahn, S., Lee, K., Mahmood, U., & Kim, H.-S. (2016). Early behavioral abnormalities and perinatal alterations of PTEN/AKT pathway in valproic acid autism model mice. *PLOS ONE, 11*, e0153298. doi:10.1371/journal.pone.0153298

4

ACETYLCHOLINE

Introduction

As outlined in Chapter 1, the neurotransmitter acetylcholine (Ach) was the first to be identified, by Lowei (1936, 1949) in his instrumental research using the frog heart. At this point it was referred to as vagal substance, but this commonly became known as acetylcholine.

The fact that this research was focused on the heart emphasises the importance of Ach outside the CNS (central nervous system). In fact, Ach is one of the core neurotransmitters at the neuromuscular junction and as such is key to the control of muscle. This is true for cardiac muscle, such as the heart, but also for striated/skeletal muscle. Ach is therefore the final neurotransmitter in a chain that allows us to enact conscious control over our muscles, and therefore act on the world.

As well as its pivotal role in the PNS (peripheral nervous system), Ach and Ach-expressing neurons play key roles within the brain. For example, Ach-expressing neurons are some of the earliest neurons to die and dysfunction in the early stages of Alzheimer's disease and Ach agonists have had some success as a therapeutic to at least slow down the progression of Alzheimer's disease. Ach neurons also play some essential roles in specific brain regions, such as the striatum, where they have been shown to be key modulators of other neurotransmitters, such as dopamine. Ach is believed to contribute to dopamine's involvement in action selection within these circuits. We will discuss dopamine in Chapter 5, but for the moment this helps to remind us that neurotransmitters rarely function in isolation. This is an important point to keep in mind as you read through this book.

So, what of the machinery behind the synthesis, packaging, degradation and neurotransmission of Ach? Well, this is no less fascinating than any of the other neurotransmitters we cover in this book. As you may have already started to realise from reading other chapters, many of these mechanisms underpinning Ach have been targeted as potential therapeutics for a range of issues, such as the previously mentioned Ach agonist used to slow the progression of Alzheimer's. These mechanisms have also been extensively researched as their function has been connected to a number of other issues, such as addiction.

In this chapter, we will first look at the mechanisms, before exploring what research has told us about the involvement of Ach, and its mechanisms, in Alzheimer's, addiction and autism spectrum disorder (ASD).

Outline of acetylcholine's synthesis, packaging and reuptake

Before we consider how Ach and its mechanistic might be altered in various atypical behavioural states, it is important to understand how the machinery functions under normal conditions. To do this we are now going to explore how Ach is synthesised, packaged for release, how it binds with receptors to transmit a signal and, finally, how it is degraded and taken back into the neuron ready for the cycle to start again.

Before we get stuck into this, it is worth thinking briefly about where Ach is expressed in the brain. Ach is expressed across a number of brain regions and by a number of different types of neurons. Key areas of expression include the pedunculopontine nucleus, where Ach neurons project out into midbrain and subcortical structures, the basal forebrain, where Ach neurons project towards frontal regions, and within the striatum, where some Ach-expressing neurons are in fact interneurons, rather than projection neurons. See Figure 4.1 for further detail of Ach neuronal expression and projections. Because of this localised expression and targeted projections, Ach plays a role in numerous different behavioural states and is a key player in modulating our behaviour. We will explore this further, later in this chapter.

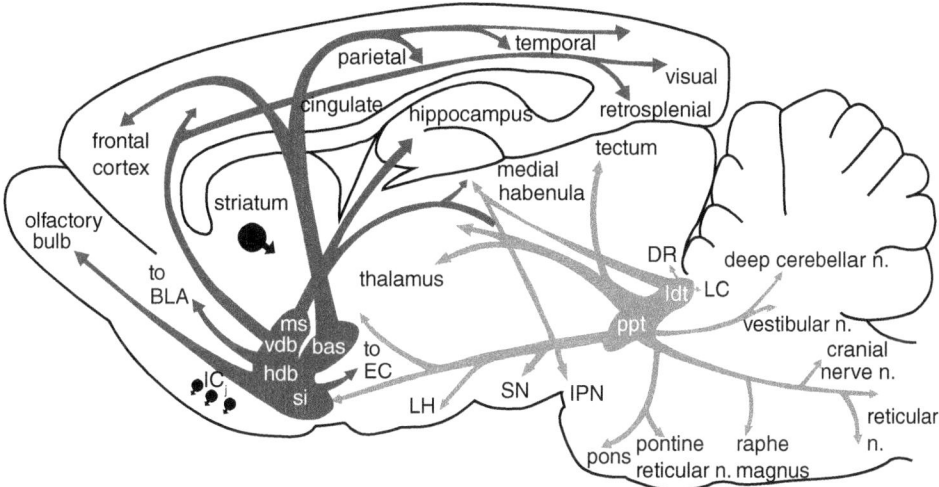

Figure 4.1 The location of cholinergic neurons and their projection networks in the brain

Note: Abbreviations: bas, nucleus basalis; BLA, basolateral amygdala; DR, dorsal raphe; EC, entorhinal cortex; hdb, horizontal diagonal band nucleus; Icj, islands of Cajella; IPN, interpeduncular nucleus; LC, locus ceruleus; ldt, laterodorsal tegmental nucleus; LH, lateral hypothalamus; ms, medial septal nucleus; ppt, pedunculopontine nucleus; si, substantia nnominate; SN, substantia nigra; vdb, vertical diagonal band nucleus.

Source: Figure and abbreviations reprinted from Woolf, N. J., & Butcher, L. L. (2011). Cholinergic systems mediate action from movement to higher consciousness. *Behavioural Brain Research*, *221*(2), 488–498. Copyright (2011), with permission from Elsevier.

Acetylcholine synthesis

Compared to some other neurotransmitters covered in this book, the synthesis of Ach is relatively straightforward. Ach is synthesised from two precursors which are partly derived from cellular metabolism. These precursors are acetyl-CoA and choline. Synthesis is a one-step process that is catalysed by the enzyme cholineacetyltrans-ferase, commonly referred to as ChAT. ChAT is most abundant in the terminal end of neurons, although some can also be found in axons. You may be thinking: if the building blocks are products of metabolism, and metabolism happens in every cell/neuron, why isn't acetylcholine expressed by all neurons? Well, the simple answer is that only certain neurons express the enzyme ChAT. This makes ChAT an excellent target if you want to label and visualise cholinergic neurons. The majority of the enzyme ChAT is free in the cytoplasm (typically referred to as soluble ChAT), although some is also found bound to the vesicles that package Ach for release. This type is typically referred to as membrane-bound ChAT. The membrane-bound ChATs suggest that there is a tight coupling between Ach synthesis and its packaging into vesicles.

Back to the precursors. Above, I suggested that the precursors for Ach synthesis are 'partly' a product of cellular metabolism. This is only part of the story and it is important to keep in mind that the precursors that form acetylcholine, namely acetyl-CoA and choline, can come from different sources. It depends on whether you are talking about the synthesis of acetylcholine in the PNS or the brain. For example, choline is synthesised in the liver and can also be taken in as a dietary precursor. So where do these two precursors come from in the brain? Well, the acetyl-CoA used for Ach synthesis, in human brains, typically comes from pyruvate. Pyruvate is involved in glucose metabolism, and is largely confined to the inner membrane of mitochondria. For further information on the complexity of this see Figure 4.2.

It is essential for pyruvate to be transported out of the mitochondria for synthesis of ACh by ChAT in the cytoplasm. Little is known about how this happens, but it is generally considered to be a key rate limiting process[1] in the synthesis of ACh. So, what about choline? This is a bit more complex than acetyl-CoA in the brain as choline can come from a number of sources. One interesting point of acetylcholine synthesis in the brain is that, although some (considered to be approximately 50%) of the choline is provided by metabolic processes, the other 50% is believed to come from the reuptake of choline as part of the recycling process in the synaptic cleft, after acetylcholine has done its job in neurotransmission. Many believe that this uptake mechanism is the main rate limiting factor in the synthesis of ACh. In fact, much evidence points to this being the case, with studies employing a high-affinity antagonist of the choline reuptake channels finding a significant reduction in the release of ACh, when the neuron is stimulated for a prolonged period. Of course,

---

[1]A rate limiting step is generally considered to be the part which is the slowest out of a chain of steps that make up a chemical reaction.

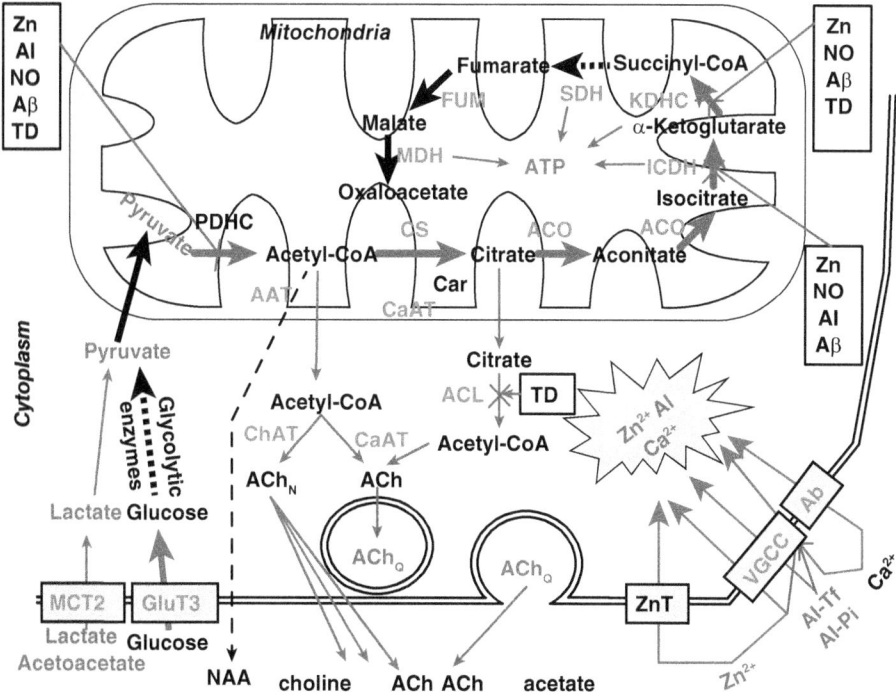

Figure 4.2 The complex mechanisms of glucose metabolism in cellular mitochondria

As part of this process, acetyl-CoA is produced as an intermediary. This is a key source of the precursor for acetylcholine synthesis within neurons.

Source: Reprinted from: Szutowicz, A. et al. (2013). Acetyl-CoA the key factor for survival or death of cholinergic neurons in course of neurodegenerative diseases. *Neurochemical Research, 38*(8), 1523–1542. Under CC BY 4.0 license.

as well as the choline being available for uptake as part of the 'recycling process', choline can also get into the brain via the blood–brain barrier. This involves specific choline transporters found in endothelial cells that are a constituent part of the blood–brain barrier. Consequently, this uptake mechanism can also transport choline from dietary intake into the brain and then into the neuron for synthesis into ACh.

The availability of precursors, and the mechanisms by which they enter the neuron, are key rate limiting factors in the synthesis of ACh. There is, however, one more rate limiter. This is acetylcholine itself, which can limit the rate of acetylcholine synthesis. It does this by acting as an allosteric antagonist at a binding site on the enzyme ChAT. This keeps acetylcholine levels within tightly-bound limits. This is worth remembering, as we are homeostatic beings. This means most of our function is optimal when kept within a tightly-controlled range. Divergence from this optimal level, by either an increase or a decrease in the availability of a neurotransmitter and/or its machinery, is likely to have distinct negative effects on the cell and the organism as a whole. This idea is at the very heart of everything that is discussed in this book as it is the guiding principle behind why alterations in neurotransmitter levels may result in atypical behaviour of the neuron and the whole organism.

Acetylcholine packaging

The vast majority of ACh in nerve terminals is packaged into vesicles for release. The mechanisms of vesicular packaging involve the vesicular transporter VAChT. This transporter is coded for by the Slc18a3 gene. The gene is located within the first few positions of the DNA that also codes for the enzyme ChAT. Therefore, when the enzyme is coded for in a cell, the vesicular transporter is also coded. This should come as little surprise as we already discussed, in the synthesis section, that ACh synthesis and packaging are tightly coupled. This is especially apparent when you consider the presence of membrane-bound ChAT on vesicles.

VAChTs share a very similar morphology to VMATs (vesicular monoamine transporters), which are discussed in other chapters in this book, focused on the monoamines dopamine and serotonin. For more detail on their structure, refer to Chapters 5 and 6. The most salient point, in terms of the anatomy/morphology of VAChTs, is that they have 12 transmembrane-spanning domains. These form a tight structure around a central binding site. The binding site is open on one side of the membrane but blocked at the other. Because ACh is cationic, it is important for there to be a PH difference between the neuronal cytoplasm and the vesicle lumen (inner space in the vesicle). This PH difference is set up by the transport of H^+ ions, via the proton pump. If you recall from Chapter 1, this is an active process and requires ATPase to provide the required energy. To then transport ACh into the vesicle lumen, as a rule, it is considered that two H^+ ions are required to be exchanged for the transport of one ACh molecule. The first H^+ ion is thought to facilitate the movement in of the ACh molecule. The second is thought to be responsible for the reorientation of the VAChT, so the binding site returns to facing the neuronal cytoplasm. This means the transporter will once again be ready for the bindings and transport of another molecule of ACh.

A point of interest with the packaging of ACh into vesicles is that, in theory, the vesicle lumen has the capacity to concentrate ACh up to 100-fold that of ACh concentrations in the neuronal cytoplasm. However, this doesn't seem to be the case, with studies suggesting that ACh levels within the vesicle lumen only seem to be concentrated at around the upper limits of ACh concentration in the neuronal cytoplasm. So, what does this mean? Well, it suggests a rather extreme limiting factor is likely to block the loading of ACh into vesicles. This is yet to be discovered.

Acetylcholine reuptake/degradation

Once released into the synapse, ACh is rapidly degraded. This largely takes place within the synapse, rather than the neurotransmitter being reabsorbed by the presynaptic neuron and/or astrocyte, before the degradation process occurs, as is the case with some other neurotransmitters covered in this book.

The enzyme acetylcholinesterase (AChE), which facilitates this degradation, is found in high concentrations within the synapse and is ranked as one of the most rapid enzymes known. This means that ACh is only in the synaptic cleft for a brief period, and concentrations rapidly reduce. AChE is a hydrolysing enzyme[2] which produces acetate and choline from hydrolysis of acetylcholine. At this point, much of the choline is taken back up by the presynaptic terminal. This is facilitated by a membrane-bound channel protein, typically referred to as ChT (choline reuptake transporter). At the same time as transporting choline back into the presynaptic terminal, ChTs also transports sodium (Na^+). Many believe this reuptake mechanism to be the main rate limiting factor in the synthesis of acetylcholine as approximately 50% of the choline used to synthesise acetylcholine comes from this recycling process. Some of you might be thinking, what happens to the acetate? There is some suggestion that this may also be recycled, but this is yet to be conclusively shown. What we do know is the acetate and acetyl-CoA produced as a result of metabolism is enough to supply the necessary precursors for further ACh production, within the terminal end. Interestingly, a number of studies have shown that this is not the case for choline, and without the presence of ChTs recycling choline from the synapse, ACh synthesis cannot be maintained. Several studies have also shown that ChTs can be found bound to synaptic vesicles. As these vesicles fuse with the presynaptic terminal end and release ACh into the synapse, the ChTs bound to the vesicle membrane fuse and become part of the presynaptic terminal end's membrane (see Figure 4.3). This produces a neat system, whereby release of ACh via vesicles also provides the mechanism by which choline can be recycled for the synthesis of ACh and the cycle to start once again.

AChE and ChTs would seem to be likely candidates for research focusing on alterations in ACh and the atypical behaviour states with which altered ACh levels are associated. Indeed, a hot research topic of recent years has been reversible anti-acetylcholinesterases, which have been trialled and approved for use as treatments for patients diagnosed with Alzheimer's disease. We will discuss these and

---

[2]This is the breakdown of a compound, facilitated by an enzyme, by the compound's interaction with a molecule of water.

their efficacy later in the chapter. ChT targeting as a potential therapeutic has also been explored, but far less so than AChEs, making the latter an exciting current area of the research literature.

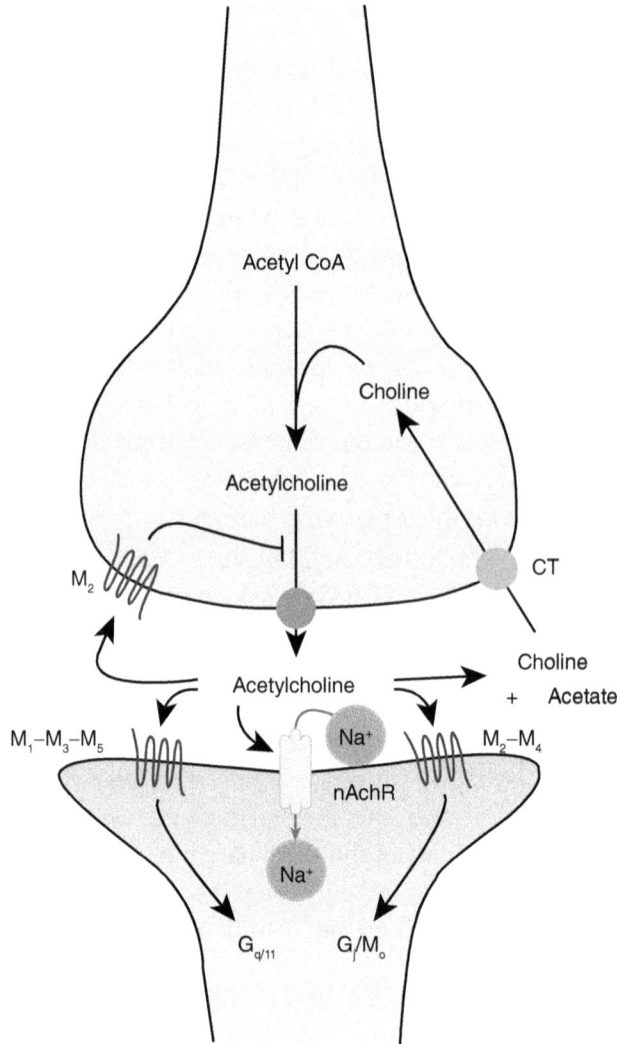

Figure 4.3 Illustration of acetylcholine binding at pre- and postsynaptic muscarinic receptors and the reuptake of choline after dissociation from the receptors

Once released from the presynaptic neuron, by binding of vesicle to the presynaptic membrane, acetylcholine binds to receptors on the postsynaptic neuron. Once the acetylcholine dissociates from these receptors, it is taken back into the presynaptic neurons by choline reuptake transporters (CTs) expressed in the membrane of the presynaptic neuron. These CTs are thought to arrive in the membrane, as the are found bound to vesicles. Therefore, as vesicles containing acetylcholine bind to the membrane for release, they also insert to CTs, ready for acetylcholine reuptake.

Source: © Smedlib / CC BY-SA 4.0

Test Yourself 4.1

From the following table select the words/phrases that are true.

Table 4.1

VMATs	acetyl-CoA and choline	reuptake transporters are only found in the presynaptic membrane	approximately 50% of choline is reuptake from the synapse and recycled
choline and acetyl-CoA are both a product of cellular metabolism	ChAT	ACh-expressing neurons are both projection neurons and interneurons	membrane potential is the most important factor in the packaging of ACh into vesicles

Answers to all Test Yourself questions are at the end of the chapter.

Outline of acetylcholine receptors

As ever, receptors for neurotransmitters are essential for neurotransmission, facilitating neuronal communication. Here we will discuss the two main classes of acetylcholine receptors and touch upon subtypes that fall under these two main classes. Like many other neurotransmitters covered in this book, there are both ionotropic and metabotropic receptors for acetylcholine. This is the main factor that distinguishes the two classes of acetylcholine receptor, with nicotinic acetylcholine receptor being ionotropic and muscarinic acetylcholine receptors being metabotropic. If you are a bit unclear about what ionotropic and metabotropic means, and how this results in different dynamic responses when the neurotransmitter binds, I recommend stopping at this point and reading the section on ionotropic and metabotropic receptors in Chapter 1.

There are several subtypes of both nicotinic and muscarinic receptor, some of which are important to us as we focus on the brain in this book, while others are less important as they are mainly found in the PNS and specifically at the neuromuscular junction.

There is only really one type of nicotinic receptor of interest when considering their effects on neurons in the brain, although there are a number when we consider the effects of muscarinic receptor activation. Muscarinic receptors have a very heterogeneous distribution pattern, with specific types being found in extremely localised brain regions and circuits. This is important to keep in mind as it suggests that different types of muscarinic receptor may contribute to specific behaviour and output controlled by specific brain regions and/or circuits.

This is now a good point to address the names of these receptors. They may seem somewhat unusual, especially compared to other neurotransmitters that have more obvious and seemingly logical names, such as $GABA_a$ and $GABA_b$. There is, however, a logic to the naming of muscarinic and nicotinic receptors, and this lies in the co-factors that can also bind to them. In the case of nicotinic receptors, it is the exogenous tertiary alkaloid nicotine and in the case of muscarinic, it is the tertiary alkaloid muscarine. Nicotine is obviously relatively well known to most of us as one of the active compounds in tobacco. However, muscarine is also an interesting compound that is highly toxic and found in a range of poisonous mushrooms.

Nicotinic receptors

Let's start with the mechanistically simpler of the two receptor classes. As mentioned, nicotinic receptors are ionotropic, therefore they are ligand gated ion channels. When acetylcholine binds to nicotinic receptors it causes a conformational shape change, which facilities the opening of the channel. This then allows both sodium and calcium to move down their electrochemical gradient into the intracellular domain of the cell.

Nicotinic receptors are typically made up of a combination of five subunits, and these types of receptors are typically said to be pentameric. The subunits include α, β, γ, δ and ϵ subunits. To make life a little easier for us, when we consider nicotinic receptors in the brain, they are believed to be made up of only α and β subunits. The fact that these different combinations exist is important for us to keep in mind because certain combinations result in much higher affinity for acetylcholine and other ligands than other combinations. This means that their dynamic effect on the neuron that expresses them depends on the subunit combination. Therefore, different brain regions and different circuits may be more affected by the local release of acetylcholine than others. The most common homeric types in the brain consist of the $\alpha7$ subunit and the most common heterometric type in the brain consists of the $\alpha4\beta2$. However, it is important to remember that there are many others, each with a distinct set of biophysical and pharmacological properties. This might make it sound like all the different 'types' of nicotinic receptors are very different, but it is important to remember that they are all ionotropic. Therefore, their dynamics, once activated, involve rapid responses with the channel typically opening within tens of microseconds, once ACh has bound. Their effects on the target neuron are therefore very rapid indeed.

Muscarinic receptors

So, now for muscarinic receptors. These are metabotropic receptors and are therefore not directly bound to ion channels. Instead, they are bound to G proteins in their intra-cellular domain, and these G proteins mediate their effect upon the target neuron when acetylcholine binds. As outlined in Chapter 1, when the signalling molecule (in this case acetylcholine) binds to these receptors it triggers a series of events that results in activation and/or deactivation of numerous intracellular secondary messengers and enzymes. Muscarinic receptors are no exception. They are generally split into two sub-classes which include M1/M3/M5 receptors and M2/M4 receptors. They are divided into these two groups as within each group they are generally considered to trigger similar intracellular signalling cascades, with the consequence generally being excitatory for M1/M3/M5 activation by acetylcholine and inhibitory for M2/M4 activation by acetylcholine. So, you may be thinking what intracellular pathways and secondary messengers does each trigger? Well, M1/M3/M5 muscarinic receptors have a $G_{q/11}$ bound proteins which, when released, can increase the activity of the enzyme PLC (phospholipase C). This results in the production of two key secondary messengers, namely IP3 (inositol 1.4.5 trisphosphate) and DAG (diacylglycerol). These secondary messengers have a number of effects on the cell: one is the activation of the enzyme PKC and another is increasing the intracellular release of Ca^{2+}, which, as we discussed in the introduction, facilitates all kinds of intracellular processes, including vesicle exocytosis and cellular proliferation. At the cell membrane, through these secondary messengers, the binding of ACh to M1/M3/M5 receptors has been shown to inhibit the M-current. This is a voltage-sensitive current that is produced by potassium efflux from the neuron. The inhibition of this by M1/M3/M5 activation results in slow depolarisation of the neuron and repetitive cell firing.

Binding of ACh at M2/M4 muscarinic receptor causes the release of the $G_{i/o}$ protein. The αi subunit of this G protein targets the enzyme adenylyl cyclase and inhibits it, consequently reducing the synthesis of the secondary messenger cAMP. This ulti-mately results in the reduced activity of PKA (protein kinase A). In addition, at the cell membrane, inwardly rectifying potassium channels (commonly referred to as GIRKs) are activated. These are activated by the βγ G-protein subunit and result in cellular hyperpolarisation. M2/M4 activation by ACh binding can also inhibit voltage gated Ca^{2+} channels, resulting in a reduction of vesicle exocytosis, among other things.

Although M1/M3/M5 and M2/M4 muscarinic receptors promote distinctly dif-ferent intracellular signalling pathways, and consequently different neuron effects when activated, there are also some commonalities. For example, both are known to upregulate MAP kinases. It is very important to keep in mind that these effects are complex, and it is not uncommon for both types of muscarinic receptors to be expressed on the same target neuron.

─Focus on research ◕─

Professor Susan Wonnacott

If you are interested in ACh receptors, especially nicotinic receptors, an excellent starting point is to search for Professor Susan Wonnacott. Professor Wonnacott has published a range of research focusing not only on the typical mechanistic and mode of action of ACh receptors, but also on their role in atypical behavioural states, especially their contribution to addiction.

A good resource to start your reading of Professor Wonnacott's research can be found via the University of Bath's research portal: https://researchportal.bath.ac.uk/en/persons/sue-wonnacott

---

Classic topics in acetylcholine research

ChAT-ChR2 mice: a 'happy accident' – what happens if we alter the amount of VAChT?

We are going to take a bit of a different approach first when discussing topics in acetylcholine research. Instead of focusing on one atypical behavioural state and considering what research tells us about the altered mechanistic, we are going to start by looking at what happens to behaviour when one of the mechanisms is altered. Specifically, we are going to explore what happens to the behaviour of animals when their levels of VAChTs increases.

This is an interesting area that came about as a result of a methodological accident of sorts. Researchers were developing a genetically-modified mouse model that expressed the channel protein channelrhodopisn (ChR2), specifically in cholinergic neurons (for more information on the importance of the protein ChR2 in current neuroscience research, see the Focus on research methods box below). They did this by inserting the code for ChR2 into the mouse DNA sequence that coded for ChAT. This inadvertently produced a repeat sequence, which resulted in the ChR2 protein being expressed in cholinergic neurons. But it also resulted in an overproduction and expression of the vesicular transporter for acetylcholine (VAChT) in these neurons. This overexpression is typically referred to as increased cholinergic tone and resulted in an animal with a hypercholinergic biological phenotype.

Initially, this was not known and in the early 2000s a number of studies with these animals reported activation of the cholinergic neurons, by stimulating the ChR2 channel protein, resulting in distinct behavioural change. But, of course, there was a huge caveat in the fact that these animals also had an increased number of VAChTs, and therefore a likely higher level of ACh ready for release into the synapse, compared to 'normal' animals. Once this was recognised, researchers began to look at the behaviour of these animals to see if the increased VAChT resulted in altered behavioural states. Multiple studies identified altered cognitive processes in these animals as well as altered motor output and motivational responses to rewarding stimuli. For example, research by Kolisnyk et al. (2016), which was one of the first studies to identify the increased cholinergic tone and expression of VAChTs in these animals, found that using a battery of behavioural and cognitive tests,[3] these animals exhibited distinct improvements in motor endurance but impaired motor learning when compared to control animals. These animals were also found to exhibit numerous cognitive deficits, including dysfunction of spatial and working memory and attentional deficits. Interestingly, other research has found contradictory improvements in the acquisition of spatial memory in these animals. For example, Nagy and Aubert (2015) found that these animals showed improvement in Morris water maze task performance, specifically exhibiting an increased number of strategies to reach the target in the water maze. On a mechanistic level, Nagy and Aubert (2015) found that this seemed to correlate with the animals exhibiting lifelong increases in the complexity of their dendritic arbours within the hippocampus.

Much of the research using these animals feeds in well with the areas we will discuss next. This will focus on the contribution of altered cholinergic mechanisms to specific atypical behaviour states, including degenerative disorders, such as Alzheimer's, addiction to drugs of abuse and developmental conditions such as autism spectrum disorder.

Focus on research methods

Optogenetics

Optogenetics has emerged over the last 15 years as one of the most influential and employed research techniques in neuroscience. It is a technique that combines genetic engineering with the use of optical technology, such as super thin fibre optics. It enables the 'activation'

(Continued)

---

[3]Including the rotarod test, the Barnes maze and the Morris water mazes.

of neurons via stimulation with a specific wavelength of light. The genetic engineering[4] element involves the insertion of the gene for a channel protein known as channelrhodopsin (ChR2). These proteins occur naturally in certain photosensitive algae and allow the flow of sodium across the membrane when triggered by light. In a neuron this would result in depolarisation. The great advantage about inserting the gene for channelrhodopisn is that, depending on where you insert it, you can target their expression to a specific population of neurons. So, in the context of the current chapter, placing the code for ChR2 directly before the DNA that codes for the VAChT will result in ChR2 expression in cholinergic neurons only. This has greatly improved our ability to be selective in our investigation of specific neuron types, which has ultimately helped us unpack the contribution of specific populations of neurons to complex behaviour and circuit-level interactions.

For an excellent outline of the technique, discussion of its research impact and consideration of its strengths and limitations I would strongly recommend the following:

Deisseroth, K. (2015). Optogenetics: 10 years of microbial opsins in neuroscience. *Nature Neuroscience*, *18*(9), 1213-1225. doi:10.1038/nn.4091

Acetylcholine and dementia

Alzheimer's disease is considered to be the most commonly occurring form of dementia, with estimates of approximately 35 million patients worldwide. It is a devastating condition both for the individual and those around them. Understanding the biological underpinnings is therefore essential. It is typified by memory loss and cognitive deficits, such as attentional deficits and difficulty organising thoughts. Interestingly, some of the first symptoms to emerge can be alterations in mood, especially increases in aggression, reduction in mood and impulsive behaviours, among many others. Alzheimer's disease is typically classed into mild and severe disease states. Towards its end state, the patient is typically left dependent on others with the inability to communicate.

You may be able to see some of the links between the research using ChAT-ChR2 mice (outlined above) and degenerative disorders such as dementia and, more specifically, Alzheimer's disease. Certainly, from the research with ChAT-ChR2 mice it is evident that alterations in acetylcholine and cholinergic neurons plays a key role in multiple types of memory, such as working and spatial memory, and aspects of

---

[4]This is not the only way to 'insert' ChR2. Many researchers have optimised the injection of ChR2 via the use of a viral vector, which transfects the cells in the locality of the injection site. This technique allows you to target specific neurons in specific regions.

memory, such as memory formation and recall. However, it is important to remember that work with these animals only suggests that increases in cholinergic tone, caused by an increased number of VAChTs, impact memory and not that it results in the diseased state of Alzheimer's disease. Although it may not do this, it is certainly compelling evidence that suggests a strong link between cholinergic tone and memory processing.

There is, however, an extensive body of research that suggests a strong connection between degeneration of cholinergic neurons, alterations in acetylcholine levels, alterations in acetylcholine's mechanisms and the pathological disease state of dementias, such as Alzheimer's disease. This connection between Alzheimer's disease and cholinergic neurons goes back to the late 1970s and early 1980s, when multiple research groups identified reductions in cholinergic markers, in post-mortem tissue samples taken from people who had received an early diagnosis of Alzheimer's disease. These reductions included various machinery, such as reductions in the enzyme ChAT that catalyses the synthesis of ACh from its precursors, and reductions in the biomarkers for the two main groups of ACh receptors, namely nicotinic and muscarinic receptors. Together, these data resulted in the formulation of the cholinergic hypothesis of Alzheimer's disease, and age-related memory dysfunction more broadly. This hypothesis is outlined in Bartus et al.'s (1982) highly influential paper, which I strongly recommend reading. Of course, this early research and the cholinergic hypothesis of Alzheimer's disease relied heavily on research with post-mortem tissue, ex vivo. This has obvious limitations and only reveals the alterations that have occurred at the end stage of the disease, rather than what we see early on. It means it is practically impossible to establish a definitive direction of effect. Based on this early research, it is just as likely that the reductions in cholinergic biomarkers could be a consequence of Alzheimer's disease rather than a cause. However, we cannot completely discount this early research as a number of these early studies did consider this issue. They managed to collect tissue from individuals who died at various stages of the illness progression, as well as taking measures of the disease severity before death. This allowed the studies, such as the work of Wilcock et al. (1982), to establish that there was a strong negative correlation between the reduction in cholinergic mechanisms, such as ChAT levels, and the severity of the dementia state. So, what these studies found is that as the severity of the dementia increased, the biomarkers for cholinergic activity decreased. By the early 2000s, techniques had developed which allowed these correlative relationships to be established in the living. This involved studies using techniques such as PET scanning.[5] These studies

---

[5]PET scans (positron emission tomography) is a technique whereby ligands that bind to certain proteins are labelled with radioactive elements. This allows you to quantify how much has bound using radiography-like techniques.

provided convincing evidence that cholinergic receptor levels reduced as the disease state progressed, suggesting that it is a likely contributory causal factor rather than consequence.

Interestingly, some cholinergic neurons seem to be lost earlier than others. Indeed, early research found that those in the basal forebrain, commonly referred to as CH4 cholinergic neurons, were some of the first to atrophy. However, this is still a hotly debated area with many researchers suggesting that alterations in the entorhinal cortex are seen before the basal forebrain. An excellent paper on this topic is that of Schmitz et al. (2016), which was also done in collaboration with the Alzheimer's Disease Neuroimaging Initiative. I suggest reading this insightful paper. The atrophy of basal forebrain cholinergic neurons is now generally considered a key biological phenotype for dementias, such as Alzheimer's disease. Why some of these cholinergic neurons degenerate earlier than others is still a somewhat unanswered question.

So, what other evidence is there that altered ACh machinery underpins memory loss/dysfunction? Well, research led by Marco Prado (2017) using animal models with disrupted VAChT, specifically in the medial forebrain bundle, found distinct reductions in working memory and aspects of spatial memory. Other studies by this group using these same animals found that the reduction of VAChT caused changes in RNA metabolism which altered key transcripts that are heavily associated with Alzheimer's disease, such as BACE1, which impacts amyloid protein and tau-protein, two of the key focal points of current research on Alzheimer's disease. We will come back to these in a moment, but what this research suggests is important to consider when contemplating the true 'cause' of Alzheimer's disease, as it suggests that, before alterations in amyloid and tau-protein comes alterations in the mechanisms of acetylcholine. It is also worth thinking back to research presented earlier in this chapter using the ChAT-ChR2 mice, as there is clear evidence of memory alterations in these animals. This provides further support for the role of altered VAChT in degenerative diseases, such as Alzheimer's disease. So, what can we conclude? Well, what we can currently say is alterations in cholinergic signalling and tone, caused by alterations in the machinery, seems to affect memory abilities and is implicated in the degeneration seen in dementias, including Alzheimer's disease. Whether they are the primary cause or simply a link in the chain is yet to be fully established.

Recently these known alterations in acetylcholine levels have been exploited as a target for rationale pharmacological therapeutics. These included numerous drugs that are reversible acetylcholinesterase inhibitors (AChE). By inhibiting AChE, these drugs essentially increase the cholinergic tone, very much as increasing the number of VAChTs does. There are a number of these drugs on the market, with more commonly used ones being Galantamine and Rivastigmine. They have unfortunately had limited success and only seem to moderately slow the progression of the disease state. This suggests that acetylcholine levels and cholinergic tone, which are mediated by mechanisms such as the VAChTs are not the key factor underpinning the degeneration of Alzheimer's. Indeed, it is important to keep in mind the treatment aetiology fallacy when considering what the efficacy of treatments tells us about the aetiology of disease states. So, even if these anti-dementia medications were more efficacious, we must not naturally assume that this means low acetylcholine levels cause Alzheimer's disease.

Because a major source of the precursors for ACh productions comes from cellular metabolism, it is likely that when there are high cellular metabolic demands, then ACh synthesis may 'lose out'. A current theory is that the link between ACh and cholinergic neuron degeneration and the dementias is neurodegenerative processes in cellular metabolism, which result in a reduced amount of the precursors for ACh synthesis. If this is the case, then alterations in ACh levels and cholinergic neuron numbers that are associated with dementias are more likely a consequence of another pathology, rather that the primary course of dementia.

The hot topics in dementia research at the moment are not focused on ACh but more centred around mutated tau-protein and beta-amyloid plaques. Discussion of these is beyond the scope of the current chapter, but we will return to them in the final chapter of this book when we discuss the contribution of gut pathology to brain disease and atypical behaviour. For now, it is best considered that these two proteins are logical primary causes of dementias such as Alzheimer's disease, as they are found in pathological forms early on in the disease state progression. It is also logical that these proteins would be involved as they damage/interfere with cellular metabolism. Consequently, they impact the production of acetylcholine, which so heavily relies on products of cellular metabolism for its synthesis. However, do keep in mind what has been outlined earlier in this chapter, as it is not a firm fact that these are the primary cause of Alzheimer's disease. Remember that there is evidence showing that altered VAChT can interfere with the production of these protein. So, the statement that ACh and cholinergic neuron loss is perhaps better thought of as related to 'disease progression' than a cause of the dementias is still a matter of debate and sustained research effort to resolve.

Acetylcholine and addiction

So, let us now consider acetylcholine and its machinery's involvement in another atypical behavioural state, namely substance use disorder. For the sake of simplicity, I will refer to this as drug addiction for the rest of this section. Drug addiction is a major issue globally. The United Nations Office on Drugs and Crime (UNODC) (www.unodc.org/unodc/en/data-and-analysis/wdr2021.html) reports that in 2018 there were an estimated quarter of a billion (269 million) people worldwide who had used drugs at least once in the previous year. There has also been a steady rise since 2006 in both the number of people who use drugs and those who have a diagnosis of substance use disorder. This report, especially Booklet 2 of the UNODC report (UNODC, 2021), also highlights the myriad of negative implications drug use can have, from the social impact to the increased risk of infections, such as hepatitis B and C. This highlights the importance of understanding the biological basis for drug addiction to aid development of rationale therapies and remediation strategies.

Several post-mortem and imaging studies identified alterations in the expression of the mechanisms behind ACh packaging and degradation, as well as ACh receptors in drug addiction. For example, research using post-mortem tissue from humans, who were chronic methamphetamine addicts, has shown that VAChT levels are elevated in this group specifically in parts of the brain, such as the dorsal striatum. These findings are not that surprising, especially when considering the known role of cholinergic neurons within the region in processing reward and modulating dopamine release, which is heavily involved in action selection. It is likely that these alterations are a result of taking the drugs and therefore may contribute to the development of an addiction. Some argue that these differences may underpin an 'addictive type trait' which could potentially result in a greater vulnerability for certain groups becoming addicts. This is difficult to prove conclusively and would require large-scale prospective studies. What we can say is that there definitely seems to be a role played by alteration of cholinergic mechanistic in drug addiction. As well as this controversy, there are also a number of studies that suggest that VAChT and ChAT level are not altered in the brains of cocaine and heroin addicts. We therefore might consider that VAChT and/or ChAT alterations may be drug-specific rather than an underpinning factor in all drug addiction.

Because of the understandable ethical limitations surrounding the research you can do with humans with a drug addiction, other methods employing animal models have been extensively utilised. One typical method using animals to explore the mechanistic of addiction is known as ICSS (intracranial self-administration). This technique involves the implantation of a stimulating electrode into various brain regions, known to be involved in reward responses and hijacked by drugs of abuse. Indeed, addiction has commonly been conceptualised as a disease of the brains reward system. Animals are then given the free choice to press a lever to activate this electrode and stimulate the brain region. Much early evidence has shown that ACh injected into the reward circuitry, such as the ventral tegmental area, increases ICSS, suggesting that ACh modulates this region and the the number of times an animal wishes to stimulate it. This has long been considered a valid model of the increased seeking of the addictive substance seen in drug addicts. However, ICSS is not the only animal paradigm used, and others, such as conditioned place preference (CPP), self-administration and reinstatement, have provided much information on the roles of ACh and its machinery in underpinning addiction. For example, CPP is used to assess the likelihood of relapse to addictive substance. Multiple studies using ACh receptor agonists have demonstrated that they can induce this conditioned place preference. Also, genetically-modified animal models that lack muscarinic ACh receptors have been shown to exhibit a reduced CPP to cocaine compared to 'typical' animals. So, what the research evidence covered so far suggests is that increased cholinergic tone is linked to the behaviour we typically associate with drug addiction.

Let's continue this line of thought by returning to the ChAT-ChR2 animals discussed earlier in this chapter. These animals have a hypercholinergic biological phenotype and therefore an increased cholinergic tone. In addition, there is a significant amount of research evidence which suggests they display a distinct and increased sensitivity to drugs of abuse, such as amphetamine and cocaine. It is not just limited to sensitivity. These animals have also been shown to exhibit increased locomotor activity to amphetamine and increased stereotypy,[6] both of which are 'typical' hallmarks of humans with a drug addiction. Studies such as Janickova et al. (2017) have also shown increased locomotor sensitisation in this mouse model to cocaine. This suggest that these alterations may well underpin drug addiction, rather than, as mentioned earlier in the chapter, simply being related to specific drugs. This research also ties back to the involvement of ACh alterations in the reward circuits with a specific focus on cholinergic interneurons in the striatum.

What about other cholinergic mechanistics? Well, one thing we have only mentioned briefly up to now is the involvement of ACh receptors. There is a vast amount of research that implicates cholinergic receptors in drug addiction. This is a complex literature but what it generally seems to suggest is that it is specific isoforms (based on their combination of subunits) of the nicotinic and muscarinic receptors that are specifically related to addiction. For example, the α7 nicotinic receptors have been shown by a number of studies to modulate the responses to morphine in the conditioned place preference paradigm. More specifically, these studies have tended to implicate alterations in these specific receptors to underpin relapse in addiction. This is an important point because relapse is one of the biggest issues that stands in the way of individuals with a drug addiction overcoming their addiction; therefore, identifying targets for pharmacological intervention, such as the α7 nicotinic receptors is highly prized. It is also important to think what this means for ACh receptors'

---

[6]Stereotypy is defined as the repetition of a behaviour without any obvious goal-directed purpose.

involvement in addiction. Remember that addiction constitutes multiple stages (see the ideas of George Koob, outlined in the section covering D3 receptors' involvement in addiction, in Chapter 5 on dopamine). Therefore, what we see here is that the ACh receptor may be implicated in one stage of addiction, but may not underpin all stages and the whole experience of addiction.

Let's now focus a little on a couple of specific abused substances, namely nicotine (in the form of tobacco) and alcohol. Following on from the previous paragraph, it is logical, based on their name, that we would expect one key cholinergic receptor group to be implicated in nicotine addiction, namely nicotinic receptors. This is indeed the case, with nicotinic receptors being key targets for nicotine, which is an allosteric agonist at these receptor sites. Again, it is the case that different isoforms of these receptors are implicated to varying degrees in tobacco and alcohol addiction. What is particularly interesting here is that a number of human genome wide association studies (GWAS) have actually identified polymorphisms in nicotinic receptor genes, which seems to convey an increased risk of cigarette/tobacco dependence. Some of the key players seem to be $\alpha3$, $\alpha5$ and $\beta4$ containing nicotinic receptors. For an excellent review of this literature, I strongly recommend the work of Brunzell, Stafford, and Dixon (2015). It is not just these GWAS that implicate subtypes of nicotinic receptors in cigarette/tobacco addiction. Research using genetically-modified animals also seems to present strong evidence. For example, research using a knockout mouse which presents with a non-functioning $\beta2$ nicotinic subunit shows that they lack key addictive responses to nicotine, such as reduced self-administration of nicotine and no conditioned place preference to nicotine administration. It is important to remember here that neurotransmitters do not work in isolation. It is particularly important in the context of addiction as much of the research evidence strongly implicates the interaction between acetylcholine and dopamine in the reinforcing effects of nicotine, and consequently nicotine addiction. Evidence, such as the research of Picciotto and colleagues (1998, 2012), suggests that nicotine mediates its rewarding effects by specifically targeting nicotinic receptors on dopaminergic terminals within the ventral striatum.[7]

So, what about alcohol addiction? Well, what is interesting is that there is a strong relationship between nicotine and alcohol addiction. It is extremely common for the substances to be co-abused. In fact, it is estimated that approximately 80–90% of alcoholics are also smokers. This was often explained using psychosocial models,

---

[7]The ventral striatum is part of the striatum, which is the main input to a group of nuclei, collectively referred to as the basal ganglia. There is a large body of research which heavily implicates the ventral striatum in reward and pleasure responses, with much research showing that drugs of abuse have their effects by hijacking this circuitry.

for example, the availability of cigarettes when drinking, etc. Evidence has emerged that the co-occurrence may not be mediated by psychosocial factors but be a consequence of shared biological alterations. Lots of pharmacological studies with nicotinic receptor antagonists (mecamylamine) show decreased/reduced ethanol intake in animals and agonists of nicotinic receptors increase ethanol intake. However, as is typically the case, there is a move in the literature towards more specific antagonists and agonists for nicotinic receptor subtypes (dependent on pentameric organisation) with a suggestion that it is not all nicotinic receptors, but certain subtypes that are likely implicated in increased ethanol intake and alcohol addiction.

So, in conclusion, much of the discussion in this section seems to present a convincing case for the involvement of ACh and its machinery in addiction. What we see is that it is perhaps not ACh *per se*, but alterations in specific mechanisms, such as alterations in the different isoforms of nicotinic receptors, that are implicated in cigarette and alcohol addiction. However, many believe that rather than AChs machinery being the key component underpinning addiction, it is a link in the chain. A good point to remember is that neurotransmitters do not have their effect in isolation and often involve complex interactions with other molecules and other neurotransmitters. ACh, in the context of addiction, is a good case in point. Much research suggests that the role of ACh in addiction is fundamentally linked to its interconnection with dopamine and dopamine release. Indeed, much research has shown that the involvement of ACh receptors is linked to their expression on dopaminergic neurons, in regions such as the VTA and subsequent modulation of dopamine release.

Novel topics in acetylcholine research

Acetylcholine and autism spectrum disorder

Autism spectrum disorder (ASD) is classified as a developmental disorder by the American Psychiatric Association's DSM-5 (2013). It is characterised by deficits in two core domains: social communication deficits and RRBIs (restricted, repetitive behaviours and interests). It might seem strange, but there is a logical connection between ASD and addiction, especially RRBIs and addiction. You may be able to make the connections if you consider the previous section on ACh and addiction. A well-known pathology that develops in many drug addicts is punding, which is a form of stereotypical repetitive behaviour that does not have a seemingly clear goal. It also involves intense interests in objects that, again, seemingly have little reward or goal value. This sounds strikingly like the types of restricted and repetitive behaviours and interests often exhibited by individuals with a diagnosis of ASD. Therefore, if ACh and its mechanisms are so heavily associated with these aspects of addiction, it seems rational that alterations in ACh mechanisms may, to some degree, underpin certain symptoms of ASD.

There is indeed emerging evidence to make this connection between ASD symptomology and ACh mechanisms. Interestingly, research going back to the late 1990s and early 2000s identified that cholinergic neurons of the basal forebrain, so heavily implicated in dementias, are also found to be altered in children with a diagnosis of ASD, with a distinct change in number, size and structure. It is not just a case that there are evident changes in the expression and morphology of cholinergic neurons. A number of early studies also reported reductions in ACh precursors in those with a diagnosis of ASD, specifically reduced levels of the precursor choline.

GWAS (genome wide association studies) implicate some of the same genes as are commonly seen in drug addiction, as mentioned in the previous section of this chapter. These genes typically code for specific cholinergic receptors and it is with these receptors that the main, emerging link between ASD and ACh lies. So, what about cholinergic receptors? Well, more recently, the research literature has started to focus more heavily on the links between the various nicotinic and muscarinic receptor subtypes and ASD. As is evident with the other atypical behavioural states we have discussed in this chapter, cholinergic receptors are becoming the key cholinergic mechanism implicated in ASD. Much of the post-mortem literature has identified reductions in M1 muscarinic receptors and various different nicotinic subunits in those individuals with a diagnosis of ASD. An interesting study by Wang et al. (2015) exposed an animal model of ASD (BTBR mice) to varying doses of nicotine. It was found that at certain doses there was a distinct reduction in repetitive behaviours and increased levels of social interaction, suggesting the involvement of nicotinic ACh receptors in modulating these core deficits in ASD.

However, other mechanisms related to ACh may also be involved in ASD. Recently, researchers have started to investigate the use of various cholinergic agonists as potential therapeutic interventions for those with a diagnosis of ASD. Two common drugs are Donepezil and Galamatine. These are AChEIs (choline acetyltransferase inhibitors), and they therefore increase cholinergic tone by inhibiting ACh degradation. They have been trialled extensively to help slow the progression of dementia. Research by Karvat and Kimchi (2014) was some of the first to show (in an animal model of ASD-like behaviour, named BTBR mice) that administration of Donepezil by i.p. injection[8] and via direct injection into the striatum results in reduced ASD-like

---

[8]i.p. injection typically involves injection into the fatty tissue of the stomach. The drug then diffuses into the vasculature and makes its way to the brain.

symptomology, including reducing cognitive rigidity and increasing social interaction. AChEIs have also been trialled in humans with a diagnosis of ASD with some evidence of reduced symptomology, although these trials have typically been relatively small scale; for example, the prospective study using Galantamine of Nicolson, Craven-Thuss, and Smith (2006) only had a sample size of 13.

So what can we conclude about the link between ASD and alterations in ACh mechanisms? Well, it seems likely that ACh plays some role in the symptomology of ASD and it seems most likely that these are mediated by ACh receptors. However, it is important to remember that ASD is a condition typified by a number of symptoms and core deficits, and currently no research definitively provides evidence for the alterrations in ACh mechanisms underpinning all these signs and symptoms. It is therefore, at the moment, a more realistic conclusion to suggest ACh is implicated in some deficits exhibited by those with an ASD diagnosis, but they are unlikely to be the only biological mechanism underpinnng the full range of symptoms.

─Test Yourself 4.2─

Find the connections. Which of the following are common factors implicated in at least two of the three atypical behaviours outlined in this chapter:

1. Alterations in ACh and the basal ganglia circuitry. The basal ganglia includes: the VTA, the striatum and parts of the basal forebrain (ventral pallidum), among other structures.
2. The potential efficacy of AChEIs for the treatment of symptoms.
3. GWAS (genome wide association studies) implicate some of the same genes, especially those genes related to ACh receptors.
4. Evidence from ChAT-ChR2 mice can be used to support the involvement of ACh machinery in the atypical behaviours.

Conclusion

The investigation of acetylcholine and its mechanism has a rich seam of research. Its synthesis, packaging, degradation, reuptake and receptors are at times a complex affair, but this is what makes these mechanisms such potentially fruitful focuses of research. It is indeed notable that ACh mechanisms are evidently implicated in a range of conditions and clinical disorders, as we have discussed in this chapter. This tells us something about the primary importance of acetylcholine to our experience. At several points throughout this chapter, I have mentioned the link between acetylcholine and the reward system. It is worth keeping these links in mind. An increasing body of research evidence suggests that ACh and cholinergic neurons are key modulators of this circuitry. As such, we would expect alterations in ACh mechanisms to underpin a myriad of different atypical behavioural states, where altered reward values and the selection of actions based on reward value is key to the behaviour. It is also important to keep in mind that ACh's mechanisms do not act in isolation. Again, the reward system is a good example of this, as the interaction between AChs (especially ACh receptors) and dopamine release is emerging as a key mechanism underpinning action selection and mediating the reward value of objects.

Test Yourself Answers

Test Yourself 4.1

Table 4.2

	acetyl-CoA and choline		approximately 50% of choline is reuptake from the synapse and recycled
choline and acetyl-CoA are both a product of cellular metabolism	ChAT	ACh expressing neurons are both projection neurons and interneurons	

Test Yourself 4.2

This is a bit of trick question because all four are implicated in the relationship between ACh and at least two of the three atypical behaviours outlined in this chapter.

1. Alterations in ACh and the basal ganglia circuitry. The basal ganglia include: the VTA, the striatum and parts of the basal forebrain (ventral pallidum), among other structures. **There are clear links with ACh in the basal ganglia and both ASD and addiction.**
2. The potential efficacy of AChEIs for the treatment of symptoms. **There are clear links, with AChEIs being trialled as treatment for both Alzheimer's disease and ASD, although there is less definitive research evidence related to their use as a treatment for ASD.**
3. GWAS (gemone wide association studies) implicate some of the same genes, especially those genes related to ACh receptors. **There are clear links with ACh receptor genes, addiction and ASD.**
4. Evidence from ChAT-ChR2 mice can be used to support the involvement of ACh machinery in the atypical behaviours. **There are clear links with the ChAT-ChR2 mice, memory alterations and drug addiction.**

References

Key information

Loewi, O. (1936). The chemical transmission of nerve action. *Nobel lecture*.

Hendrickson, L., Guildford, M., & Tapper, A. (2013). Neuronal nicotinic acetylcholine receptors: Common molecular substrates of nicotine and alcohol dependence. *Frontiers in Psychiatry, 4*(29). doi:10.3389/fpsyt.2013.00029

Arvidsson, U., Riedl, M., Elde, R., & Meister, B. (1997). Vesicular acetylcholine transporter (VAChT) protein: a novel and unique marker for cholinergic neurons in the central and peripheral nervous systems. *J Comp Neurol, 378*(4), 454–467.

Inazu, M. (2019). Functional expression of choline transporters in the blood-brain barrier. *Nutrients, 11*(10), 2265. doi:10.3390/nu11102265

Pahud, G., Salem, N., van de Goor, J., Medilanski, J., Pellegrinelli, N., & Eder-Colli, L. (1998). Study of subcellular localization of membrane-bound choline acetyltransferase in Drosophila central nervous system and its association with membranes. *Eur J Neurosci, 10*(5), 1644–1653. doi:10.1046/j.1460-9568.1998.00177.x

Carroll, P. T. (1994). Membrane-bound choline-O-acetyltransferase in rat hippocampal tissue is associated with synaptic vesicles. *Brain Research, 633*(1), 112–118. doi:https://doi.org/10.1016/0006-8993(94)91529-6

Fisher, S. K., & Wonnacott, S. (2012). Chapter 13 - Acetylcholine. In S. T. Brady, G. J. Siegel, R. W. Albers, & D. L. Price (Eds.), *Basic Neurochemistry (Eighth Edition)* (pp. 258–282). New York: Academic Press.

Ferguson, S. M., & Blakely, R. D. (2004). The choline transporter resurfaces: new roles for synaptic vesicles? *Mol Interv, 4*(1), 22–37. doi:10.1124/mi.4.1.22

CHaT ChR2 mice – a 'happy accident'

Janickova, H., Prado, V. F., Prado, M. A. M., El Mestikawy, S., & Bernard, V. (2017). Vesicular acetylcholine transporter (VAChT) over-expression induces major modifications of striatal cholinergic interneuron morphology and function. *Journal of Neurochemistry, 142*(6), 857–875. doi:10.1111/jnc.14105

Deisseroth, K. (2015). Optogenetics: 10 years of microbial opsins in neuroscience. *Nature Neuroscience, 18*(9), 1213–1225. doi:10.1038/nn.4091

Acetylcholine and addiction

American Psychiatric Association. (2013). Diagnostic and statistical manual of mental disorders (5th ed.) (DSM-5). Washington, DC: APA.

Hendrickson, L., Guildford, M., & Tapper, A. (2013). Neuronal nicotinic acetylcholine receptors: common molecular substrates of nicotine and alcohol dependence. *Frontiers in Psychiatry, 4*(29). doi:10.3389/fpsyt.2013.00029

Janickova, H., Prado, V. F., Prado, M. A. M., El Mestikawy, S., & Bernard, V. (2017). Vesicular acetylcholine transporter (VAChT) over-expression induces major modifications of striatal cholinergic interneuron morphology and function. *Journal of Neurochemistry, 142*(6), 857–875. doi:10.1111/jnc.14105

Picciotto, M. R., Zoli, M., Rimondini, R., Léna, C., Marubio, L. M., Pich, E. M., . . . Changeux, J.-P. (1998). Acetylcholine receptors containing the β2 subunit are involved in the reinforcing properties of nicotine. *Nature, 391*(6663), 173–177. doi:10.1038/34413

Picciotto, M. R., Higley, M. J., & Mineur, Y. S. (2012). Acetylcholine as a neuromodulator: cholinergic signaling shapes nervous system function and behavior. *Neuron, 76*(1), 116–129. doi:10.1016/j.neuron.2012.08.036

Wright, V., Georgiou, P., Bailey, A., Heal, D., Bailey, C., & Wonnacott, S. (2018). Inhibition of alpha7 nicotinic receptors in the ventral hippocampus selectively attenuates reinstatement of morphine-conditioned place preference and associated changes in AMPA receptor binding: Alpha7 nAChRs in ventral hippocampus mediate reinstatement of morphine-CPP. *Addict Biol, 24*. doi:10.1111/adb.12624

Leonard, S., Mexal, S., & Freedman, R. (2007). Smoking, genetics and schizophrenia: Evidence for self medication. *Journal of Dual Diagnosis, 3*(3–4), 43–59. doi:10.1300/J374v03n03_05

Crittenden, J. R., Lacey, C. J., Lee, T., Bowden, H. A., & Graybiel, A. M. (2014). Severe drug-induced repetitive behaviors and striatal overexpression of VAChT in ChAT-ChR2-EYFP BAC transgenic mice. *Front Neural Circuits, 8*, 57. Retrieved from www.frontiersin.org/article/10.3389/fncir.2014.00057

Siegal, D., Erickson, J., Varoqui, H., Ang, L., Kalasinsky, K. S., Peretti, F. J., . . . Kish, S. J. (2004). Brain vesicular acetylcholine transporter in human users of drugs of abuse. *Synapse, 52*(4), 223–232. doi:10.1002/syn.20020

Gould, R. W., Duke, A. N., & Nader, M. A. (2014). PET studies in nonhuman primate models of cocaine abuse: translational research related to vulnerability and neuroadaptations. *Neuropharmacology, 84*, 138–151. doi:10.1016/j.neuropharm.2013.02.004

Williams, M. J., & Adinoff, B. (2008). The role of acetylcholine in cocaine addiction. *Neuropsychopharmacology : official publication of the American College of Neuropsychopharmacology, 33*(8), 1779–1797. doi:10.1038/sj.npp.1301585

Redgrave, P., & Horrell, R. I. (1976). Potentiation of central reward by localised perfusion of acetylcholine and 5-hydroxytryptamine. *Nature, 262*(5566), 305–307. doi:10.1038/262305a0

Singh, J., Desiraju, T., & Raju, T. R. (1997). Cholinergic and GABAergic modulation of self-stimulation of lateral hypothalamus and ventral tegmentum: effects of carbachol, atropine, bicuculline, and picrotoxin. *Physiol Behav, 61*(3), 411–418. doi:10.1016/s0031-9384(96)00452-0

Siegal, D., Erickson, J., Varoqui, H., Ang, L., Kalasinsky, K. S., Peretti, F. J., . . . Kish, S. J. (2004). Brain vesicular acetylcholine transporter in human users of drugs of abuse. *Synapse, 52*(4), 223–232. doi:10.1002/syn.20020

Fink-Jensen, A., Fedorova, I., Wörtwein, G., Woldbye, D. P., Rasmussen, T., Thomsen, M., . . . Basile, A. (2003). Role for M5 muscarinic acetylcholine receptors in cocaine addiction. *J Neurosci Res, 74*(1), 91–96. doi:10.1002/jnr.10728

Brunzell, D. H., Stafford, A. M., & Dixon, C. I. (2015). Nicotinic receptor contributions to smoking: insights from human studies and animal models. *Curr Addict Rep, 2*(1), 33–46. doi:10.1007/s40429-015-0042-2

Acetylcholine and dementia

Ferreira-Vieira, T. H., Guimaraes, I. M., Silva, F. R., & Ribeiro, F. M. (2016). Alzheimer's disease: Targeting the cholinergic system. *Current Neuropharmacology, 14*(1), 101–115. doi:10.2174/1570159x13666150716165726

Prado, V. F., Janickova, H., Al-Onaizi, M. A., & Prado, M. A. M. (2017). Cholinergic circuits in cognitive flexibility. *Neuroscience, 345*, 130–141. doi:https://doi.org/10.1016/j.neuroscience.2016.09.013

Kolisnyk, B., Al-Onaizi, M., Soreq, L., Barbash, S., Bekenstein, U., Haberman, N., . . . Prado, M. (2016). Cholinergic surveillance over hippocampal RNA metabolism and Alzheimer's-like pathology. *Cerebral Cortex, 27*, 1–15. doi:10.1093/cercor/bhw177

Szutowicz, A., Bielarczyk, H., Jankowska-Kulawy, A., Pawełczyk, T., & Ronowska, A. (2013). Acetyl-CoA the key factor for survival or death of cholinergic neurons in course of neurodegenerative diseases. *Neurochemical Research, 38*(8), 1523–1542. doi:10.1007/s11064-013-1060-x

Schliebs, R., & Arendt, T. (2006). The significance of the cholinergic system in the brain during aging and in Alzheimer's disease. *Journal of Neural Transmission (Vienna, Austria : 1996), 113*(11), 1625–1644. doi:10.1007/s00702-006-0579-2

Wilcock, G. K., Esiri, M. M., Bowen, D. M., & Smith, C. C. (1982). Alzheimer's disease. Correlation of cortical choline acetyltransferase activity with the severity of dementia and histological abnormalities. *J Neurol Sci, 57*(2–3), 407–417. doi:10.1016/0022-510x(82)90045-4

Grothe, M. J., Ewers, M., Krause, B., Heinsen, H., & Teipel, S. J. (2014). Basal forebrain atrophy and cortical amyloid deposition in nondemented elderly subjects. *Alzheimers Dement, 10*(5 Suppl), S344–353. doi:10.1016/j.jalz.2013.09.011

Al-Onaizi, M. A., Parfitt, G. M., Kolisnyk, B., Law, C. S., Guzman, M. S., Barros, D. M., . . . Prado, V. F. (2017). Regulation of cognitive processing by hippocampal cholinergic tone. *Cereb Cortex, 27*(2), 1615–1628. doi:10.1093/cercor/bhv349

Bartus, R. T., Dean, R. L., Beer, B., & Lippa, A. S. (1982). The cholinergic hypothesis of geriatric memory dysfunction. *Science, 217*(4558), 408. doi:10.1126/science.7046051

Marucci, G., Buccioni, M., Ben, D. D., Lambertucci, C., Volpini, R., & Amenta, F. (2021). Efficacy of acetylcholinesterase inhibitors in Alzheimer's disease. *Neuropharmacology, 190,* 108352. doi:https://doi.org/10.1016/j.neuropharm.2020.108352

Bowen, D. M., Smith, C. B., White, P., & Davison, A. N. (1976). Neurotransmitter-related enzymes and indices of hypoxia in senile dementia and other abiotrophies. *Brain, 99*(3), 459–496. doi:10.1093/brain/99.3.459

Schmitz, T. W., Nathan Spreng, R., Weiner, M. W., Aisen, P., Petersen, R., Jack, C. R., . . . The Alzheimer's Disease Neuroimaging, I. (2016). Basal forebrain degeneration precedes and predicts the cortical spread of Alzheimer's pathology. *Nature Communications, 7*(1), 13249. doi:10.1038/ncomms13249

Nagy, P. M., & Aubert, I. (2015). Overexpression of the vesicular acetylcholine transporter enhances dendritic complexity of adult-born hippocampal neurons and improves acquisition of spatial memory during aging. *Neurobiology of Aging, 36*(5), 1881–1889. doi:https://doi.org/10.1016/j.neurobiolaging.2015.02.021

Acetylcholine and ASD

Marotta, R., Risoleo, M. C., Messina, G., Parisi, L., Carotenuto, M., Vetri, L., & Roccella, M. (2020). The Neurochemistry of autism. *Brain Sciences, 10*(3). doi:10.3390/brainsci10030163

Fasano, A., Barra, A., Nicosia, P., Rinaldi, F., Bria, P., Bentivoglio, A. R., & Tonioni, F. (2008). Cocaine addiction: from habits to stereotypical-repetitive behaviors and punding. *Drug Alcohol Depend, 96*(1-2), 178–182. doi:10.1016/j.drugalcdep.2008.02.005

Deutsch, S. I., Urbano, M. R., Neumann, S. A., Burket, J. A., & Katz, E. (2010). Cholinergic abnormalities in autism: is there a rationale for selective nicotinic agonist interventions? *Clin Neuropharmacol, 33*(3), 114–120. doi:10.1097/WNF.0b013e3181d6f7ad

Martin-Ruiz, C. M., Lee, M., Perry, R. H., Baumann, M., Court, J. A., & Perry, E. K. (2004). Molecular analysis of nicotinic receptor expression in autism. *Brain Res Mol Brain Res, 123*(1-2), 81–90. doi:10.1016/j.molbrainres.2004.01.003

Karvat, G., & Kimchi, T. (2014). Acetylcholine elevation relieves cognitive rigidity and social deficiency in a mouse model of autism. *Neuropsychopharmacology, 39*(4), 831–840. doi:10.1038/npp.2013.274

Kemper, T. L., & Bauman, M. (1998). Neuropathology of Infantile Autism. *Journal of Neuropathology & Experimental Neurology, 57*(7), 645–652. doi:10.1097/00005072-199807000-00001

Takechi, K., Suemaru, K., Kiyoi, T., Tanaka, A., & Araki, H. (2016). The α4β2 nicotinic acetylcholine receptor modulates autism-like behavioral and motor abnormalities in pentylenetetrazol-kindled mice. *European Journal of Pharmacology, 775*, 57–66. doi:https://doi.org/10.1016/j.ejphar.2016.02.021

Wang, L., Almeida, L. E. F., Spornick, N. A., Kenyon, N., Kamimura, S., Khaibullina, A., . . . Quezado, Z. M. N. (2015). Modulation of social deficits and repetitive behaviors in a mouse model of autism: the role of the nicotinic cholinergic system. *Psychopharmacology, 232*(23), 4303–4316. doi:10.1007/s00213-015-4058-z

Nicolson, R., Craven-Thuss, B., & Smith, J. (2006). A prospective, open-label trial of galantamine in autistic disorder. *J Child Adolesc Psychopharmacol, 16*(5), 621–629. doi:10.1089/cap.2006.16.621

Ghaleiha, A., Ghyasvand, M., Mohammadi, M. R., Farokhnia, M., Yadegari, N., Tabrizi, M., . . . Akhondzadeh, S. (2014). Galantamine efficacy and tolerability as an augmentative therapy in autistic children: A randomized, double-blind, placebo-controlled trial. *J Psychopharmacol, 28*(7), 677–685. doi:10.1177/0269881113508830

5

DOPAMINE

Introduction

For at least the last 50 years, the dopamine system in the brain has been identified as playing a key role in multiple behaviours and functions, from controlling motor function, selecting appropriate actions based on experience of reward, to mood and perception. Of course, this also means that when the dopaminergic system dysfunctions, it results in alterations to motor function, action selection, mood and perception. Indeed, dysfunction of the brain's dopamine system is firmly embedded in the literature on degenerative disorders, such as Parkinson's disease and psychiatric illnesses, such as major depressive disorder (MDD), schizophrenia and substance use disorder. In this chapter we will first look at the machinery of dopamine synthesis, packaging and reuptake before taking a closer look at how dysfunction of this machinery is implicated in various atypical states.

Outline of dopamine's synthesis, packaging and reuptake

Like any system, the dopamine system relies on various workers to help put the molecule together (synthesis), get it ready for delivery (packaging) and take back any returns (reuptake). All these constituent workers, which include proteins and enzymes, are essential for the smooth running of the system. Therefore, if one of them dysfunctions, this has implications for the whole system. It often results in alterations to the amount of dopamine that can be produced, released and reuptaken. Many disorders, such as schizophrenia, are classically characterised as disorders typified by imbalances in the levels of dopamine. However, there can also be alterations in how the receptors for dopamine function, which can have massive implications for its functional effect. We will discuss this in more detail later in the chapter.

Dopamine's synthesis

Dopamine belongs to a family of compounds known as catecholamines. In Figure 5.1, you can see the structure of dopamine, which is very similar to the structure of serotonin (see Figure 6.1 in Chapter 6). Their similarity is because they are related and part of the broad group of neurotransmitters referred to as monoamines. This group also includes epinephrine and norepinephrine, which are beyond the scope of this book. However, as a point of interest, dopamine is an intermediate compound[1] in the biosynthesis of epinephrine and norepinephrine.

---

[1]This means dopamine is formed as part of the chain of reaction that results, finally, in the production of epinephrine and norepinephrine.

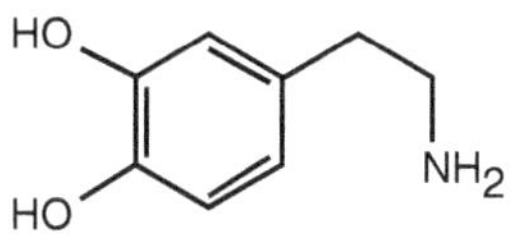

Figure 5.1   The molecular structure of dopamine

As an amine, this structure is very similar to that of other amines, such as serotonin.

Probably one of the main reasons why dopamine has garnered such research interest over the last half century is that, in the mammalian brain, there are only a few distinct nuclei (these are large collections of neurons) that contain dopamine.[2] The two main nuclei in the mammalian brain are the ventral tegmental area (VTA) and the substantia nigra pars compacta (SNc), which you can see in Figure 5.2. We will revisit these when we discuss the role of dopamine in disorders and behaviour later on in the chapter. Dopaminergic neurons in these nuclei send axonal projections in a rather complex pattern to many other locations in the brain. One specific target is the striatum, which is a part of the basal ganglia and is heavily implicated in reward processes and motor control.

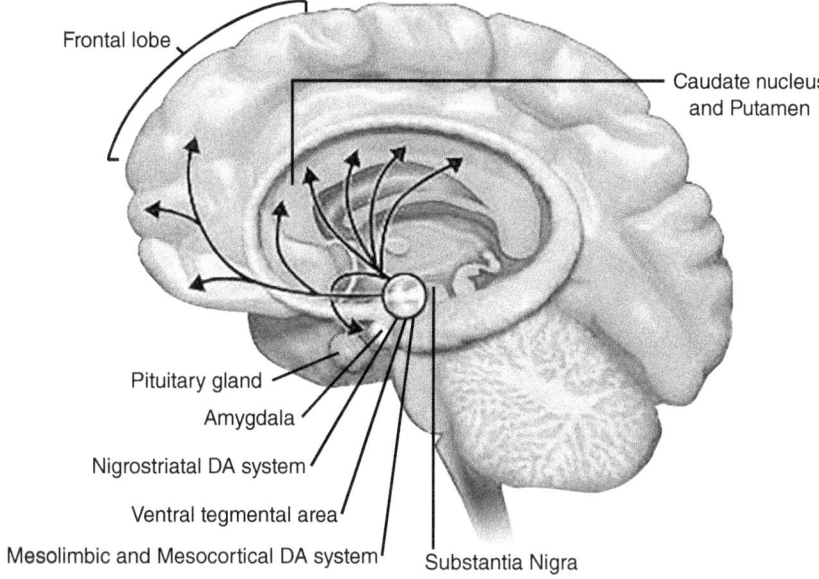

Figure 5.2   The location of dopaminergic neurons and their projection networks in the brain

Dopaminergic neurons are largely located in two specific nuclei, namely the substantia nigra pars compacta and the ventral tegmental area. These neurons project to a number of other territories in the brain, including the cortex and striatum.

Source: © Bruce Blaus / CC BY-SA 4.0

---

[2]Although this is true to some degree, recent research has shown that there are in fact dopaminergic neurons expressed within other brain regions and that they are likely interneurons.

Hopefully based on what was mentioned in the previous section, you are starting to make connections between dopamine, brain regions and its roles in behaviour.

So how is it synthesised? The starting point for catecholamine synthesis is the essential amino acid phenylalanine. Being an essential amino acid means that it cannot be synthesised in the mammalian body, therefore it must be ingested as part of our diet. Common foods that we get it from include meat, fish, cheese and lentils. The enzyme phenylalanine hydroxylase (removal of an oxygen and hydrogen molecule) can then catalyse the production of tyrosine. For many, tyrosine is considered the starting point of dopamine synthesis as this is a non-essential amino acid, which can be ingested as part of a balanced diet, but can also by formed in the liver by hydroxylation of phenylalanine. Once we have tyrosine, two more steps and two more enzymes are needed before we get dopamine. It is worth considering that, like any other systems, the more parts there are to the system the more potential areas there are for things to go wrong.

At this point, synthesis is now taking place in the nerve terminal. In the first step, tyrosine is hydrolysed to dihydroxyphenylalanine (DOPA) by the enzyme tyrosine hydroxylase. This is one of the most important enzymes in the biosynthesis of catecholamines as it is the rate limiting enzyme, essentially controlling the speed at which dopamine, and the other catecholamines, are produced. It is also important as it is only expressed in dopaminergic neurons in the brain, and is therefore a key target for researchers wanting to visualise these neurons. The acronym DOPA might be familiar to some of you as the synthetic form of it, L-DOPA, is a common treatment used to control tremors and rigidity of movement, which are key symptoms of degenerative motor disorders, such as Parkinson's disease.

Test Yourself 5.1

In the late 1960s, reports came out that patients who were being treated with L-DOPA were reporting some of the classic symptomology of schizophrenia, such as hallucinations and delusions. What would this mean for the role of dopamine in both Parkinson's patients and patients with schizophrenia?

1.   Both schizophrenia and Parkinson's disease are caused by low levels of dopamine.
2.   Both schizophrenia and Parkinson's disease are caused by high levels of dopamine.
3.   Schizophrenia is likely a result of increased levels of dopamine, while Parkinson's disease is likely a result of low levels of dopamine.
4.   Schizophrenia is likely a result of decreased levels of dopamine, while Parkinson's disease is likely a result of high levels of dopamine.
5.   All of the above.
6.   None of the above.

Answers to all Test Yourself questions are at the end of the chapter

Once DOPA is formed, the final step involves the enzyme dopa decarboxylase, removing the carboxyl group to produce dopamine for DOPA. These steps can be seen in Figure 5.3. As mentioned earlier, it is important to remember that this is not necessarily the end of the process for catecholamine synthesis. Remember, dopamine is only one of three major catecholamines, including norepinephrine and epinephrine. However, at this point, dopamine is now ready to perform its role as a neuromodulator.

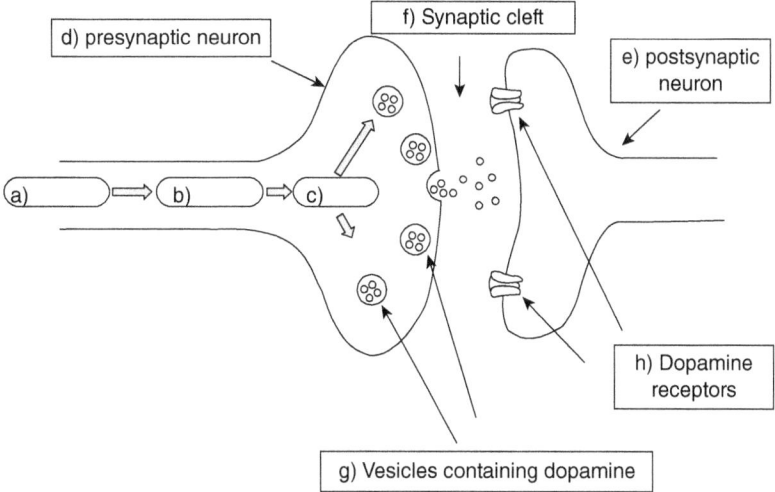

Figure 5.3   The two-step process of dopamine synthesis

Dopamine is synthesised in a two-step process from the precursor tyrosine and L-DOPA in the terminal end of the presynaptic neuron. (a) = tyrosine, (b) L-DOPA, (c) dopamine.

Source: Adapted by permission from Springer Nature: Youdim, M., Edmondson, D., & Tipton, K. (2006). The therapeutic potential of monoamine oxidase inhibitors. *Nature Reviews Neuroscience, 7*, 295–309. Copyright (2006).

Dopamine packaging

Dopamine, free in the cytoplasm, is highly neurotoxic. It is therefore essential that once it is synthesised it is rapidly packaged for release, by exocytosis, out of the neuron and into the synapse. This involves the loading of the transport packages, known as vesicles. For a refresher on this, look back to Chapter 1 of this book. The membrane transporters that package dopamine into vesicles are called vesicular monoamine transporters (VMATs). This tells us something interesting, namely that these transporters package other 'amines', as well as dopamine, into the vesicles. Therefore, the vesicles that hold, and eventually merge with the synaptic membrane to release dopamine into the synapse, also release other molecules. One of these other molecules is a monoamine, serotonin, which is the subject of the next chapter. Let's take a minute to consider what this might mean on a biological and behavioural level. First, on a biological level, it means that these two molecules are often co-released. Second, on a behavioural level, it means that any alteration in behaviour known to be affected by changes in dopamine levels may also, at least partly, be due to changes in serotonin levels. This reveals an important fact about neurotransmitters, which I emphasise at many points in this book: namely, they do not work in isolation. Certainly, many psychiatric disorders, such as major depressive disorder, are characterised by alterations in both serotonin and dopamine transmission. These VMATs may therefore be a common molecular mechanism underpinning behavioural dysfunction.

As well as these vesicles containing other 'amines' there is also significant evidence that suggests that the **synthesis** of the other catecholamines (norepinephrine and epinephrine) takes place inside these vesicles as they are known to express the hydrolysing enzyme dopamine β hydroxylase, which converts dopamine to norepinephrine.

Back to VMATs

VMATs are part of a superfamily of membrane transporters known as TEXANs (toxin extruding antiporters). These TEXANs are a family of bacterial transporters that remove toxins from within cells and exchange them for extracellular protons that are located in the extracellular part of the cell membrane, as illustrated in Figure 5.4. VMATs evolved from this superfamily, which gives you some appreciation of why they transport dopamine, which is known to be highly neurotoxic.

There are two key types of VMAT found in mammals. These are VMAT1 and VMAT2. They are both coded for by different genes. Historically, there was some contention over whether VMAT1 was found in the central nervous system, with the majority of VMAT1 expression being in the peripheral endocrine system. Recent evidence suggests that VMAT1s are indeed expressed in the mammalian brain, and alterations have profound effects on behaviour, including links to anxiety disorder and bipolar depression. VMAT2s have long been known to be expressed in the mammalian central nervous system, specifically in monoaminergic neurons, such as those that are dopaminergic. VMAT2 plays a key role in keeping catecholamines ready for release from neuronal terminals upon depolarisation of the presynaptic neuron. Recent evidence has suggested a strong relationship between alterations in VMAT2s and schizophrenia.

Once dopamine has been packaged into the vesicles by the VMATs it is ready for release. This happens upon the arrival of an action potential at the synapse, whereby calcium channels open, calcium flows into the presynaptic terminal and this promotes the synaptic vesicles to move and merge with the synaptic membrane, resulting in the release of molecules, such as dopamine, into the synapse. At this

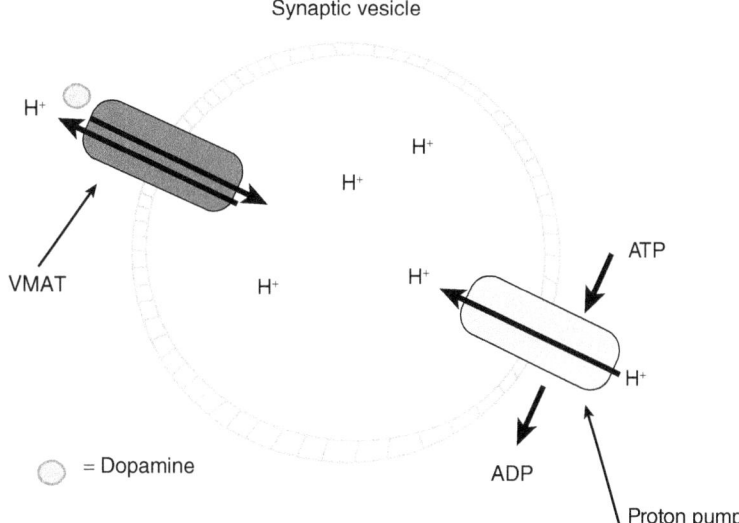

Synaptic vesicle

H⁺

ATP

VMAT

H⁺

ADP

○ = Dopamine

Proton pump

Figure 5.4 Illustration of a vesicle packaging dopamine with the typical channels embedded in its membrane

One point of note is that VMATs help package other monoamines as well as dopamine. Therefore, upon fusion of the vesicle with the membrane, and release of its content into the synapse, it is likely that other neurotransmitters are also released as well as dopamine.

Source: Adapted from: Chaudhry, F. A., Edwards, R. H., & Fonnum, F. (2008). Vesicular neurotransmitter transporters as targets for endogenous and exogenous toxic substances. *Annual Review of Pharmacology and Toxicology, 48*(1), 277–301. doi:10.1146/annurev.pharmtox.46.120604.141146

point, neuromodulators such as dopamine diffuse across the synaptic cleft and bind to their receptor targets on the presynaptic membrane. The receptors are extremely important and we will return to the brief, but first we will look at dopamine reuptake into the presynaptic neurons.

Dopamine reuptake

Once dopamine has been released into the synapse it readily diffuses across the cleft and binds with its target receptors. Eventually the dopamine dissociates from these receptors and a job needs to be done to 'mop it up'. This job is done by a transmembrane protein called DAT (dopamine transporter). This is part of a family of transmembrane proteins that remove cathecholamines from the synaptic cleft and bring them back into the presynaptic cell. Each of these transporters is specific for a unique catecholamine. In the case of DAT, it is dopamine. The DAT is like many other reuptake transporters we have discussed in other chapters, as it is an active sodium-dependent transport process, which moves dopamine against a concentration gradient. Remember, dopamine is highly neurotoxic, so once it has been returned from the synaptic cleft into the presynaptic cell, it needs to be rapidly broken down. The process of breaking down dopamine, once it is back in the presynaptic neuron, is called catabolism and requires two enzymes, namely monoamine oxidase (MAO) and catechol O-methyltransferase (COMT).

As mentioned earlier, lots of research suggests that several psychiatric disorders, including schizophrenia and affective disorders, are at least partly caused by alterations in dopamine levels. DATs are a key part of the system, regulating extracellular and intracellular dopamine levels, and dysfunction of DATs has been heavily implicated in the aforementioned psychiatric disorders.

Outline of dopamine receptors

So now is time to look at how dopamine has its effects on the postsynaptic neuron. Once dopamine is released from the presynaptic neurons it diffuses across this synapse and binds to specific receptors, with the classic analogy being that of a lock and key. There are a number of dopamine receptors and it is important to remember that as well as these receptors being located on the postsynaptic neuron, some are also expressed on the presynaptic neuron.

Dopamine receptors are generally considered in two broad classes, referred to as D1-like receptors and D2-like receptors. The D1-like receptors are made up of dopamine D1 and D5 receptors, while the D2-like receptors include D1, D3 and D4. Many of the other neurotransmitters we have discussed, and will discuss, in this book have both ionotropic and metabotropic receptors, which have different effects on the neuron across a different time course. Dopamine is an exception as all its receptors are metabotropic and are therefore all considered to be GPCRs (G-protein-coupled receptors). As a rule, it is generally considered that D1-like receptors are excitatory in their effects, while D2-like receptors are inhibitory. Once bound to the receptor, dopamine triggers a complex cascade of intracellular secondary messengers, as can be seen in Figure 5.5. For a more in-depth discussion of these intracellular processes, refer to Chapter 1 of this book. The key secondary messenger altered as a result of dopamine binding to D1-like and D2-like receptors is cAMP. This is increased in concentration as a result of dopamine binding to D1-like receptors and reduced in concentration as a result of dopamine binding to D2-like receptors. Activation of both receptors also has effects on several other intracellular mechanisms, such as alterations in the state of potassium and calcium channels. Again, as a general rule, D1-like receptors are found on the postsynaptic neuron, while D2-like receptors are found on the presynaptic neuron. This has led to the belief that D2-like receptors act as auto receptors. Auto receptors are essentially a negative feedback loop. They act in the signal transduction pathway to reduce any further release of the neurotransmitter, in this case dopamine. So, when dopamine levels in the synapse get too high, this is the switch turning off any further release.

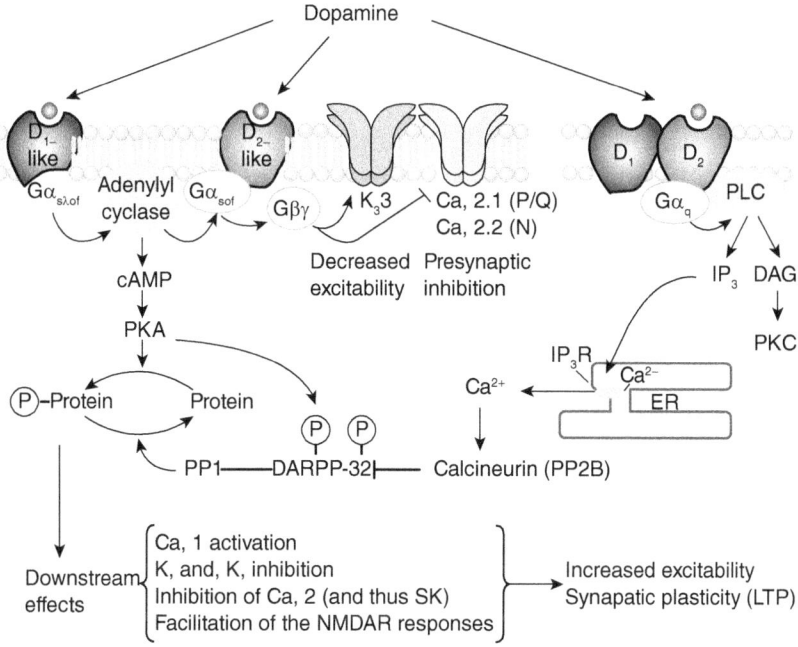

Figure 5.5 The complex intracellular cascades modulated as a consequence of dopamine bindings at D1-like and D2-like receptors

Both D1-like and D2-like dopamine receptors are metabotropic. This means that they are not directly coupled to ion channels and binding of dopamine to the receptors results in the dissociation of specific, intracellularly-bound G proteins, which, once dissociated, modulate a number of proteins to produce a myriad of intracellular effects. This includes the modulation of enzymes, such as adenylyl cyclase and the modulation of potassium channels.

Source: Savica, R., & Benarroch, E. E. (2014). Dopamine receptor signaling in the forebrain. *Neurology, 83*(8), 758. doi: https://doi.org/10.1212/WNL.0000000000000719. Reproduced by permission from Wolters Kluwer Health, Inc.

It is worth noting that there are other receptors that have affinity for dopamine, such as the trace amine associated receptors (TAARs). These are also GPCRs, but they are beyond the scope of this book. As well as these TAARs, further factors relating to D1-like and D2-like receptors complicate the picture when it comes to dopamine receptors. In a simplistic way, we often conceive of dopamine receptors as consisting of these two families that are separate and distinct, although recent research has shown that D1-like and D2-like receptors can form what are known as heteromeric complexes. These functional heterodimers can activate a distinct intracellular secondary messenger signalling pathway compared to D1-like and D2-like alone. For example, activation of a D1/D2 receptor heterodimer has been shown to result in the release of BDNF (brain-derived neurotrophic factor), which has been heavily associated with aspects of drugs addiction and abuse (Perreault et al., 2014).

Classic topics in dopamine research

As discussed earlier, dopaminergic projections in the brain are specific and target unique structures, with most of the dopaminergic input to these regions emanating from the VTA or the SNc. One key region innervated heavily by dopaminergic neurons from both the SNc and the VTA is the striatum. The striatum is one of the main input nuclei of a collection of structures that make up the basal ganglia. Dorsal regions in the striatum are heavily associated with control of voluntary movement, while ventral regions in the striatum have had a long association with reward and pleasure. As dopamine projects into these regions and modulates neurons within the region, it therefore makes sense that dopamine and dopamine alteration may well affect the behaviour in which these regions are known to be involved. Indeed, there is a long history of research on the contribution of dopamine to voluntary movement, with much evidence that loss of dopaminergic projections, into the dorsal striatum, contribute to the pathology seen in patients with movement disorders, such as Parkinson's disease. Similarly, alteration in dopaminergic input into the ventral striatum has been shown to underly much behavioural pathology associated with changes in reward value and pleasure, such as substance use disorder. In the following sections we will explore this in more detail.

Test Yourself 5.2

In the next section we will focus on the role alterations in the machinery that underpin dopamine synthesis, storage and reuptake, as well as dopamine receptors, play in behaviour. It is therefore essential that you are confident with this machinery and know all the key terms.

From the following list select the key terms related to dopamine and put them under the headings of: Synthesis, Storage, Reuptake or Receptors.

Table 5.1

Synthesis	Storage	Reuptake	receptors
D1-like	D2-like	COMT	RER
VMAT	GAD	SNc	VTA
5HT	MAO	VGAT	Muscarinic

Synthesis	Storage	Reuptake	receptors
DOPA	MGLUr	NK1r	MORs
L-DOPA	DAT	Metabotropic	TH
VaChT	Nicotinic	NOS	SERT

Dopamine and reward learning

For many years, dopamine has been heavily implicated in reward learning. Reward learning can take several forms but generally involves the biasing of future action based on the expectations of certain outcomes. Early research posited a broader catecholamine hypothesis of reward, where the key molecule was noradrenaline. It is important to remember that at this time dopamine was thought of as simply a precursor to noradrenaline and not a neurotransmitter in itself. Evidence eventually mounted up against noradrenaline as the main player in reward. This was largely a result of advancements in pharmacology as a host of specific agonists and antagonists were developed/discovered that didn't affect catecholamines in general, but would target the individual subtypes, such as dopamine, serotonin and noradrenaline. Much of Roy Wise's pivotal research signed the death papers for the noradrenaline hypothesis of reward, as he convincingly showed in a set of anatomical studies (Corbett & Wise, 1980) that animals would not work to self-stimulate the noradrenaline centres in the brain, but would to self-stimulate the dopaminergic centres (ventral tegmental area and substantia nigra par compacta).

As mentioned above, at this time in the late 1970s, evidence was gathering that dopamine was indeed a neurotransmitter in its own right and that targeted pharmacological modulation of dopamine seemed to dramatically alter reward-related behaviour, independent of noradrenaline. Specifically, dopamine antagonists seemed to alter animal behaviour in a way that implied a 'devaluation of reward rather than an impairment of performance capacity' (Wise, 2008, p.3). This mounting evidence led Roy Wise (1978) to propose the anhedonia hypothesis of dopamine, which suggested that dopamine played a key role in the subjective experience of pleasure associated with positive rewards. Therefore, if dopamine was blocked from having its biological effects, it would reduce the relative 'pleasure' received from a rewarding stimulus and reduce the amount the stimulus sought. Over the years, the anhedonia hypothesis of dopamine's involvement in reward has largely been disproven, although it is still contested to some degree. Certainly, research that uses modern scanning techniques with humans shows a strong correlation between subjective expressions of pleasures and the occupancy of dopamine receptors in ventral striatal territories, such as the nucleus accumbens. Interestingly, this research suggests a specific increase in the occupancy of D2 receptors as subjective pleasure increases, perhaps helping to refine the contribution that dopamine plays, specifically in the 'pleasure' aspects of reward.

Recent research has seen many moves towards arguing for one of two other theories implicating dopamine in reward. One suggests that dopamine acts to enhance the salience of stimuli in the environment and to motivate the pursuit of reward; the other suggests that dopamine acts as a time stamp marking the likely occurrence of a predicted reward and therefore producing an associative link. Berridge (2007a) succinctly sums this up as a debate between dopamine's involvement in: 'liking', 'wanting' or 'learning', respectively. We will discuss the idea that dopamine as a time stamp (learning) in due course. It is worth noting at this point that there is much evidence that suggests different brain circuits are involved in 'liking' and 'wanting', as illustrated in Figure 5.6. Therefore, it seems reasonable to consider that dopamine may play different roles in these aspects of behaviour.

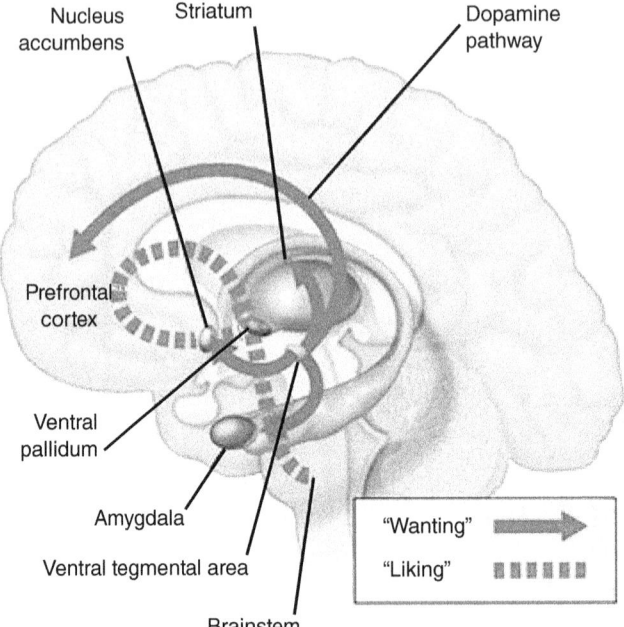

Figure 5.6 Circuits and brain regions identified as involved in either 'liking' or 'wanting' behaviours

Notice dopaminergic projections innervate some of these regions, such as the nucleus accumbens, but do not directly innervate others, such as the ventral pallidum.

Source: Gaskin, S. (2021) *Behavioral neuroscience: essentials and beyond*. Thousand Oaks, CA: Sage.

First, Berridge and colleagues have employed some interesting methods to try to tease apart the role of dopamine in 'liking' versus 'wanting'. Fundamentally, they argue that all vertebrates, from rodents to primates, have evolutionarily conserved facial gestures, such as those seen in Figure 5.7, that suggest 'liked' substances (such as sweet tastes) and 'disliked' substances (such as bitter tastes). These 'liking' responses can therefore be quantified in a paradigm commonly referred to as the taste reactivity paradigm. At the same time, the amount of 'wanting' can be easily quantified by giving the animal (normally rodents) free access to the substance and measuring how much they take of it in a set time.

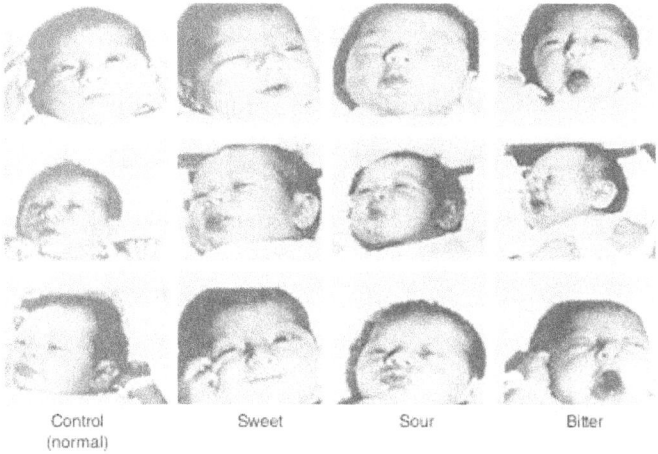

Control (normal)　　　Sweet　　　Sour　　　Bitter

Figure 5.7　Evolutionary conserved orofacial responses produced in response to 'pleasurable' solutions, such as sucrose, and aversive/unpleasant solutions, such as bitter quinine

Source: Steiner, J. E. (1977). Facial expressions of the neonate infant indicating the hedonics of food related chemical stimuli. In *Taste and development: the genesis of sweet preference*. DHEW Publication No. NIH 77–1068. Bethesda, MD: United States Department of Health, Education, and Welfare, pp. 173–189.

Early studies by Berridge and colleagues involved lesioning of the dopaminergic projections into the ventral striatal territories. These lesions and loss of dopaminergic innervation produced no alterations in animals' 'liking' responses. Further research using neuroleptics, essentially to block dopamine receptors, showed the same lack of effect. As we moved into the early 2000s, the developments of transgenic animals[3] resulted in a hyperdopaminergic mouse line. This was caused by the genetic knockdown of the dopamine transporter gene. These animals again

---

[3]A transgenic animal is one where a gene, or part of a gene, has been inserted into the animal's genome by a researcher.

showed no alterations in their 'liking' responses to natural rewards but, crucially, they did show alterations in the 'wanting' of the natural reward (Peciña et al., 2003). These findings are not just constrained to animal research. In fact, much work with humans also suggests a stronger association between dopamine and 'wanting' than between dopamine and 'liking'. One good example focuses on a specific subgroup of patients diagnosed with the motor degenerative disease Parkinson's disease (PD). This subgroup displays increased compulsive behaviours around taking/'wanting' drugs and increased levels of gambling behaviour. These patients are said to have dopamine dysregulation syndrome (DDS) and will typically report higher 'wanting' but not altered 'liking', which correlates strongly with the amount of free dopamine, after taking the medication to help control the symptoms of Parkinson's disease, L-DOPA.

But what about the role of dopamine in learning? As mentioned previously, the role of reward is to bias future actions and therefore produce learning. A popular theory of dopamine that has arisen is often referred to as the reward prediction error (RPE). The RPE centres around the notion that dopamine essentially acts as a time stamp, which identifies the condition where something rewarding has happened and links this to a stimulus that is associated with the arrival of the reward.

Focus on research 🔍

Schultz, dopamine and the behaviourists - links across the century

Fundamentally, what Schultz and colleagues (1997, 2004, 2016) proposed directly links back to the early 1900s and the work of the behaviourists, such as Edward Thorndike. In his law of effect, Thorndike talks about the notion that a stimulus acts as a stamp, donating the likelihood of reward. In many ways, Schultz and others have appropriated this terminology and attributed the stamp to a biological mechanism, namely alterations in dopaminergic neuron firing and, ultimately, dopamine release.

To fully understand the work of Schultz and colleagues, and others, researching the notion that dopamine acts as a time stamp for reward, you must have a reasonable grasp of some of the basic concepts of behaviourist dogma, as illustrated in Figure 5.8.

For example, much of the research that focuses on the idea of dopamine as a time stamp employs classical conditioning/associative learning paradigms. In these paradigms, you have an unconditioned stimulus (UCS), which is the stimulus that will eventually be associated with the reward. Once it is associated with the reward, it is renamed as the conditioned stimulus (CS).

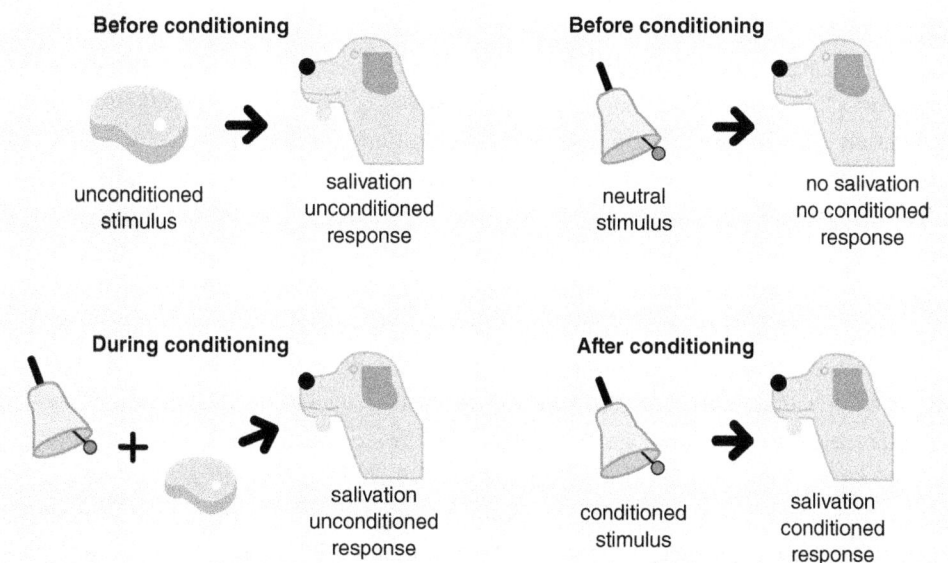

Figure 5.8   An illustration of the classic Pavlovian conditioning paradigm

The classic Pavlovian conditioning paradigm is where an unconditioned stimulus (food), which produces an unconditioned response (salivation), is paired with a neutral stimulus (sound of bell). After multiple pairings, the neutral stimulus becomes conditioned and evokes the salivatory response without the need for presenting the food.

Source: Higgs, S., Cooper, A., & Lee, J. (2020). *Biological psychology* (2nd ed.). London: Sage.

The evidence to support the reward prediction error theory of dopamine's involvement in reward learning initially came from the work of Schultz and colleagues (1997, 2004, 2016). This work involved recording the electrical activity of dopaminergic neurons as primates completed classical associative learning tasks. So, how did they do this? First, they presented the animal with an unconditioned stimulus (UCS), quickly followed by a reward (typically a pleasurable sweet solution, fruit juice). It was noted at this point that the dopaminergic neurons increased in firing directly after the reward. After several pairings of the UCS and the reward the increase in dopaminergic neuron firing happened directly after the UCS. At this point, the UCS is said to become a conditioned stimulus (CS). After this, Schultz and colleagues ran trials where the CS was presented but the reward was omitted. Something very interesting happened here as when a predicted reward was expected, there was a sudden pause in the firing of the dopamine neurons. This has been taken as evidence that alterations in the firing rate of dopaminergic neurons, and the subsequent release of dopamine, act as a time stamp identifying when a reward is expected (based on cues) and, importantly, also acting as a time stamp to identify the absences of reward, the RPE.

It would make sense that if Schultz and colleagues' idea that dopamine is a teaching signal is correct, then if dopamine levels were altered, the ability to predict and ultimately learn from reward might also be altered. Indeed, there is evidence to support this notion. One intriguing study carried out by Zaghloul et al. (2009) used patients (diagnosed with Parkinson's disease) undergoing deep brain stimulation surgery. Using multelectrode arrays, the researchers recorded from the substantia nigra pars compacta (SNc) in these patients, while they completed a learning task where the reward was 'theoretical' financial gains. SNc 'dopaminergic' neurons were seen to increase firing in reponse to unpredicted rewards. This suggests, as the primate studies of Schultz and colleagues do, that dopaminergic neuron activity might provide a key stamp highlighting the differences between predicted and actual outcome. There is also a myriad of animal studies that show, using dopamine agonists, antagonists and receptor blockade, that alterations to dopamine signalling either before or after various learning paradigms, such as a Pavlovian condition, can significantly disrupt performance in a learning task and/or consolidation of the task. For example, dopamine agonists given directly after a Pavlovian learning task greatly increase an animal's willingness to work for a conditioned stimulus, presumably strengthening the association (Everitt & Robbins, 2005).

The idea that dopamine may act as a time stamp, facilitating reward learning and acting as a prediction error mechanism, is appealing, although there is reasonable evidence that suggests learning can happen 'normally' without the necessity of dopamine. One good example of this involves the use of mutant mice that lack the enzyme tyrosine hydroxylase and therefore do not synthesise dopamine. Although these mice show acute Parkinsonian-type symptoms, they have been shown to be able to effectively learn by choosing a spout that delivered a preferable substance/natural reward (sucrose solution) over one that simply provided water.

It is an appealing notion that dopamine plays an integral underpinning role in our experience of reward and ultimately learning. Indeed, if you consider dopamine in this broad sense, then it fits neatly with many of the atypical behaviours, disorders and conditions that are often typified by alterations in dopamine synthesis, release and signalling, from motor disorders, such as Parkinson's disease, to substance use disorder and, more broadly, addiction. In the following sections we will discuss the role of dopamine and alterations in dopamine in the context of schizophrenia, Parkinson's disease and drug addiction.

Dopamine and schizophrenia

Schizophrenia is a complex disorder characterised by many different symptoms during an active phase. It can generally be seen as characterised by a loss of touch with reality, whereby the individual has difficulty distinguishing between what is real and unreal. This is generally referred to as psychosis. Symptoms are grouped into three main categories, which are: positive, negative and disorganised symptoms.

1. Positive symptoms include hallucinations (sensing things that are not there) and delusions (false beliefs that persist despite the presence of disconfirmatory evidence).
2. Negative symptoms include a loss or blunting of emotion responses, a reduced ability to plan and a reduced ability to speak.
3. Disorganised symptoms include problems with logically planning and organising language, a good example being word salads, where sentences often form as an incoherent jumble of words.

For a more thorough outline of the diagnosis of schizophrenia, consult the American Psychiatric Association's website (www.psychiatry.org/patients-families/schizophrenia/what-is-schizophrenia), the American Psychiatric Association's diagnostic manual, known as the DSM-5 (APA, 2013) or the *International Classification of Diseases* (ICD-11), which is another diagnostic manual, but published by the World Health Organisation (WHO) (2019).

The dopamine hypothesis of schizophrenia has been influential since the early 1960s. In its earliest guise, it postulated that excessive dopaminergic neurotransmission was the main biological feature of schizophrenia. This was first postulated after the discovery that Chlorpromazine, which was originally developed as a sedative, had antipsychotic effects. It was discovered that Chlorpromazine administration resulted in a significant upshift in dopamine metabolites in the brain but did not alter dopamine concentrations. Other sources of evidence to support the idea that excessive dopaminergic transmission underpinned schizophrenia arrived in the shape of amphetamine psychosis. Large amounts of amphetamine can produce a state of psychosis very similar to that seen in patients with paranoid schizophrenia, and when antipsychotics, such as Chlorpromazine, are given to those with amphetamine psychosis, it largely eradicates this psychosis. So, what does this tell us about dopamine and schizophrenia? Well, amphetamine is well known to dramatically increase the release of amines, such as dopamine. Therefore, the assumption was that amphetamine psychosis was also caused by excessive dopaminergic transmission. Further evidence to support the role of excessive dopamine in schizophrenia came from patients diagnosed with Parkinson's disease. Later in this chapter we will discuss the role of dopamine in Parkinson's disease, which is characterised be a dramatic loss of dopaminergic neurons. Interestingly, this provided some support for the excessive dopamine theory of schizophrenia, as when the dopamine precursor levodopa (L-DOPA) is used to treat PD, it reduces the symptomology of PD but has the side effect of producing schizophrenic symptomology. As the theory was refined, there was an increasing focus on the idea that excessive dopaminergic transmission is caused by alterations in the density or affinity of the receptors for dopamine in those diagnosed with schizophrenia. Indeed, in an elegant review of the literature, Kestler, Walker, and Vega (2001) concluded that when controlling for methodological differences and differences in sample characteristics, there was still a moderate effect, with an increased density of postsynaptic, D2-like dopamine receptors seen in those diagnosed with schizophrenia.

However, over the course of the subsequent 10–20 years little consistent research evidence emerged definitively implicating excessive dopaminergic transmission as the core aetiology (or cause) of schizophrenia. It is worth remembering that schizophrenia is a complex disorder, characterised by an array of different symptomology. Indeed, this early research often neglected this important point, focusing on schizophrenia as though it was one disorder, rather than the reality, which is a disorder characterised by a collection of various subtypes with multiple different symptomologies.

──Test Yourself 5.3──

Post-mortem investigation: Much of the research carried out during these years consisted of using methods to quantify the amount of dopamine and its metabolites in post-mortem (after death) tissue from patients who had been diagnosed with schizophrenia during their lives. There are major issues with this method that might undermine the validity of these results and provide some explanation as to why the research was so inconsistent. Can you think of at least one?

Suggested answers are provided at the end of the chapter.

What became apparent in the literature is that there was a seeming difference in the concentrations of dopamine, depending on the brain region. This could provide some of the explanation for why there was so much inconsistency in the early research, as regional variation was rarely considered. It became clear that there was a reasonably clear picture of **hyperdopaminergia** (high levels of dopamine transmission) in subcortical structures, especially the basal ganglia, and **hypodopaminergia** (low levels of dopamine transmission) in cortical regions, such as the prefrontal cortex. Linked to this was the emerging evidence that D1-like and D2-like dopamine receptors were unevenly distributed in the brain. Large-scale meta-analysis, such as Kestler, Walker, and Vega's (2001) research, show modest increases specifically of D2/D3 receptors in the striatum, while D1 receptors in this region seem unaltered. However, differences in D1 receptors have been identified in the literature, with these mainly being in the cortex. The direction of alteration is less consistent, with some studies reporting an increase in D1 receptors in the cortex, while others report decreases. Again, one major confound here is the issue of causation. Certainly, it could be possible that rather than these dopamine receptor changes causing schizophrenia, they are a response to schizophrenia, in the sense that they may be a response to altered dopamine levels, with upregulation and downregulation of receptors being well known to be dependent on the levels of the neurotransmitters in a region.

Genetics and the molecular mechanisms of dopamine synthesis storage and reuptake

One major advance over recent years is the rise of studies involving genetic manipulation and gene targeting. Schizophrenia, for a long time, has been, at least to some degree, considered to have a genetic aetiology. Research in this area provides some reasonable support for the role of dopamine. In fact, many of the gene variants associated with schizophrenia are key players in dopaminergic synthesis, storage and signalling. One good example is the COMT gene, which programmes for an enzyme essential for the degradation of dopamine. When the chromosome that carries this gene is deleted, it results in a number of symptoms, including those associated with a schizophrenia diagnosis and psychosis. An important point here is that deletion of this gene results in several symptoms, some of which are less 'typical' of schizophrenia, so although it may play some part, it certainly isn't the only contributor. Other well-researched contributors include the DISC1 gene, which recent research heavily implicated in regulating dopamine function, and the DRD3 gene, which is responsible for the synthesis of D3 receptors. In fact, the reality with the genetic base of schizophrenia is that it is very likely to be polygenic. For a good review of this see Sullivan (2005).

Gene editing has also allowed us to develop many animal models of schizophrenia, one of which is the DAT knockout mouse. These mice mimic many of the behavioural signs of schizophrenia, including psychomotor agitation, prepulse inhibition and stereotyped behaviours. However, no animal model can fully replicate all the symptomology of schizophrenia, with some symptoms, such as the core symptoms of delusions and hallucinations, being difficult, if not impossible, to measure objectively in animals.

The reality is that with genetics and schizophrenia gene variants are only likely to account for a small number of cases of schizophrenia. Other factors, such as environment, seem likely to play a key role in the aetiology of schizophrenia. However, this does not preclude dopamine from still being a key player in the aetiology of schizophrenia, for example one environmental factor related to the onset of schizophrenia is stress. Stress has also been shown to elevate dopamine release. Therefore, in this instance dopamine alteration may well mediate the onset of schizophrenic symptoms.

Other biogenic amines and schizophrenia

As ever, it is essential not to think that the research presented in these pages is the only evidence out there. For example, there are several lines of argument that implicate other biogenic amines in schizophrenia as well as dopamine. One example is serotonin. Many of the radioligand studies mentioned above that focus on the expression of dopamine receptors can actually be considered in the context of serotonin, as many of these radioligands are believed also to bind to serotonin receptors. Therefore, the picture we see in the studies looking at dopamine receptors in schizophrenia may at least partly be due to alterations in serotonin receptors as well. Indeed, emerging research suggests the involvement in specific subgroups of serotonin, known as 5HT2A receptors. One line of evidence comes from the effectiveness of Clozapine, an atypical antipsychotic for the treatment of schizophrenia. Clozapine is a highly selective 5HT2A antagonist, so the line of argument goes that if it treats the symptoms, then it must contribute to the cause. Although this does provide some support, it is somewhat of a false argument, as many drug therapies alleviate symptoms but don't treat the cause. For example, if you broke your arm and you were given pain killers, the reduction in pain wouldn't mean that your arm was fixed.

Dopamine and Parkinson's disease

Parkinson's disease is a devastating motor degenerative disorder that is characterised by several motor and non-motor deficits. The key motor deficits include postural rigidity, a resting tremor and Bradykinesia (slowness of movements). Parkinson's disease is a degenerative disorder, so its occurrence is positively correlated with age. One of the main biological phenotypes in Parkinson's disease is a dramatic reduction of dopamine and dopaminergic neurons. Indeed, the main therapy, which proves extremely effective at controlling the motor deficits, is the dopamine precursor Levodopa (L-DOPA).

As previously mentioned, one of the main regions that dopaminergic neurons project from is the SNc. These projections heavily innervate the striatum, specifically the dorsal striatum. For many years the function of the striatum was considered to be solely a motor control structure. Recent research suggests its functions are much more varied, although it is still maintained that one of its core functions is the control of motor output. As such, it makes a lot of sense that dopamine, and dysfunction of the dopamine supply to this region, might underly pathology associated with motor control, such as Parkinson's disease.

Some of the earliest links between dopamine loss and Parkinsonian symptoms came accidently when, in the early 1980s, a novel synthetic heroin started making its way onto the streets of various towns in California. Several individuals who had taken this drug presented with many of the typical motor symptoms of Parkinson's disease, such as rigidity and tremors, as well as some of the non-motor features, such as deficits in higher cognitive functioning. These individuals also responded strongly and rapidly to the dopamine precursor L-DOPA. It later became apparent that the active ingredient in this synthetic heroin (MPTP) seemed to lead to specific neuronal loss of the dopaminergic neurons in the SNc – the very same ones that project extensively into the dorsal striatum. On a molecular level, this is because MPTPs metabolite MPP+ is readily taken up by the dopamine membrane uptake transporters found on dopaminergic neurons in the SNc. Once inside the neuron, MPP+ is highly neurotoxic, leading to their death. Some of you might be wondering why dopaminergic neurons in other regions and other cells that express dopamine transporters are not affected? This is a very sensible question and one with no definitive answer, although it is thought that other dopaminergic neurons, for example in the ventral tegmental area, may be protected by other molecules they express, such as the calcium-binding protein calbindin, or by a lower affinity of MPP+ at their dopamine reuptake sites.

This evidence led to the reasonable assumption that 'normal' Parkinson's disease, not induced by MPTP ingestion, may be induced by specific loss of these dopaminergic neurons. So, the next question is what normal mechanism could dysfunction that would lead to loss of these neurons, and ultimately Parkinson's disease? Well, as we've previously mentioned, dopamine is highly neurotoxic, and neurons do their best to avoid the build-up of dopamine intracellularly by packaging dopamine into vesicles. It would therefore be fair to suggest that alteration in VMATs, the vesicular transporters responsible for dopamine's packaging, may lead to neuronal cell death and Parkinsonian symptomology. There are several lines of evidence to support this conclusion. First, early research using immunohistochemical labelling of post-mortem tissue, from human patients diagnosed with Parkinson's disease, displayed reduced VMAT2 levels compared to control human tissue, especially in the striatum. This suggests that reduced VMAT levels, and therefore cytotoxic build-up of intracellular dopamine, may underpin Parkinson's disease. However, post-mortem research has its limitations, most notably in establishing a direction of effect. For example, here we cannot definitively say if reduced VMAT levels caused Parkinson's disease in these patients or if reduced VMAT levels are a result of Parkinson's disease.

Next, there is evidence from pharmacological studies that inhibiting VMAT function produces symptomology of Parkinson's disease. Specifically, a drug called Reserpine, which was originally formulated as an anti-hypertensive drug, is known to be a

VMAT inhibitor. Research has shown that when Reserpine is administered to animals you see many of the typical motor deficits seen in human patients with a Parkinson's diagnosis. Interestingly, research by Skalisz et al. (2002) has shown convincing evidence that administration of Reserpine not only produces motor deficits akin to human Parkinson's disease in animals, but also induces alterations in mood, specifically anhedonia, which are known non-motor deficits seen in the early stages of Parkinson's disease.

Although pharmacological manipulation is an improvement over post-mortem research, providing a more convincing case for causality, probably the best line of evidence for alterations in VMAT expression/function underpinning Parkinson's disease comes from research using transgenic animals. So, what happens when you genetically manipulate the expression of VMATs? Well, the simple answer is when you knock out VMATs by deleting the VMAT gene(s), you see the death of the animals within a few days post-partum (after birth). It is therefore obvious that a full 'knockout' model is not viable. However, other genetic manipulations can be done to reduce the expression levels of VMAT, while keeping the animals viable. One such animal is the KA1 line, which was developed through gene targeting. This produced a mouse that expressed only 5% of the 'normal' levels of VMAT2. Multiple studies have shown this animal to exhibit a behaviour phenotype, much like human Parkinson's disease patients. However, there is one major caveat with this model, namely that the reduction in VMAT expression was also accompanied by a loss in α-synuclein. Alterations in α-synuclein are another heavily researched factor considered likely to contribute to Parkinson's disease pathology. Therefore, using this model, it is difficult to tease apart the role VMAT alteration and/or α-synuclein may play in Parkinson's disease pathology.

These KA1 mice were an excellent starting point and, through careful breeding, researchers managed to develop a line that had the low expression levels of VMAT2 while still expressing α-synuclein. These mice are referred to as the VMAT-deficient mice. These animals show many of the biological and behavioural phenotypes of human Parkinson's disease, such as numerous changes in dopaminergic release within the striatum and significant deficits in motor coordination, which appeared at ages that corresponded well with the typical age of symptom expression in human patients. These animals also show something that many animal models of Parkinson's disease don't, namely, early alterations in hedonic responses, which present before the classic motor deficits. Despite all this, there is no one mouse model that faithfully reproduces all the symptomology seen in human patients with Parkinson's disease, so we cannot be conclusive. However, it seems likely that alterations in VMAT functionality is a good contender for playing a causal role in Parkinson's disease.

As we get narrower and narrower in our focus as molecular biologists, we can sometimes forget to raise our head up from the microscope and look at the bigger picture. Indeed, it is important to remember that as well as loss of dopaminergic neurons, Parkinson's disease is also underpinned by alterations in other neurons and neurotransmitters. For example, there is a known loss of serotonergic neurons in the raphe nuclei of patients with Parkinson's disease, and known degeneration of the norepinephrine neurons of the locus coeruleus. However, if we think again for a moment about what underpins all these neurotransmitters, we come to the revelation that they are all biogenic amines, and therefore are all packaged into vesicles by VMATs. This still is not the full picture and it is essential to consider the role of other factors, such as Lewy bodies in Parkinson's pathology. Lewy bodies are abnormal protein deposits. They consist of the protein α-synuclein, which plays a key role in 'healthy' neuronal functioning. However, when it binds and forms Lewy bodies it fundamentally stops normal functioning of the brain and neurons. Recent research has provided convincing evidence for a link between abnormal α-synuclein in the gut and the development of Parkinson's disease pathology, with abnormal gut α-synuclein making its way to the brain and contributing to dopaminergic cell death. So, what we see here is that dopaminergic cell death could be a result of another underlying pathology. Thus, it is not the cause of Parkinson's disease *per se*, but is a symptom of the real underlying cause. This serves to emphasise the complexity of the molecular pathology underpinning Parkinson's disease and should aid to remind you that at the very forefront of knowledge, what we think is right is, at best, only likely to be part of the answer.

Novel topics in dopamine research

D3 receptors and drug addiction

It should be no surprise, considering dopamine's role in shaping reward, that dopamine has been heavily implicated in the addiction potential of many drugs of abuse. The general belief is that drugs of abuse hijack the reward circuitry, of which dopaminergic projections, especially into the ventral striatum, are a key component.

Dopamine receptors have been traditionally classified as D1-like and D2-like, with D1-like consisting of D1 and D5 receptors, while D2-like consist of D2, D3 and D4 receptors. Because of this simplification, many specific roles that D5, D3 and D4 receptors play in behaviour have been overlooked, with the general assumption being that they play the same active role as D1, for D5 and D2 for D3 and D4. Of course, this simplification wasn't done simply as a way of reducing the complexity; it was largely done due to a lack of specific agonists and antagonists for D5, D3 and D4. However, over recent years there has been significant progress made in this regard, and a wider selection of agonists and antagonists have been discovered that allow us to specifically target the individual dopamine receptors. The rise of gene editing and targeted optogenetic approaches has also helped. A good example of what this increased ability to target specific receptors reveals is the rise of D3 receptors as a major player in addiction. D3 receptors are much more sparsely distributed than their category lead D2 receptors, although their distribution gives clues to their potential role. Indeed, D3 receptors are well known to be localised to basal ganglia nuclei, such as the ventral striatum, which, as previously discussed, is known to be involved in reward learning and pleasure response. It is therefore a logical suggestion to consider that alterations in D3 receptors, within this region, may play a key role in pleasure and reward modulation, and may be affected when drugs of abuse 'hijack' this circuitry.

Early research, such as the work of Staley and Mash (1996), used post-mortem tissue from those who had been a victim of a cocaine overdose. This research used radioligands to visualise D3 receptors and found that there was a significantly higher density of D3 receptors, specifically localised to regions of the striatum, with a distinctly higher density in the nucleus accumbens region of the ventral striatum, in those who had died of cocaine overdose compared to control tissue from apparently drug-free individuals. This research fits well with what we know of the nucleus accumbens as a region involved in the modulation of the pleasure aspects of reward, and suggests that density of D3 receptors in this region may contribute to addiction. It is worth remembering at this point the major caveat of post-mortem research, namely that we cannot fully establish the direction of effect. Although it might seem like D3 density contributes to addiction, it could be that other processes that contribute to addiction result in a change in D3 receptors.

Although there are several other studies supporting the findings of Staley and Mash (1996), the increase in D3 receptor density is not a consistent finding across all addictions. Many other studies have found the opposite, with lower levels of D3 receptors, or no difference in expression between the 'drug addict' and 'non-drug addict' groups. A good example of this type of research focused on smokers as the 'drug addict' group. The research seems to show a much more complex picture than that with cocaine addicts. The researchers found a significant reduction in D3 receptors in smokers and a negative correlation between the number of cigarettes smoked a week and the number of D3 receptors.

Drug addiction isn't a unitary concept

In common parlance, we may think of drug addiction as though it is one concept. However, there are generally considered to be 'stages' of addiction:

1. The first is initiation: This is when the individual acquires the drug addiction. It is typified by generally increased intake of the drug over a period.
2. The second is maintenance: this is typified by a compulsive seeking and intake of the drug.
3. The third is withdrawal/relapse/negative affect.

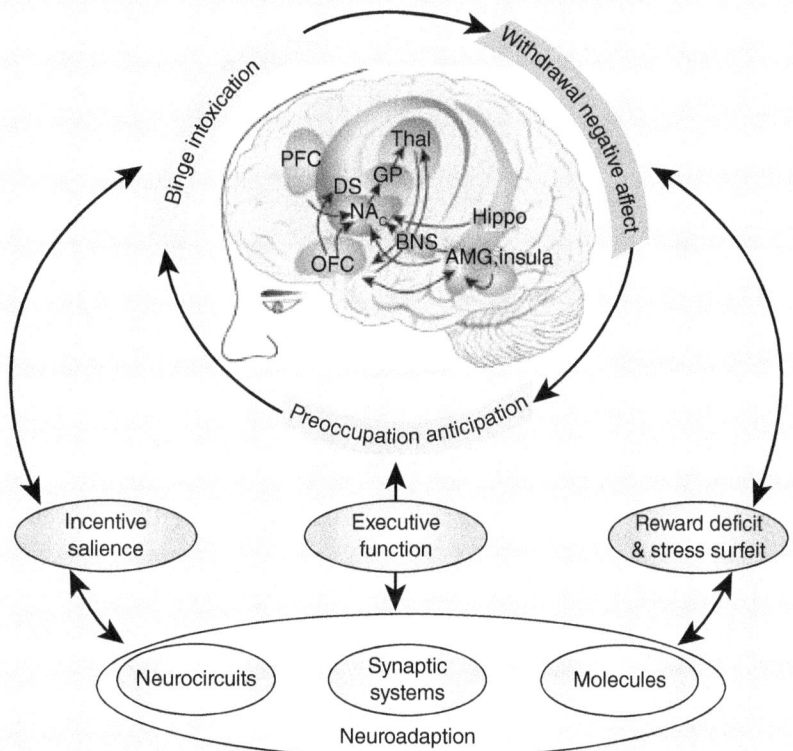

Figure 5.9 Addiction as a three-stage process, with each process correlated with the involvement of specific brain regions

Addiction is conceived as consisting of three stages, each of which involves multiple distinct brain regions. Binge intoxication and withdrawal stages are thought to be mainly governed by subcortical regions, such as the nucleus accumbens, while preoccupation and anticipation involve both subcortical and cortical regions.

Source: Reprinted by permission from Springer Nature: Wise, R. A., & Koob, G. F. (2014). The development and maintenance of drug addiction. *Neuropsychopharmacology, 39*(2), 254–262. Copyright (2014).

(Continued)

So, why does this matter? Well, when we consider the role of neurotransmitters and their receptors, such as D3 receptors, in addiction, we need to consider the stage at which the human or animal model being used is. This is important as the results we find will likely differ between the stages. For example, if D3 receptors increase during initiation, but don't alter during maintenance or withdrawal, then we can draw the conclusion that D3 receptors are involved in the formation of addiction, and therefore any pharmacological intervention to be used with 'addicts' will only be useful at this point. However, if we find that D3 receptors alter during withdrawal, and this is the factor that shapes how 'hard' a withdrawal the individual suffers, then the pharmacological intervention may well be useful for those trying to break the cycle of drug addiction.

George Koob is a pre-eminent researcher within the addiction literature. Koob adds an extra layer to the 'stages' of addiction as he also believes that these different stages are mediated by different brain regions and brain circuits, as shown in Figure 5.9. There is certainly a fair amount of literature which supports this, with the basal ganglia circuitry being key to the initial stages of binge intoxication, cortical regions being key to anticipation and preoccupation (maintenance) and amygdaloid regions (with a bit of the ventral striatum) being key players in the withdrawal and negative affect stage.

Much animal research has been conducted on this topic, with a key focus on the use of specific agonists and antagonists during the initiation and maintenance stages of drug addiction. One such example is the early work of Caine and Koob (1993). In their research they implanted cannulas into mice and trained them to self-administer cocaine through these cannulas. Once the animals were self-administering cocaine, they began to add a specific D3 receptor agonist at the same time as cocaine. What they found is that this increased the intervals between cocaine self-administration, but, crucially, it didn't stop the mice self-administering the cocaine. This was considered an intriguing finding and was taken as suggesting that agonism of D3 receptors increases the 'potency' (reward value) of the cocaine, and therefore it increased the interval between self-administration as the 'hit' was greater. This conclusion is supported by the self-stimulation literature,[4] where cocaine administration increases the amount an animal will self-stimulate, and D3 antagonists, administered at the same time as cocaine, have been seen to reduce the number of times an animal

---

[4]This is the phenomenon whereby animals will actively press a lever to stimulate, via an implanted electrode, certain parts of the brain. Key regions for self-stimulation are the ventral striatum and the medial prefrontal bundle.

will self-stimulate their brain. However, although this all seems to suggest that D3 receptors are involved in the rewarding value of cocaine, and therefore the maintenance of the addiction, research that has given animals cocaine to self-administer on a minimal fixed ratio, which required only a small number of responses to get the cocaine, while manipulating the D3 receptors with agonists and antagonists has been surprisingly unclear in its results. For example, many studies have found that when the fixed ratio is low, D3 antagonist and agonist do not seem to affect the amount the animal will self-administer. This suggests that it is only when the animal must work hard for the cocaine reward that D3 receptors mediate the response. Therefore, we can perhaps see D3 receptors as being involved in incentivising the behaviour, rather than initiating the addiction.

Gene editing and transgenic animal models have really started to clarify some of these details. This is no truer than in the case of D3 receptors' involvement in drug addiction. Much research has indicated that animals show a distinct increase in both drug-seeking and drug-taking behaviour when D3 receptor expression has been genetically 'knocked out'. One particularly good example is the research of Zhan et al. (2018). This study is particularly good as its design considers the various 'stages' of addiction previously considered. They found that D3-knockout animals learned to self-administer heroin faster and self-administered more of it, suggesting a role of D3 receptors in the initiation of addiction. They also found that the D3-knockout animals would work harder to obtain a heroin reward than control animals, persisting much longer in the pursuit of the drug in a progressive ratio reinforcement task, suggesting that alterations in D3 receptors play some role in the maintenance and compulsive seeking stage of addiction. Finally, they found that the D3-knockout animals showed a higher seeking behaviour compared to controls during extinction training (this is where the 'reward' – heroin in this case – is removed) and showed significantly increased responses when they were challenged with heroin after extinction, suggesting that D3 receptors are also involved in the relapse and reuptake of an addiction.

Despite all this, much research evidence points towards other mechanisms related to dopamine and, in many cases, other mechanisms that are not dopamine-dependent that modulate 'drug addiction'. A good example comes from the laboratory of Christian Lüscher (Creed et al., 2016), whose research implicates alterations in the signalling of GABAergic projection neurons, specifically in the ventral pallidum, in 'stages' of addiction. As ever, one thing to always keep in mind is that the brain is a complex system, and it is unlikely that answers will boil down to the involvement of one neurotransmitter or one receptor in the myriad of behaviours its circuitry controls.

Focus on research

Influential reseachers

T. Celeste Napier is a pre-eminent researcher in key areas that appear throughout this chapter. Over the years she has heavily published core papers on the effects of dopamine, and its agonists and antagonists, in various regions, including basal ganglia territories, such as the dorsal and ventral striatum. As discussed in this chapter, these regions, and dopamine alterations, are heavily implicated in reward-related behaviour, drug addiction and Parkinson's disease. Her current research is still heavily entwined with these regions and these disorders, with a particular focus on a subset of Parkinsonian patients that display increased drug abuse and impulse control disorders, which may be linked to the known alteration in dopaminergic neuron function within those diagnosed with Parkinson's disease.

There are many good papers, both reviews and primary research, from the lab of T. Celeste Napier, but here are a couple, relevant to this chapter, for your further reading:

Napier, T. C., Kirby, A., & Persons, A. L. (2020). The role of dopamine pharmacotherapy and addiction-like behaviors in Parkinson's disease. *Progress in Neuro-Psychopharmacology and Biological Psychiatry, 102.* doi:10.1016/j.pnpbp.2020.109942

Napier, T. C., & Persons, A. L. (2019). Pharmacological insights into impulsive-compulsive spectrum disorders associated with dopaminergic therapy. *The European Journal of Neuroscience, 50*(3), 2492–2502. doi:10.1111/ejn.14177

Conclusion

Dopamine and its molecular mechanism are well studied, especially in the context of behaviour related to reward. It seems extremely likely that this is indeed dopamine's core function, providing a putative time stamp informing us when an expected reward is present, but also, with its absence, identifying for the organism when an expected reward does not appear. Alteration in dopamine's ability to do this, especially in the reward circuitry, seems likely to underpin its role in a myriad of atypical behavioural states, such as addiction and schizophrenia. In addition, dopamine's machinery is still very much seen as a key player and treatment target for a range of degenerative motor disorders, such as Parkinson's disease. Despite all this, it is essential to remember that, as a monoamine, dopamine and its machinery interacts with several other neurotransmitters, such as serotonin, which is also a monoamine and will be focused on in the next chapter. This adds a level of complexity to our understanding of dopamine and should start you thinking about how much of what we know about dopamine may also, at least partially, involve other neurotransmitters and their mechanisms.

Test Yourself Answers

Test Yourself 5.1

Point 3 - this is the historical view of the role of dopamine in schizophrenia and Parkinson's disease; however, it is important to consider that more recent research suggests that the relationship is much more complicated than this and, certainly for schizophrenia, there is plenty of evidence that suggests the imbalance in dopamine levels is brain region dependent.

Test Yourself 5.2

Table 5.2

Synthesis	Storage	Reuptake	Receptors
TH	VMAT	DAT	D1-like
DOPA		COMT	D2-like
VTA		MAO	Metabotropic

Test Yourself 5.3

a. Many of these patients had spent at least some time being treated with antipsychotic drugs, such as Chlorpromazine and Haloperidol, which were trying to 'normalise' dopamine transmission levels.

b. The definition of schizophrenia in these studies was often very broad and seldom was consideration given to the 'type' of schizophrenia the patients had presented with before death.

c. There were a lot of inconsistencies in the methods used to store and process the tissue post-mortem, which could have resulted in degradation of the tissue.

References

Key information

Tobias, J. A. Y., & Merlis, S. (1970). Levodopa and Schizophrenia. *Jama, 211*(11), 1857–1857. doi:10.1001/jama.1970.03170110063025

Eiden, L. E., & Weihe, E. (2011). VMAT2: a dynamic regulator of brain monoaminergic neuronal function interacting with drugs of abuse. *Ann N Y Acad Sci, 1216*, 86–98. doi:10.1111/j.1749-6632.2010.05906.x

Huot, P., & Parent, A. (2007). Dopaminergic neurons intrinsic to the striatum. *J Neurochem, 101*(6), 1441–1447. doi:10.1111/j.1471-4159.2006.04430.x

Jaber, M., Jones, S., Giros, B., & Caron, M. G. (1997). The dopamine transporter: a crucial component regulating dopamine transmission. *Mov Disord*, *12*(5), 629–633. doi:10.1002/mds.870120502

Perreault, M. L., Hasbi, A., O'Dowd, B. F., & George, S. R. (2014). Heteromeric Dopamine Receptor Signaling Complexes: Emerging Neurobiology and Disease Relevance. *Neuropsychopharmacology*, *39*(1), 156–168. doi:10.1038/npp.2013.148

The classics

American Psychiatric Association. (2013). Diagnostic and statistical manual of mental disorders (5th ed.) (DSM-5). Washington, DC: APA.

World Health Organisation. (2019). International classification of diseases (11th revision) (ICD-11). Geneva: WHO.

Schultz, W., Dayan, P., & Montague, P. R. (1997). A neural substrate of prediction and reward. *Science*, *275*(5306), 1593–1599. doi:10.1126/science.275.5306.1593

Schultz, W. (2004). Neural coding of basic reward terms of animal learning theory, game theory, microeconomics and behavioural ecology. *Curr Opin Neurobiol*, *14*(2), 139–147. doi:10.1016/j.conb.2004.03.017

Schultz, W. (2016). Dopamine reward prediction error coding. *Dialogues Clin Neurosci*, *18*(1), 23–32. Retrieved from www.ncbi.nlm.nih.gov/pubmed/27069377

Wise, R. A. (1978). Catecholamine theories of reward: a critical review. *Brain Res*, *152*(2), 215–247. doi:10.1016/0006-8993(78)90253-6

Corbett, D., & Wise, R. A. (1980). Intracranial self-stimulation in relation to the ascending dopaminergic systems of the midbrain: a moveable electrode mapping study. *Brain Res*, *185*(1), 1–15. doi:10.1016/0006-8993(80)90666-6

Cannon, C. M., & Palmiter, R. D. (2003). Reward without dopamine. *J Neurosci*, *23*(34), 10827–10831. doi:10.1523/jneurosci.23-34-10827.2003

Everitt, B. J., & Robbins, T. W. (2005). Neural systems of reinforcement for drug addiction: from actions to habits to compulsion. *Nature Neuroscience*, *8*(11), 1481–1489. doi:10.1038/nn1579

Kestler, L. P., Walker, E., & Vega, E. M. (2001). Dopamine receptors in the brains of schizophrenia patients: a meta-analysis of the findings. *Behav Pharmacol*, *12*(5), 355–371.

Wise, R. A. (2008). Dopamine and reward: the anhedonia hypothesis 30 years on. *Neurotoxicity Research*, *14*(2-3), 169–183. doi:10.1007/BF03033808

Berridge, K. C. (2007a). The debate over dopamine's role in reward: the case for incentive salience. *Psychopharmacology*, *191*(3), 391–431.

Peciña, S., Cagniard, B., Berridge, K. C., Aldridge, J. W., & Zhuang, X. (2003). Hyperdopaminergic mutant mice have higher "wanting" but not "liking" for sweet rewards. *J Neurosci, 23*(28), 9395–9402. doi:10.1523/jneurosci.23-28-09395.2003

Langston, J. W. (2017). The MPTP Story. *Journal of Parkinson's Disease, 7*(s1), S11–S19. doi:10.3233/JPD-179006

Miller, G. W., Erickson, J. D., Perez, J. T., Penland, S. N., Mash, D. C., Rye, D. B., & Levey, A. I. (1999). Immunochemical analysis of vesicular monoamine transporter (VMAT2) protein in Parkinson's disease. *Exp Neurol, 156*(1), 138–148. doi:10.1006/exnr.1998.7008

Skalisz, L. L., Beijamini, V., Joca, S. L., Vital, M. A. B. F., Da Cunha, C., & Andreatini, R. (2002). Evaluation of the face validity of reserpine administration as an animal model of depression -Parkinson's disease association. *Progress in Neuro-Psychopharmacology and Biological Psychiatry, 26*(5), 879 -883. doi:https://doi.org/10.1016/S0278-5846(01)00333-5

Taylor, T. N., Caudle, W. M., & Miller, G. W. (2011). VMAT2-Deficient Mice Display Nigral and Extranigral Pathology and Motor and Nonmotor Symptoms of Parkinson's Disease. *Parkinsons Dis, 2011*, 124165. doi:10.4061/2011/124165

Kestler, L. P., Walker, E., & Vega, E. M. (2001). Dopamine receptors in the brains of schizophrenia patients: a meta-analysis of the findings. *Behav Pharmacol, 12*(5), 355–371.

Howes, O. D., & Kapur, S. (2009). The dopamine hypothesis of schizophrenia: version III--the final common pathway. *Schizophr Bull, 35*(3), 549–562. doi:10.1093/schbul/sbp006

Seeman, P., & Kapur, S. (2000). Schizophrenia: more dopamine, more D2 receptors. *Proceedings of the National Academy of Sciences of the United States of America, 97*(14), 7673–7675. doi:10.1073/pnas.97.14.7673

Sullivan, P. F. (2005). The Genetics of Schizophrenia. *PLOS Medicine, 2*(7), e212. doi:10.1371/journal.pmed.0020212

Schmidt, C. J., Sorensen, S. M., Kehne, J. H., Carr, A. A., & Palfreyman, M. G. (1995). The role of 5-HT2A receptors in antipsychotic activity. *Life Sci, 56*(25), 2209–2222. doi:10.1016/0024-3205(95)00210-w

Volkow, N. D., Wang, G. J., Fowler, J. S., Logan, J., Gatley, S. J., Wong, C., . . . Pappas, N. R. (1999). Reinforcing effects of psychostimulants in humans are associated with increases in brain dopamine and occupancy of D(2) receptors. *J Pharmacol Exp Ther, 291*(1), 409–415.

Evans, A. H., Pavese, N., Lawrence, A. D., Tai, Y. F., Appel, S., Doder, M., . . . Piccini, P. (2006). Compulsive drug use linked to sensitized ventral striatal dopamine transmission. *Ann Neurol, 59*(5), 852–858. doi:10.1002/ana.20822

Zaghloul, K. A., Blanco, J. A., Weidemann, C. T., McGill, K., Jaggi, J. L., Baltuch, G. H., & Kahana, M. J. (2009). Human substantia nigra neurons encode unexpected financial rewards. *Science*, *323*(5920), 1496–1499. doi:10.1126/science.1167342

Hot topics

Lohoff, F. W. (2010). Genetic variants in the vesicular monoamine transporter 1 (VMAT1/SLC18A1) and neuropsychiatric disorders. *Methods Mol Biol*, *637*, 165–180. doi:10.1007/978-1-60761-700-6_9

Multani, P. K., Hodge, R., Estévez, M. A., Abel, T., Kung, H., Alter, M., . . . Lohoff, F. W. (2013). VMAT1 deletion causes neuronal loss in the hippocampus and neurocognitive deficits in spatial discrimination. *Neuroscience*, *232*, 32–44. doi:10.1016/j.neuroscience.2012.11.023

Perreault, M. L., Hasbi, A., O'Dowd, B. F., & George, S. R. (2014). Heteromeric Dopamine Receptor Signaling Complexes: Emerging Neurobiology and Disease Relevance. *Neuropsychopharmacology*, *39*(1), 156–168. doi:10.1038/npp.2013.148

Kim, S., Kwon, S. H., Kam, T. I., Panicker, N., Karuppagounder, S. S., Lee, S., . . . Ko, H. S. (2019). Transneuronal Propagation of Pathologic alpha-Synuclein from the Gut to the Brain Models Parkinson's Disease. *Neuron*, *103*(4), 627–641.e627. doi:10.1016/j.neuron.2019.05.035

Heidbreder, C. A., Gardner, E. L., Xi, Z.-X., Thanos, P. K., Mugnaini, M., Hagan, J. J., & Ashby, C. R. (2005). The role of central dopamine D3 receptors in drug addiction: a review of pharmacological evidence. *Brain Research Reviews*, *49*(1), 77–105. doi:https://doi.org/10.1016/j.brainresrev.2004.12.033

Wise, R. A., & Koob, G. F. (2014b). The development and maintenance of drug addiction. *Neuropsychopharmacology : official publication of the American College of Neuropsychopharmacology*, *39*(2), 254–262. doi:10.1038/npp.2013.261

Czermak, C., Lehofer, M., Wagner, E. M., Prietl, B., Gorkiewicz, G., Lemonis, L., . . . Liebmann, P. M. (2004). Reduced dopamine D3 receptor expression in blood lymphocytes of smokers is negatively correlated with daily number of smoked cigarettes: a peripheral correlate of dopaminergic alterations in smokers. *Nicotine Tob Res*, *6*(1), 49–54. doi:10.1080/14622200310001656858

Staley, J. K., & Mash, D. C. (1996). Adaptive increase in D3 dopamine receptors in the brain reward circuits of human cocaine fatalities. *J Neurosci*, *16*(19), 6100–6106. doi:10.1523/jneurosci.16-19-06100.1996

Caine, S. B., & Koob, G. F. (1993). Modulation of cocaine self-administration in the rat through D-3 dopamine receptors. *Science, 260*(5115), 1814–1816. doi:10.1126/science.8099761

Zhan, J., Jordan, C. J., Bi, G.-H., He, X.-H., Gardner, E. L., Wang, Y.-L., & Xi, Z.-X. (2018). Genetic deletion of the dopamine D3 receptor increases vulnerability to heroin in mice. *Neuropharmacology, 141*, 11–20. doi:10.1016/j.neuropharm.2018.08.016

Creed, M., Ntamati, N. R., Chandra, R., Lobo, M. K., & Luscher, C. (2016). Convergence of Reinforcing and Anhedonic Cocaine Effects in the Ventral Pallidum. *Neuron.* doi:10.1016/j.neuron.2016.09.001

Napier, T. C., & Persons, A. L. (2019). Pharmacological insights into impulsive-compulsive spectrum disorders associated with dopaminergic therapy. *European Journal of Neuroscience, 50*(3), 2492–2502. doi:10.1111/ejn.14177

Napier, T. C., Kirby, A., & Persons, A. L. (2020). The role of dopamine pharmacotherapy and addiction-like behaviors in Parkinson's disease. *Progress in Neuro-Psychopharmacology and Biological Psychiatry, 102.* doi:10.1016/j.pnpbp.2020.109942

6

SEROTONIN

Introduction

Serotonin is one of the most well-known neurotransmitters in popular culture. One of the main reasons for this is the fact that alterations in serotonin expression and function have been implicated in multiple behaviours and clinical disorders. In fact, some go as far as to suggest that it is involved in almost every physical function. Serotonin is especially prominent in the mood disorders literature and is part of the hotly debated 'monoamine' theories of depression. Many researchers, since the 1960s, have suggested that serotonin is a key player in mood states and affective disorders, including major depressive disorder. This was a key turning point in the psychiatric treatment of mood disorders, leading to a shift towards biological treatments as a key component of an intervention, aimed at helping reduce the symptomology of psychiatric disorders. As such, serotonin and the drugs, such as Prozac, that act on its function, have become ingrained in the collective consciousness. In this chapter we will discuss the underpinning molecular mechanism of serotonin's synthesis, packaging, reuptake and receptors before considering the roles that these molecular mechanisms might play in a range of atypical behaviours, including their classic role in 'mood' disorders and their role in typical and atypical sleep.

Outline of serotonin's synthesis, packaging and reuptake

As discussed in the previous chapter, there is a vast 'machinery', and a number of key workers, that help contribute to neurotransmitters' synthesis, packaging, reuptake and the transmission of their effects. This is also true for serotonin. Alterations at any point in this machine or its key workers can therefore contribute to biological pathology and changes in our behaviour. It is therefore absolutely essential to fully understand the workers so we can identify the possible source of alterations in output.

Serotonin is found in various locations in our bodies, not just in the central nervous system. In fact, the majority of serotonin in our body is found in the enteric nervous system, which plays a key role in controlling the function of our GI (gastrointestinal) tract. We will return to its role in the GI tract in Chapter 10. In the current chapter we will largely be focusing on its role within the brain, and how alterations within the brain may result in behaviour changes. But first, more about the machinery.

Serotonin's synthesis

As mentioned in the introduction, serotonin belongs to a large family of chemicals known as monoamines (see Figure 6.1). This broad group includes various molecules, some of which are close relatives of serotonin, while others are distant cousins. The broad group includes dopamine (see Chapter 5 for more on this) and epinephrine, which can be considered as cousins, and melatonin, which can very much be considered a closer relative, perhaps akin to a sibling.

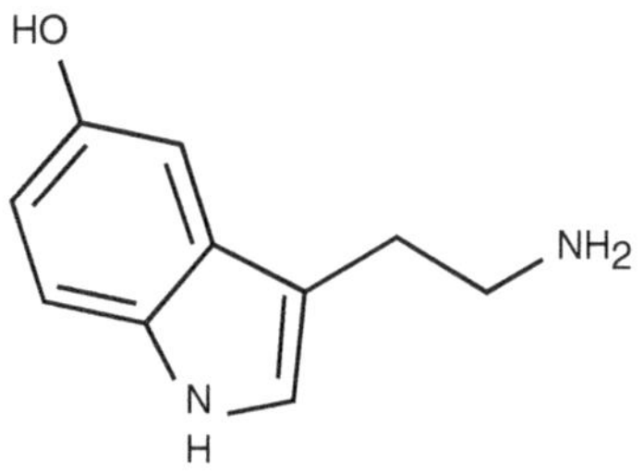

Figure 6.1 Molecular structure of serotonin

If you take a moment to look back to the dopamine chapter you will see the similarity of serotonin with other monoamines.

So where is serotonin synthesised in the brain? Well, the vast majority of serotonin in the brain comes from a small region that is considered part of the reticular formation in the brain stem. This region is known as the raphe nuclei (see Figure 6.2). Projections from the raphe nuclei (expressing serotonin) innervate several different regions in the mammalian brain, including parts of the basal ganglia, such as the striatum, which, as we discussed in Chapter 5 on dopamine, are heavily implicated in reward and pleasure responses. Other regions innervated by these projections include the amygdala, the hypothalamus, the hippocampus and the cortex.

An interesting side point to consider is that anything in the brain's stem and, more broadly, the lower and central regions of the brain, is evolutionarily conserved. What this means is that they have been around in our brain for a long time, so they are likely to be involved in modulating behaviour that is essential for our survival. A good example of this is that these serotoninergic neurons in the raphe nuclei are found in lizards (*Anolis sagrei*). Recent research suggests that they are integral to sex and aggression behaviour in these animals – two behaviours that will help the animal survive and pass on its genes. Maybe we are not as far away as we think from these lizards!

So, much like dopamine, probably one of the main reasons serotonin has garnered research interest over the last half-century is that, in the mammalian brain, there are only a few distinct nuclei, in this case the raphe nuclei, that express serotonin[1] and their projections are targeted and localised, suggesting that they may play a modulatory role in the behaviour these brain regions shape. We will begin to unpack this later on in this chapter.

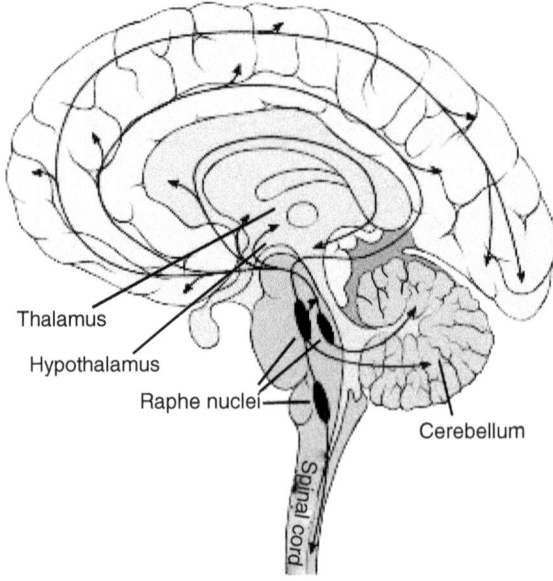

Figure 6.2 The raphe nuclei and serotonergic projections from this region into other territories

The vast majority of serotonin synthesised in the brain is synthesised by serotonergic neurons of the raphe nuclei. These serotonergic neurons then project to numerous brain regions, including cortical territories and subcortical regions, such as the limbic system and cerebellum.

Source: © Patrick J. Lynch and C. Carl Jaffe and modified by S. Jähnichen / CC BY 2.5

So how is it synthesised? The starting point, like the majority of the monoamines, is one of the common amino acids. For serotonin this is tryptophan. Tryptophan gets into the brain across the blood–brain barrier. It cannot readily diffuse and requires an active transport mechanism. Tryptophan must compete for this mechanism with other dietary amino acids. We will return to this in later parts of this chapter when we discuss tryptophan dietary depletion research, which has been widely used to investigate the effects of low levels of serotonin in numerous behaviour states, including mood disorders and sleep.

---

[1]Although this is true to some degree, recent research has shown that there are in fact serotonergic neurons expressed within other brain regions and that they are likely interneurons.

Once tryptophan has made its way into the brain, the next step is to load up the neurons as this is where the simple two-step process to produce serotonin takes place. Again, this is not a straightforward matter of diffusion but involves tryptophan being taken into neurons by a sodium-dependent uptake mechanism. Once inside the neuron, the first step is catalysed by the enzyme tryptophan hydroxylase, which removes the OH group in a process called hydrolysis to produce 5-hydroxytryptophan. This is a very important enzyme as it is considered the main rate limiting step in the synthesis of serotonin. What this means is that any alterations in this enzyme, either in terms of the amount or its ability to function, will have a dramatic effect on the subsequent expression of serotonin. Finally, 5-hydroxytryptophan is the decarboxylated (removal of a COOH group) by the enzyme AAADC (aromatic L-amino acid decarboxylase) to form serotonin, which is commonly referred to as 5-HT (5-Hydroxytryptamine).

Serotonin packaging

Serotonin is packaged into vesicles by VMATs. Cast your mind back to Chapter 5 and you will recall that this stands for vesicular monoamine transporter. Therefore these transporters aid the packaging, for release, of all the major monoamines. I won't cover this again here, but if you need a refresher have a look back to Chapter 5. Before you do so, perhaps have a go at the following quiz and see what you remember.

Test Yourself 6.1

1. How many keep types of VMATs are expressed in the mammal?
2. Which VMAT has been known to be expressed in neurons of the brain for the longest?
3. Identify at least one atypical behaviour in which alterations in VMAT function is implicated.
4. Which neurotransmitters are believed to be partly synthesised within the vesicle after their precursors are transported in by VMATs?
5. VMATs belong to a family of transporters known as TEXANs. What does this stand for?
6. What does the fact VMATs evolved from TEXANs tell us about monoamines?

Answers to all Test Yourself questions are at the end of the chapter.

It is easy to think that once serotonin is packaged into these vesicles it remains waiting for the signal to arrive at the synapse for the vesicle's merger with the cell membrane, and release into the synapse. This is not the full picture and the packaging of serotonin is somewhat of a dynamic process, with constant leakage of serotonin from the vesicle back into the neuron's cytoplasm, and reuptake via VMATs into the vesicle. The process by which VMAT transports serotonin into the vesicles is called an 'active transport' mechanism. This is because it requires an input of energy. As you can see in Figure 6.3, both VMATs and enzyme exist in the vesicle's membranes. This enzyme converts ATP (adenosine triphosphate) into ADP (adenosine diphosphate), providing the energy required to actively transport hydrogen ions into the vesicle. This sets up a proton gradient in the vesicle membrane, which fundamentally powers the import of serotonin by VMATs. Because this is such a dynamic process, you can perhaps start to get the picture of how dysfunction of VMATs will rapidly result in a build-up of serotonin outside the vesicle, resulting in a toxic environment for the neuron.

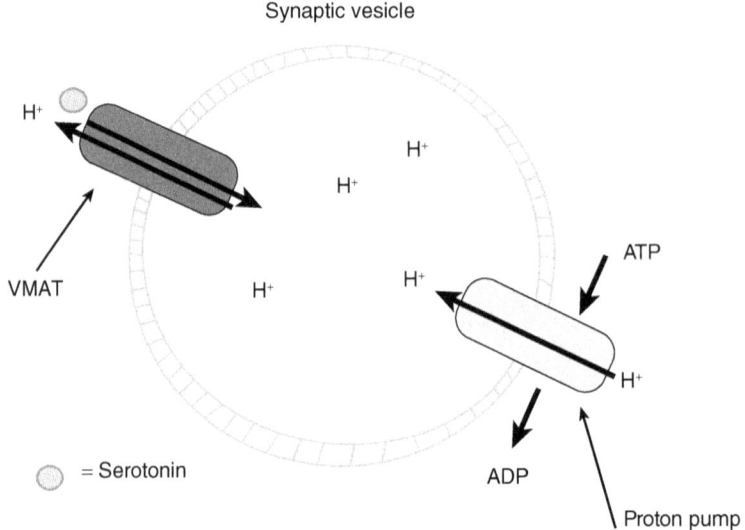

Figure 6.3   Illustration of a vesicle packaging serotonin with the typical channels embedded in its membrane

As was the case for dopamine packaging, the channel protein that facilitates the movement of serotonin into the vesicle lumen is the vesicular monoamine transporter.

Source: Adapted from: Chaudhry, F. A., Edwards, R. H., & Fonnum, F. (2008). Vesicular neurotransmitter transporters as targets for endogenous and exogenous toxic substances. *Annual Review of Pharmacology and Toxicology, 48*(1), 277–301. doi:10.1146/annurev.pharmtox.46.120604.141146

Serotonin reuptake

Once serotonin has been released into the synapse and has had its effects at its receptor sites, the time comes for it to be brought back into the neuron and repackaged, awaiting future action potentials to arrive and the cycle to start again. This process of reuptake is managed by the membrane transporter SERT. SERT is a member of the neurotransmitter sodium symporters (NSS), which include the transporters for dopamine, serotonin, norepinephrine, glycine and GABA. They all function in a very similar way but are each specific for their unique neurotransmitter. They are known as symporters as they simultaneously transport molecules in and out of the cell. In the case of SERT, it simultaneously transports sodium ions, chloride ions and serotonin into the neuron while transporting potassium out. It manages to bring serotonin by harnessing the sodium gradient across the cell membrane, which effectively means serotonin is 'dragged' in with sodium against serotonin's concentration gradient. You might be thinking why does potassium leave? This is a good question and is essential to the neuron as it means the whole process is electroneutral. Thus, this process should not cause any overall change in the neuron's membrane potential, which otherwise could trigger further serotonin release. As previously mentioned, monoamines are cytotoxic once back inside the neuron, so it is essential that, once intracellular, serotonin is rapidly broken down into its constituent parts. This is done by the enzyme monoamine oxidase (MAO), which is located in the mitochondrial membrane inside the neuron. MAO has two isoforms, one is known as MAO-A, which seems to have the greatest affinity for serotonin. The other is known as MAO-B and is generally considered to be more selective for other amines, such as benzylamine. This seems straightforward and you would expect that if MAO-A is selective for serotonin, then serotonergic neurons of the brain would express this type. However, this is not completely the case and most research seems to show that the cell bodies of serotoninergic neurons in the raphe nucleus predominantly express MAO-B. The projections and the terminal ends of these raphe nuclei neurons seem to express the type we'd expect, namely MAO-A. This is somewhat perplexing, and the reason is not entirely resolved.

These reuptake mechanisms have been extensively researched and are a popular target for many of the main drug therapies related to affective disorders and anxiety disorders. We will cover this in more depth later in this chapter.

─ Focus on concepts ☁ ────────────────

The interaction between neurotransmitter machinery – SERT and nNOS

It is important to keep in mind throughout this book that neurotransmitters and their molecular mechanisms do not act in isolation. In fact, as we have already seen in this chapter,

(Continued)

there is a lot of interplay between the mechanisms controlling the synthesis, packaging and reuptake of serotonin and dopamine.

This interplay between neurotransmitters goes beyond those that share the same family. One good example of this is the interplay between serotonin reuptake transporters (SERT) and the enzyme that is key to the synthesis of neuronal nitric oxide (nitric oxide synthase (nNOS)), which is a fascinating 'gaseous' neurotransmitter we focus on in Chapter 9. The enzyme nNOS is able to bind onto an intracellular part of SERT. When it does this, it dramatically slows down the amount of SERT in the neuron's plasma membrane and therefore, as a result, reduces the amount of serotonin taken back up into the presynaptic neuron. This means there is more serotonin left in the synapse. Why is this interesting? Well, a significant number of the drugs used to treat depression are what is known as selective serotonin reuptake inhibitors (SSRIs). These drugs directly reduce the function of SERTs and therefore result in more serotonin in the synapse. Fundamentally, bindings of nNOS seem to have the same effect. nNOS could therefore be a novel and intriguing new target for antidepressants, and perhaps one that has fewer of the traditional side effects that are associated with SSRIs.

Further reading

Garthwaite, J. (2007). Neuronal nitric oxide synthase and the serotonin transporter get harmonious. *Proceedings of the National Academy of Sciences, 104*(19), 7739. doi:10.1073/pnas.0702508104

Joca, S. R. L., Sartim, A. G., Roncalho, A. L., Diniz, C. F. A., & Wegener, G. (2019). Nitric oxide signalling and antidepressant action revisited. *Cell and Tissue Research, 377*(1), 45–58. doi:10.1007/s00441-018-02987-4

Outline of serotonin receptors (ionotropic and metabotropic)

As is the case with the majority of neurotransmitters, there are several different receptor types specific to serotonin. In fact, there are generally considered to be seven types, and within each of these there are various further subtypes, with around 14 receptor subtypes in total. Much of the recent research evidence linking serotonin receptors to 'atypical behaviour' focuses on these various different subtypes. It is important to keep in mind, when you are doing your wider reading, that the older literature may have used less specific methods to investigate the role of serotonin receptors, so any conclusions you read may be contradictory to more recent research. This is largely due to progression in techniques and the development/availability of more and more specific antagonists and agonists in pharmacological research, and radioligands in imaging studies, which can now target the subtypes of serotonin receptors. We will look at research in this area later in this chapter. For now, let's pick apart a little bit of the detail on the subtypes.

The main types are labelled one to seven with serotonin three receptors (5HT3R) standing out on a mechanistic level from the other six. As the title of the section implies, the reason is that they are the only serotonin receptor that is known to be ionotropic. If you can think back to previous chapters, especially Chapter 1, you will remember that this means a lot for the effects on the neuron, the main thing being that activation of this type of channel causes rapid onset and rapidly reversible electrical responses. Before we go any further, this is an excellent opportunity to remind yourself about the main differences between ionotropic and metabotropic receptors by having a go at the following task.

Test Yourself 6.2

Ionotropic vs metabotropic

Grab a piece of paper and draw a table with two columns. Label one column ionotropic and the other column metabotropic. Now place the following terms and phrases under the correct headings.

Table 6.1

Generally postsynaptic	Long latency	Binding site is combined with the channel	Short latency	Pre- and postsynaptic
Slow responses	G-protein-coupled, involving secondary messengers	Binding site is separate from the channel	Second messenger independent	Rapid response, typically 10–50 ms

Hopefully, you are pretty confident about the main difference between ionotropic and metabotropic receptors. Now we're are going to look more deeply at metabotropic serotonin receptors, and the intracellular effects their activation has on secondary messenger pathways.

As mentioned above, 5HT3 receptors are distinct as they are the only known ionotropic serotonin receptor. This means the other six receptor types are all G-protein-coupled receptors or, in other words, metabotropic. Although the six may all be metabotropic, activation of each doesn't result in the same effect on the neuron. This is because the different types either activate different secondary messenger systems or have opposing effects on the secondary messenger systems. One distinct difference that splits these metabotropic receptors into two groups is the effect they have on the enzyme adenylyl cyclase (AC) and, ultimately, the levels of the secondary messenger cAMP (cyclic adenosine monophosphate). Activation of 5HT4, 6 and 7 receptors causes a confirmation change (shape change) in the coupled G proteins that results in activation of AC and increased levels of cAMP, while 5HT1 and 5 receptor activation causes a chain of events that results in inhibition of AC and decreases levels of cAMP. cAMP has its main effect on the enzyme protein kinase A (PKA), which is key to the regulation of 'energy' with the neuron in the form of sugar, glycogen and lipid metabolism. If we were broadly to say what this means for the neuron, we would say 5HT4, 6 and 7 activation results in excitation, while 5HT1 and 5 activation results in inhibition. It is important to remember that this is very much a simplification and that the intracellular pathways modulated by activation of both these types of receptors has all manner of effects on gene expression, resulting in diverse effects on the neuron, such as synaptogenesis and synthesis of enzymes, which are essential to produce neurotransmitters.

For those of you with a keen eye, you may be thinking 'we're missing one'. Indeed, we are. The last type of metabotropic serotonin receptor is 5HT2R. I have left this till last as this type stands out from the rest. Although they are G-protein-coupled, their activation, and the subsequent activation of the G protein, results in three different

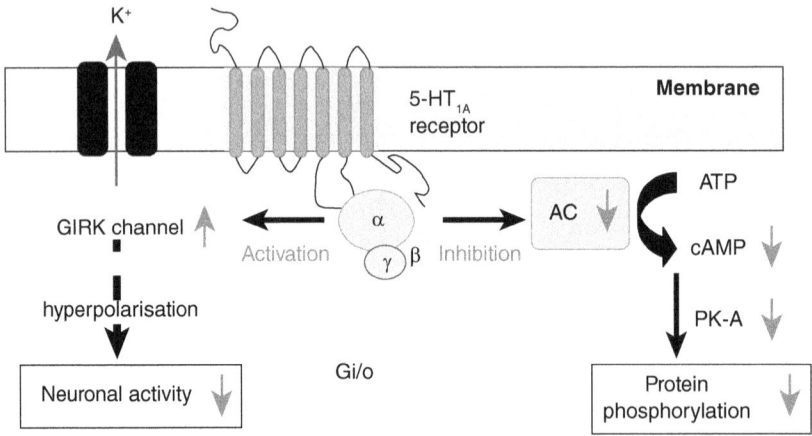

Figure 6.4   The intracellular processes modulated by the binding of serotonin at 5HT1A receptors

The specific serotonin receptor subtype 5HT1A has general inhibitory effects on the expressing neuron. This is largely mediated by the dissociation of the bound G protein, which activated K+ channels, resulting in neuronal hyperpolarisation. In addition, binding of serotonin to the 5HT1A receptor, and the subsequent dissociation of the G protein, also results in downregulation of the enzyme adenylyl cyclase (AC). This results in the downregulation of a myriad of other intracellular processes, such as the production of cAMP.

Source: Figure reprinted from: Shimizu, S., & Ohno, Y. (2013). Improving the treatment of Parkinson's disease: a novel approach by modulating 5-HT1A receptors[J]. *Aging and Disease, 4*(1), 1–13.

types of secondary messengers, namely IP (inositol phosphate), PIP2 (phosphoinositol bisphosphate) and DAG (diacylglycerol). This all results in increases in calcium levels within the neuron and activation of protein kinase C (PKC), which plays a fundamental job in regulating a wide variety of proteins within the neuron.

But what about the subtypes? As mentioned at the start of this section, there are seven types of serotonin receptors and 14 subtypes. Not all the receptor types have subtypes. In fact, it is 5HT1, 2 and 5 receptors that include subtypes, with 5HT1 having the most at six (5HT1a, b, c, d, e, f). One specific subtype we will come back to in relation to its involvement in affective disorders is 5HT1a receptors. Activation of these essentially leads to hyperpolarisation of the neuron (reduced chance of firing) through activation of GIRK channels, which can be seen in Figure 6.4. This is important, especially when we consider these subtypes of receptors are commonly found

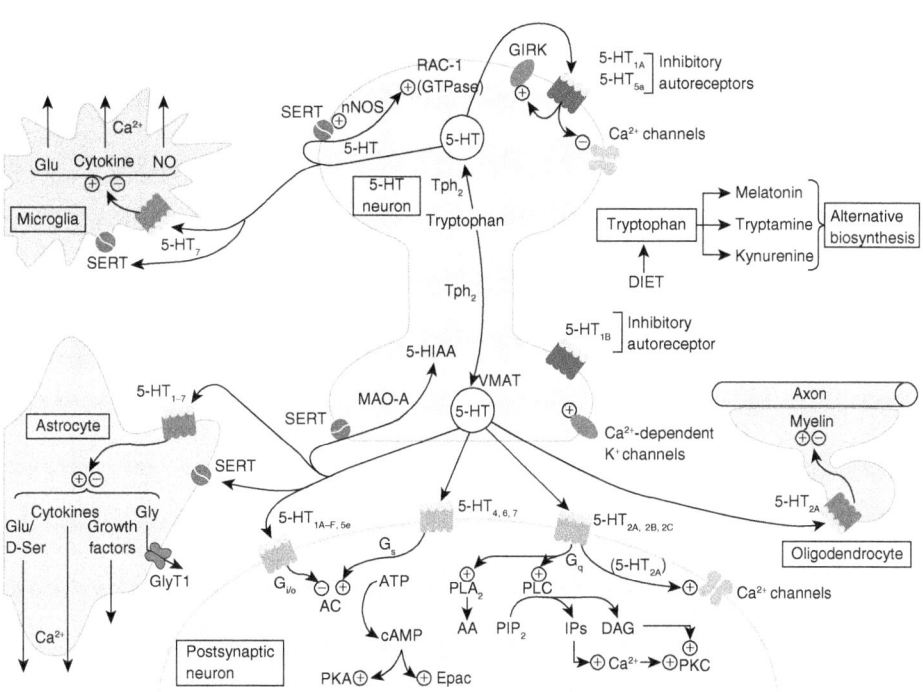

Figure 6.5 The complex mechanisms by which serotonin receptors have their effects at pre- and postsynaptic locations as well as beyond the neuron in other cells, such as astrocytes, oligodendrocytes and cytokines

Source: Reprinted from: Millan, M. J., Marin, P., Bockaert, J., & Mannoury la Cour, C. (2008). Signaling at G-protein-coupled serotonin receptors: recent advances and future research directions. *Trends in Pharmacological Sciences, 29*(9), 454–464. Copyright (2008), with permission from Elsevier.

as auto receptors on the presynaptic cell body. What this means is that activation of these, on the presynaptic neuron, results in a reduced probability of any further serotonin release. I mention this now as it is worth keeping in mind when we discuss research that implicates them in affective disorders, later in this chapter.

Unfortunately, this is a complex topic, as can be seen in Figure 6.5, and a whole book could be written on the intracellular signalling pathways of G-protein-coupled serotonin receptors. However, if you wish to read more, the article by Millan et al. (2008) is an excellent starting point.

Classic topics in serotonin research

Serotonin and mood disorders

As mentioned previously in this chapter, serotonin is part of the group of neurotransmitters called monoamines. This group also includes dopamine, epinephrine and norepinephrine. In the early 1960s, monoamines began to emerge in the literature as a commonly cited factor in mood disorders. The most common theory back then was that the monoamine norepinephrine was the key player, and that alterations in norepinephrine's expression and function were the key biological factors underpinning mood disorders. As the 1960s rolled on it became clear that there was another contender for the throne, in the form of serotonin.

Serotonin is a monoamine, but it is a member of a different subgroup from norepinephrine. Norepinephrine belongs to the subfamily catecholamines, while serotonin belongs to the subfamily of indoleamines. This can often get confusing as the textbooks will typically refer to the broad 'monoamine' hypothesis of mood disorders. In this chapter, we will specifically focus on the 'monoamine' hypothesis as it relates to serotonin.

So why serotonin? Well, let's start with a little bit of history. Much of the early evidence came for the effects of pharmacology treatments. These early treatments were generally being trialled as therapies for other issues but turned out to help reduce depressive symptomology. For examples, drug classes such as monoamine oxidase inhibitors and tricyclic antidepressants were found in the 1950s and 1960s to significantly reduce the negative symptomology of depression and, more broadly, to elevate mood. However, these drugs were non-selective, in the sense that they altered the machinery that underpinned the degradation and reuptake of all monoamines, and, as discussed earlier, monoamines take several forms. The real evidence that seemed to specifically implicate serotonin alterations in depression symptomology was the rise of selective serotonin inhibitors as antidepressant medications. They became prevalent in the research literature in the late 1970s, and by the 1980s the first one was being marketed as an antidepressant medication. The therapies include the drug Fluoxetine (typical trade name: Prozac). It essentially works by blocking serotonin reuptake and therefore increasing the amount of serotonin in the synapse. Based on the fact that these drugs are 'effective' at alleviating the symptomology of depression, you might think that it is very clearly the case that low serotonin levels must therefore cause depression. However, this is not necessarily the case, and is often referred to as the treatment aetiology fallacy. What this means is that just because a biological alteration reduces symptoms doesn't mean that it is the thing that caused the issue.

Precursor loading experiments

Some of the best early research that added solidity to the serotonin hypothesis of depression came from an experimental method known as precursor loading or precursor deletion. This method works on the principle that alterations in the precursor to serotonin, tryptophan, produces changes in the amount of serotonin available in the brain. Therefore, serotonin levels can be altered by altering the levels of dietary tryptophan. This is a little more complicated than simply adding or subtracting the amount of tryptophan in someone's diet because, as discussed earlier in this chapter, tryptophan competes for active transport across the blood–brain barrier with other large neutral amino acids (LNAA), such as tyrosine. So when doing these studies, the researchers have to carefully balance the amount of tryptophan relative to other LNAAs. For example, increased tryptophan will not have an effect on the amount of serotonin if the individual also has a diet high in other LNAAs. The method therefore requires alteration in dietary tryptophan and other LNAAs to ensure increase and decrease in free tryptophan in the brain and the resulting increases or decreases in serotonin.

The tryptophan deletion studies of Delgado and colleagues (1990) are often cited in the literature. In these studies, patients who were in remission (recovery) after experiencing an episode of clinically diagnosed major depression consented to having a low tryptophan diet followed by a protein drink high in LNAAs (other than tryptophan) the day after. This meant tryptophan was low, due to diet restrictions, but further lowered, as a ratio, because of the increased levels of other LNAAs. This resulted in a reduction of free tryptophan levels by an average of 91% in the participants. The effects of the low tryptophan diet and drink were dramatic, with 14 out of the 21 participants experiencing a depressive relapse followed by a gradual (over the course of two days) return to remission. This was the case even though they continued to take their antidepressant medication. These studies are criticised to some degree because the patients were also on a course of antidepressant medication.

Some have therefore postulated that the depressive relapses may have been due to an interaction between the tryptophan deletion and the medication. However, other studies, with participants who had previously suffered episodes of major depression but were recovered and antidepressant free, seem to find very similar results to those of Delgado and colleagues.

As further support for the role of serotonin levels in mood disorders, there are a significant number of precursor loading studies (increase the tryptophan) that also suggest that increased levels of tryptophan (remember this is a ratio, relative to the level of other LNAAs) and subsequent increase in serotonin elevates mood in clinical populations. Although there are several studies that show elevation of tryptophan to reduce depressive symptomology, there are also many studies that suggest there is no effect of a tryptophan diet on depressive symptomology. Furthermore, when we move to a non-clinical population, there is even less evidence to suggest tryptophan loading improves 'mood' state, with most studies using 'healthy' volunteers, suggesting no alterations in mood caused by alteration in tryptophan levels. This has led to many people proposing that there may be a 'vulnerable' group who are more responsive to tryptophan alterations in terms of their subsequent effect on mood state.

An interesting avenue of research, which may provide some insight into the likely 'vulnerable group' comes from research on emotional processing biases in patients diagnosed with depression. There is significant evidence which suggests that a bias towards identifying negative responses in others, such as facial gestures denoting disgust, and a bias away from identifying positive responses in others, such as facial gestures denoting happiness, may be a cognitive factor underpinning depressive states. Tryptophan loading studies have found that this type of emotional processing bias seems to be altered when tryptophan levels increase. For example, Murphy et al.'s (2006) study found that tryptophan loading increased the identification of positive facial responses and reduced the identification of negative facial gestures as well as reducing overall attentional focus on negative words. This might mean, therefore, that a 'vulnerable group' that has a negative emotional processing bias may be more susceptible to alterations of tryptophan in the diet, and decreasing tryptophan may result in altered mood to the level of a clinical diagnosis of mood disorder, while increasing tryptophan in the diet may be more efficacious for those with this negative emotional processing bias.

Another important point to keep in mind is that many studies suggest alterations in tryptophan (and subsequent serotonin levels) result in changes to other behaviour and cognitive states beyond 'mood'. For example, research shows that tryptophan loading alters sleep and alertness. This is interesting as these two factors are often also altered in patients diagnosed with major depressive disorder. Therefore, what we might be seeing is that serotonin levels alter general alertness and sleep levels, which in turn result in a shift in general mood and affective state. So we might suggest that it is not serotonin levels *per se* that cause the 'depressive state', but it is mediated by sleep and general alertness. Indeed, it is perhaps easy to see how alteration in alertness might be identified as diminished interest in pleasurable activities, which is a key diagnostic criterion for major depression, especially if you were focused on measuring 'depression'.

The machinery of serotonin: reuptake and message transmission

Serotonin reuptake transporters

Although not completely clear, a reasonably convincing amount of evidence seems to suggest some role for altered serotonin levels in mood disorders. If we now start to focus on the machinery of serotonin's neuronal reuptake and 'message transmission', we come across another area of research that heavily implicates alterations in serotonin signalling in mood disorders.

A large body of research focuses on alterations in the serotonin reuptake transporter (SERT) and mood disorders. Indeed, these are a very logical biological focus, especially when you consider that some of the most effective antidepressants are selective serotonin reuptake inhibitors. The main method used in this area of research is autoradiographic labelling. It employs a chemical which is known to bind with the biological target, which is tagged with a radioactive element. Like an x-ray, when film is exposed to this 'labelled' tissue it will give us an image and reveal the relative density of our biological target, in this case reuptake transporters.

Much research in this area has employed these radioligands to label post-mortem tissue from individuals who had been diagnosed with a mood disorder during their life and, in most cases, had died as a result of suicide. Early results were greatly inconsistent, with some studies showing increased levels of reuptake transporters, others decreased expression of reuptake transporters and yet others suggesting no difference in reuptake transporter expression levels between those with a clinical mood disorder diagnosis and non-clinical control brain tissue. I have chosen the phrase 'reuptake transporter' carefully in this paragraph as one of the major caveats of this early research was the use of rather non-specific radioligands. The most common one used was [3H]imipramine, which is known to bind with both SERTs and norepinephrine reuptake transporters. This may be one of the reasons for inconsistencies in the literature.

Things started to get a little clearer when more specific radioligands were used in the early 2000s. One of the more specific radioligands commonly used was [3H] cyanoimipramine. In studies employing this radioligand, we see either a significant reduction in SERT expression in those with mood disorder diagnosis or no change compared to control. Another factor that may contribute to the inconsistency in the literature is the brain region focused on. Most of these studies look at the prefrontal cortex, although some look at other cortical regions, while others focus on the hippocampus and the region where serotonergic projections emanate from, namely the dorsal raphe nuclei. An intriguing study that highlights the complexity of brain region was carried out by Mann et al. (2000), who found that reductions in SERT expression seemed to be specifically localised to ventral frontal cortical regions in those who had died by suicide, while dorsal frontal cortical regions were reduced in SERT expression in those with a mood disorder diagnosis who had not seemingly died by suicide.

Serotonin receptors

We are often led to believe that the main factor that underpins mood disorders is the level of serotonin in the brain. Indeed, it is worth considering that much of what we have covered about serotonin's involvement in mood disorders boils down to factors that alter serotonin levels seem to alter mood state. However, many people believe that serotonin levels *per se* are not the main issue underpinning mood disorder, but rather it is the relative expression and function of serotonin receptors. If you cast your mind back to the receptor section of this chapter, you will remember that serotonin receptor types are many, with further subtypes, resulting in approximately 14 in total. Unfortunately, discussing each one and its relative contribution to mood disorders is a book within itself. Instead, we will now focus on one of the most heavily implicated in mood disorders, namely $5HT_{1a}$ receptors. However, please do keep in mind that almost all the subtypes have been implicated in some shape or form in mood disorders.

Much like the involvement of SERT levels in mood disorders, autoradiographic methods have been commonly used to investigate the relative expression of serotonin receptors in the brain tissue. Early studies in this area typically focused on individuals who had died by suicide, with often little documentary information about their mental health. However, as we moved into the later 1990s the research tended to have a more rigorous approach and included groups that had documented mood disorder diagnosis during their life. As with the SERT autoradiographic binding literature, the results are mixed, but most studies either find no difference in expression levels of $5HT_{1a}$ receptors compared to controls or an increase in $5HT_{1a}$ expression, especially in prefrontal cortical regions and the hippocampus.

As we moved into the later 1990s and early 2000s, considerable development was made in the field of PET (positron emission tomography) imaging.

This technique involves injecting receptor agonists/antagonists (work with/ work against) tagged a radioactive marker into the bloodstream of participants, which makes their way to the brain via the vasculature. The technique allowed researchers to investigate the expression of serotonin receptors in vivo (in the living). Interestingly, the early research using this technique, such as that by Drevets et al. (2007), typically found the opposite of the post-mortem autoradiographic studies. Namely, that $5HT_{1a}$ levels (or, more specifically, the binding to these receptors) were reduced in the brain of those diagnosed with mood disorders, especially in hippocampal and cortical regions. Unusually for this research area, there is a strong consensus in the literature that lower $5HT_{1a}$ receptor binding, and therefore lower receptor levels, is the case in those diagnosed with a mood disorder.

Although all this research is intriguing and reveals a complex relationship between SERT, serotonin receptor expression and mood disorders, there are a couple of important issues that need to be considered. One problem with much of this imaging research is that it simply tells us about the amount of SERTs and 5HT receptors in the brain but tells us little about how they function. This is an important point and may explain some of the contradictions in the imaging literature. It is important to remember that just because the numbers are 'normal' doesn't mean that they are functioning correctly. Another point is that although much of this research shows altered levels of SERT and/or serotonin receptors in those diagnosed with mood disorders, it cannot establish, conclusively, if these altered levels cause the mood disorder or are a consequence of the mood disorder. This is where animal research comes in as it allows us to establish more direct causal relationships by measuring 'depressive symptomology', making a change (such as reducing the $5HT_{1a}$ receptor levels), then measuring 'depressive symptomology' again.

One example of a study that does this was carried out by Richardson-Jones and colleagues (2010). They used a genetically-modified mouse model that resulted in either reduced expression of the $5HT_{1a}$ auto receptor or increased expression of the $5HT_{1a}$ auto receptor. Those animals with the higher expression of the $5HT_{1a}$ auto receptor showed increased depressive behaviour, including less resilience to stress. Crucially, this study also showed that alteration of $5HT_{1a}$ receptors causes a change in raphe nuclei neuronal firing rate but didn't alter the basal levels of free serotonin. This provides pretty convincing evidence for the involvement of neuronal activity and $5HT_{1a}$ auto receptors in mood disorders, but less of a role for simply altered baseline serotonin levels, providing a firm suggestion that serotonin levels *per se* are not key to depressive symptomology but to the neuronal machinery underpinning its effects.

Serotonin and eating disorders

Serotonin has been heavily implicated in homeostasis.[2] One simple example of homeostasis is the body's ability to maintain its core temperature within a very narrow band around 37°C. Changes by a small degree either up or down can result in dramatic loss of biological function and ultimately death. Another key homeostatic mechanism that is essential to maintaining our body in a healthy state, and ultimately helping us survive, is the intake of food.

It would therefore seem reasonable that any atypical behaviour related to the intake of food may well be mediated by alterations in serotonin and its molecular machinery. Indeed, the idea that alterations in serotonergic tone (the normal level of serotonin in the body) underpin eating disorders has been a popular idea for many years and there is a reasonable amount of research evidence that suggests this may well be the case. We will come back to this research shortly, but first it would seem pertinent to clarify what eating disorders are.

There are several different types of eating disorder, but the two most known and understood are anorexia nervosa and bulimia nervosa. People generally think that eating disorders are typified by an individual's disinterest or dislike of food, and a popular myth is that eating disorders are caused by a lack of desire to ingest food. In fact, nothing could be further from the truth. Eating disorders are often typified by a strong desire (to the point of obsession) for food. Despite this, those diagnosed with anorexia nervosa control and restrict their intake of food in the pursuit of 'thinness'. This is done to the point of extreme malnutrition and, in many cases, organ failure and death.

Anorexia nervosa is often considered to be caused by psychosocial factors, with a rich literature focusing on the role media and cultural perceptions of 'thin' as desirable. This suggestion seems very reasonable. Hoek and van Hoeken (2003) identified that the prevalence of eating disorders increased through the twentieth century up to the 1970s, which correlates with the rise in the 'thin' ideal in western society. Nevertheless, as mentioned above, there is considerable evidence to suggest at least some contribution of underlying biological alterations, with a key player being serotonin.

Early research focused on measuring the basal levels of serotonin in those diagnosed with or recovering from anorexia nervosa. This research generally found reduced levels of serotonin metabolites (byproducts of its use) in the cerebrospinal

---

[2]Homeostasis is the body's ability to control key biological characteristics within a narrow, optimal, range in order to keep us healthy and all biological processes functioning correctly.

fluid (CSF) of those with anorexia nervosa. However, many researchers realised that there is somewhat of an issue with this, as malnutrition is known to alter serotonin levels, so many researchers started to focus on those with a diagnosis who had 'recovered' (were at normal weight and had a good/stable diet). With these individuals, researchers commonly found an increased level of serotonin metabolites, compared to controls, suggesting in fact that high levels of serotonin may be more likely to contribute as a 'causal' factor in anorexia nervosa. It is important to consider a couple of things about these findings. One is that measuring the metabolite levels in the CSF is an indirect measure, and therefore may not fully represent what is actually happening with serotonin. Second, much of this research is conducted with those who have already received a diagnosis of anorexia nervosa, even if they have 'recovered'. It may be that these alterations are a result of other factors that may be the 'real' cause of developing anorexia nervosa. We may see altered serotonin as more of a consequence of the disorder than a causal factor. Even if this is the case, it can still be viewed with interest as it may be an integral factor that limits recovery. Certainly, researchers such as Kaye, Fudge, and Paulus (2009) proposed the influential theory that serotonin alterations may perpetuate the negative cycle, resulting in chronic illness and some of the key dysfunctional thinking that typifies anorexia nervosa, such as the obsession with food and the common denial of the illness, which many diagnosed with anorexia exhibit. So, perhaps what we should really be thinking here is: 'Is serotonin involved in perpetuating anorexia nervosa?' rather than 'Does serotonin cause anorexia nervosa?'

Neuroimaging studies of SERT and serotonin receptors

Positron emission tomography (PET) has been a common technique used of recent years to investigate the machinery underpinning serotonin function in the brain, and how it may relate to anorexia nervosa. The technique allows us to visualise and quantify the levels of various cellular receptors and transport mechanisms by attaching a radioligand to a molecule that is known to bind with the receptors or transporters.

There is a pretty convincing line of evidence for alterations in the expression of 5HT1A and 5HT2A receptors in those either diagnosed with or recovering from anorexia nervosa. The generally consistent finding seems to be that there is an increased binding potential[3] of the 5HT1A receptor and a reduced binding potential for the

---

[3]Binding potential is often used as a proxy for the expression of the receptor. Therefore, increased binding potential is commonly considered to be related to increased expression of the receptor and reduced binding potential the opposite.

5HT2A receptor (especially in cortical regions) in patients diagnosed with anorexia nervosa. Although there is reasonable consistency in these findings in the literature, it is often hard to conclusively interpret the findings as different studies often focus on different brain regions. Thus, differences may, to some degree, be region-specific.

There is also a pretty convincing line of evidence suggesting that alterations in SERT (the serotonin transporter) may also be evident in those diagnosed with an eating disorder, although the literature seems to suggest SERT alterations are more common in bulimia nervosa rather than anorexia nervosa. For example, Kaye et al. (2005) only found significantly reduced binding to SERTs in patients recovering from bulimia nervosa and binge-type anorexia nervosa. Indeed, SERT reductions seem to be more specifically linked to a subtype of behaviour associated with both bulimia nervosa and anorexia nervosa, namely bingeing behaviour. One example of this is the research of Kuikka et al. (2001), who found reductions in SERT binding for those with a specific diagnosis of binge eating disorder. This raises an important issue, namely that eating disorders are complex disorders with multiple alterations in behaviour and cognition. It is therefore important to consider that alterations in serotonin may be linked to specific alterations in certain behaviour and/or cognition rather than typifying the 'whole' disorder.

Animal models

As we have discussed in this section, much of the evidence for the involvement of serotonin in anorexia nervosa comes from patients who already have a diagnosis or those who are recovering. This limits our ability to suggest that it is involved in the aetiology of the disease. In order to try to resolve this issue animal models (normally rodents) of anorexia nervosa have been developed. A common model used is what is called the activity-based anorexia model (ABA). This model involves restricting the diet of animals while giving them free access to an exercise wheel. It leads to self-starvation in the animal and hyperactivity, which are both commonly seen in patients with anorexia nervosa. Typical research with the animal model involves administration of various pharmacological agents and observation of weight, feeding and hyperactivity. There are very mixed and somewhat inconclusive results employing pharmacological manipulation of serotonin in these models. For example, ABA rats that were administered a 5HT1A agonist showed a reduction in weight loss, but it crucially didn't stop weight loss (Atchley & Eckel, 2006). Also, other research, from the same group, using a more general serotonin agonist seemed to promote more rapid weight loss and increased hyperactivity in these ABA animals. These results seem contradictory, although it is important to think back to our discussion on 5HT receptors earlier in the chapter. This is because 5HT1A receptors commonly act as auto receptors, reducing the further release of serotonin. Therefore, the fact 5HT1A agonist reduced weight loss is consistent with general serotonin agonist perpetuating weight loss and increasing activity.

It is important to be aware that the ABA model is only one animal model of several in use. For a more detailed description of other models, see van Gestel, Kostrzewa, Adan, and Janhunen (2014).

It is essential to remember that eating disorders in general, and anorexia nervosa specifically (as it has been the main focus of this section), are extremely complex disorders with a correspondingly complex aetiology. It seems likely that psychosocial and environmental factors, as well as serotonin and its molecular machinery, play roles in anorexia nervosa's aetiology. Therefore, as I have repeated many times throughout this book, it is essential to remember that serotonin is only one piece of the jigsaw, and to fully represent a complex disorder, we must have a complex and entwined theory as to its aetiology.

Test Yourself 6.3

Complete the following box by ticking the sections that are appropriate.

Table 6.2

	Increased in anorexia nervosa	Decreased in anorexia nervosa	No change in anorexia nervosa	More relevant to other eating disorders
Serotonin overall				
SERT				
5HT1A				
5HT2A				

Serotonin and sleep

Before we discuss serotonin's involvement in sleep it is important to clarify that 'normal' sleep is made up of several components and the amount of these components naturally varies as you age. As we move into a sleep cycle, we can split the stages of sleep into two components: REM (rapid eye movement) sleep and NREM (non-rapid eye movement) sleep. REM is typically associated with dreaming and a 'typical' adult has approximately six episodes of REM per night. NREM is anything that isn't REM.

In Figure 6.6, NREM sleep is stages 1, 2, 3 and 4, with stage 4 typically being referred to as 'deep sleep'. Again, as you can see in Figure 6.6, the 'typical' adult has two episodes of stage 4 sleep a night, at the beginning of the night, and this gets shallower as the night progresses. Sleep can also be spilt into components based upon the types of electrical activity happening in the brain at various points. For example, in Figure 6.6, you'll see that anything below stage 2 is typically referred to as slow-wave sleep as the electrical activity of neurons in the brain changes its pattern.

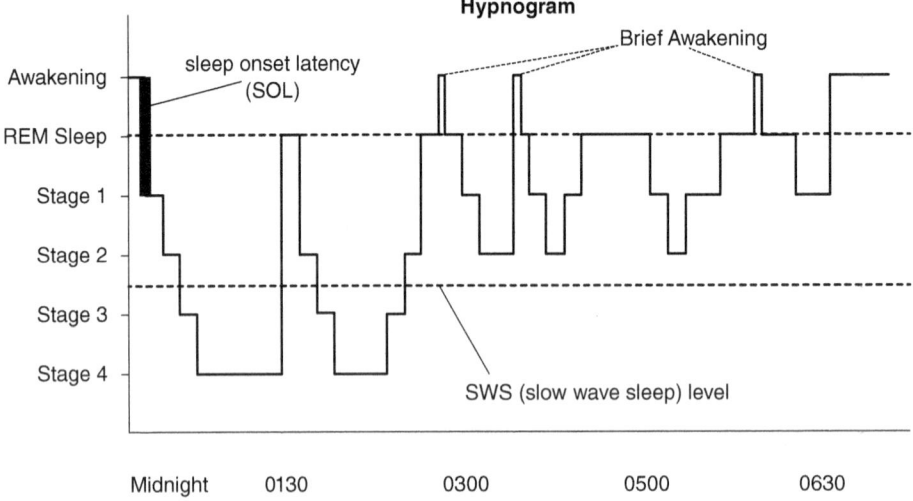

Figure 6.6   The stages of the sleep cycle during a typical night

During a typical night's sleep for an adult several stages of sleep will be moved through. As the night progresses the cycles become briefer and generally do not reach the depth (stage 3 or 4) that are seen in the first 90–120 minutes of sleep. Each of these stages of sleep is characterised by distinct electrical brain activity, such as the illustrated slow waves seen in stage 2–3.

Source: © RazerM / CC BY-SA 2.5

As already mentioned in the previous section on mood disorders, there is a rich seam of research that implicates serotonin, and its machinery, in alterations in sleep states. An important point to note on a biological level is that serotonin is a precursor to melatonin. Melatonin synthesis (from serotonin) is triggered in the pineal gland by falling light levels, and as melatonin has historically been referred to as 'nature's sleep draft', it should come as no surprise that serotonin should play some role in sleep and 'wakefulness'. So, what role might we expect it to play? Well, one of the negative side effects typically reported by patients taking SSRIs to manage mood disorders is insomnia (a reduced ability to sleep or an increase in disrupted sleep). This would suggest that greater levels of 'free' serotonin increase alertness and reduce the ability to sleep. As we will see in the next few pages, the research is complex and somewhat contradictory, although as we get closer to the present day it does seem that much of the research supports the function of serotonin in both sleep and 'wakefulness'.

Much of the early research focused on injecting serotonin into vertebrates and observing the effects on their behaviour. These injections were typically intravenous. Therefore, much like tryptophan loading studies, they would have diffuse effects across all brain regions. These studies typically found initial increases in alertness followed by a stereotyped drowsiness. These biphasic responses were thought to be mediated by serotonin's effects in different brain regions. The way researchers clarified this issue was by using animal models where various brain regions had been dissected out. For example, Koella & Czicman (1966) colleagues carried out a brainstem transection and found that, after this, only the arousal effects of serotonin were seen. It suggested that arousal effects are mediated by regions above the brain stem and that the 'sleep-inducing' effects of serotonin application are mediated by lower regions and brain stem-specific regions.

As the research progressed it became clear that there was a specific 'core' region that was the source of most of the serotonergic innervation in the brain, namely the neurons of the raphe nuclei. These neurons project out of the raphe nuclei and innervate many brain regions, including the striatum, hypothalamus, amygdala and hippocampus. Again, lesion studies were performed on this area, with research suggesting reductions in sleep as a result of its lesion. It led many to propose the hypothesis that the dorsal raphe serotoninergic projections, and increased serotonin levels as a result of their activity, were essential for 'inducing sleep'.

However, there were a few key caveats with these studies. The first is that serotoninergic neurons are essential for the regulation of our body temperature. Therefore, the loss of these neurons resulted in hypothermia, causing a reduction in sleep. This was demonstrated because sleep was not reduced when the animals were in a warm environment (Murray, Buchanan, & Richerson, 2015). Another issue with these studies is that in lower-order animals, such as mice, the reduction in sleep, due to lesions of the dorsal raphe nuclei serotonergic neurons, was not consistently found. In fact, several key studies seemed to suggest that although animals with raphe nuclei lesions displayed hyperactivity, it did not affect their sleep level. One intriguing study conducted by Cespuglio and colleagues (1976) invovled reducing the activity of raphe nuclei serotonergic neurons in cats by cooling to +10°C. This seemed to produce increases in wakefulness when done as the animals slept, but seemed to induce sleep states if the animal was awake. This led to the suggestion that serotonin may act as a switching mechanism moving between sleep to wake and moving between the stages of sleep, as outlined in Figure 6.6.

As we move forward to more 'modern' research, one key target region for raphe nuclei serotonergic projections, the ventrolateral preoptic area (VLPO), has become important to our understanding of sleep and serotonin's involvement. The VLPO is part of the hypothalamus (see Figure 6.7 for more detail on its anatomical location). Denoyer and colleagues (1989), among others, carried out early research focusing on the VLPO. These studies involved injecting various inhibitors of the machinery responsible for the synthesis of serotonin into animal models. It was found that these chemicals, which blocked the synthesis of serotonin by inhibiting the enzyme tryptophan hydroxylase, could induce insomnia. Subsequent injections of serotonin into the VLPO of these animals could reverse the insomnia and reinstate normal levels of slow wave sleep and REM (rapid eye movement) sleep. It seems to imply that blockade of serotonin is responsible for an increased inability to sleep/wakefulness, which fits well with most of the research covered above. However, the pharmacological agent used to do these experiments also have complex effects on other monoamines, including norepinephrine and dopamine, so the research was certainly far from conclusive.

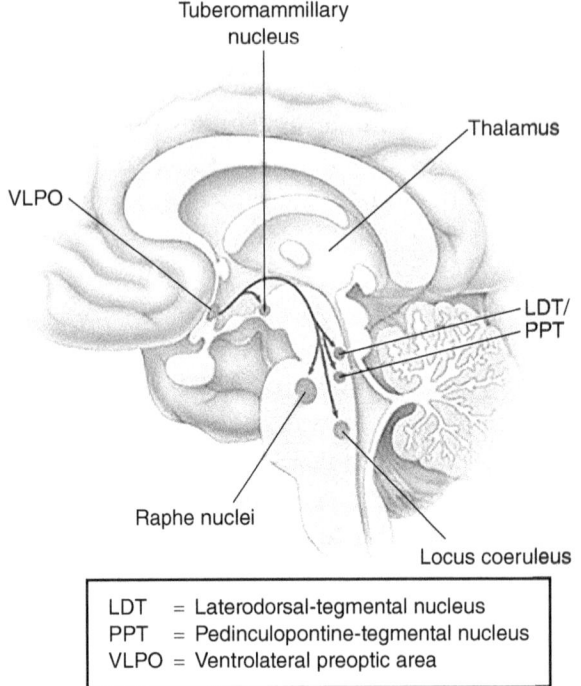

LDT	= Laterodorsal-tegmental nucleus
PPT	= Pedinculopontine-tegmental nucleus
VLPO	= Ventrolateral preoptic area

Figure 6.7   The ventrolateral preoptic area (VLPO)

The hypothalamus contains several subregions including the VLPO. It is found dorsal to the raphe nuclei and ventro-rostrally to the thalamus

Source: Gaskin, S. (2021). *Behavioral neuroscience: essentials and beyond*. Thousand Oaks, CA: Sage.

As we get to current research, the role of the VLPO has become intriguingly important, although the beliefs that low levels of serotonin induce wakefulness and high levels induce sleep changed. This occurred as the technical ability to record the electrical activity of neuron populations in vivo became prominent in the literature.[4] Indeed, the literature moved very much towards the belief that serotonergic neuronal activity (in the raphe nuclei) increases during 'wakefulness' and reduces sequentially as we move into sleep through REM and into NREM. Evidence to support the role these neurons play in the 'wakefulness' inducing effects of serotonin increased dramatically. Today, the weight of research seems to suggest that serotonergic neuronal activity (dorsal raphe nuclei) is associated with wakefulness/alertness. It makes a lot more sense if you think back to our previous discussion about melatonin inducing sleep. This is because melatonin is synthesised from serotonin, so as melatonin increases serotonin should decrease as we move into sleep and, conversely, as we move out of sleep, we would expect melatonin to break down, resulting in increased levels of serotonin. However, this is still not definitively the case and research on serotonin's involvement in sleep is still very much in flux, with some recording studies showing that specific subpopulations of serotonergic neurons in the raphe nuclei increase firing during the stages of sleep.

The effects of serotonin (released from raphe nuclei neurons) is probably largely mediated by two things: the 'type of neuron' that the serotonergic neurons target and the type of serotonin receptors that these 'target' neurons express.

Novel topics in serotonin and sleep promoting neurons

So, what are the neurons that are targeted by these serotonergic projections from the raphe nuclei? This is a very complex question to answer as these projections innervate many brain regions. However, one important target we have already mentioned, in the context of sleep, is the VLPO. In this region, serotonergic projections are known to innervate what have become known as sleep-promoting neurons, as well as targeting other neurons that also innervate these sleep-promoting neurons within the VLPO.

These, sleep-promoting neurons in the VLPO are considered to be divided into two types based upon their response to serotonin, with approximately 44% being

---

[4]These electrophysiological techniques assume that the alteration in electrical activity infers alterations in the release of neurotransmitters. In this case, alterations in serotoninergic neurons of the raphe nuclei result in an altered release probability for serotonin in the regions these neurons target.

inhibited by serotonin (type I) and 56% being excited by serotonin (type II). Sangare and colleagues (2016) think that these two types essentially act as a switching mechanism, with type II neurons being involved in sleep akin to a bouncer deciding if sleep can happen, and type I neurons being involved in executing the action and allowing you to go through the door to sleep.

It is worth remembering that like all science that is at the cutting edge, the ideas about sleep-promoting neurons are likely to change and, at best, will be adjusted and clarified and, at worst, be disproven.

Serotonin receptors and sleep

The second factor that will mediate the role of serotonergic projections on sleep are the type of serotonin receptors expressed by their targets.

Serotonin receptors seems to be involved both in the transition between awake to sleep states and the number of various stages within sleep. For example, there is some research that suggests $5HT_2$ receptors vary in their activity between awake and sleep stages, suggesting that they may be involved in instating the circadian rhythm of the sleep/wake cycle. Other research has implicated $5HT_{1B}$ receptors in the transition between awake and sleeping states, with $5HT_{1B}$ agonists increasing the awake state and reducing the amount of slow-wave sleep in rats. Looking into the stages of sleep, research has implicated the $5HT_{1A}$ receptors, specifically in REM sleep, with studies using mutant mice that do not express these receptors engaging in more REM sleep but having unchanged levels of slow-wave sleep. This has also been found when specific $5HT_{1A}$ agonists have been injected directly into the raphe nuclei, suggesting that these effects may be mediated by 5HT1A receptors' role as a presynaptic auto receptor, essentially 'switching off' the further release of serotonin.

If serotonin receptors are differentially affected in sleep and the transition between sleep and waking states, then we may expect for them to be differentially distributed on the two different types of sleep-promoting neurons. Indeed, Sangare and colleague's (2016) data does seem to suggest this, with $5HT_1$ A, B, D and F being expressed on type 1 sleep-promoting neurons, while $5-HT_{2A}$-C, 4 and 7 were found to be equally expressed on both type I and type II sleep-promoting neurons within the VLPO.

Although all this evidence seems to present as convincing evidence for serotonin alterations to be involved in sleep, it is important to remember that there are several other key players in the form of other neurotransmitters and neurohormones. For example, two of the big players in recent years are the neurohormones orexin and hypercretin (see the papers by Ursin (2002) and Scammell, Arrigoni, and Lipton (2017) for further introductions to these). In addition, there is significant evidence for the involvement of GABAergic neurons in various components of sleep, especially REM sleep. This all serves to emphasise that the mechanisms of sleep are complex, and although serotonin and its machinery may play some role, it is only one of many that contribute to a 'good night's sleep'.

Conclusion

Serotonin is without doubt a complex neurotransmitter, from its multitude of different receptor subtypes and the complex interaction between its machinery and that of other neurotransmitters, to the inconclusive and sometimes contradictory findings of research implicating its involvement in behaviour such as sleep, eating and mood. In our next chapter we will move away from what could be viewed as the more 'traditional' neurotransmitters to look at several molecules that act as neurotransmitters but are typically referred to as neuropeptides. In the following chapters you will see that the mechanisms and machinery we have so far covered in this book do not necessarily hold true for all neurotransmitters.

▬Test Yourself Answers▬

Test Yourself 6.1

1. Two.
2. VMAT2.
3. Schizophrenia, major depressive disorder, anxiety disorders.
4. Epinephrine and norepinephrine.
5. Toxin extruding antiporters.
6. It suggests monoamines are toxic within the neuron.

Test Yourself 6.2

Table 6.3

Generally postsynaptic (ionotropic)	Long latency (metabotropic)	Binding site is combined with the channel (ionotropic)	Short latency (ionotropic) long	Pre- and postsynaptic (metabotropic)
Slow responses (metabotropic)	G-protein-coupled, involving secondary messengers (metabotropic)	Binding site is separate from the channel (metabotropic)	Second messenger independent (ionotropic)	Rapid response, typically 10–50 ms (ionotropic)

(Continued)

Test Yourself 6.3

Table 6.4

	Increased in anorexia nervosa	Decreased in anorexia nervosa	No change in anorexia nervosa	More relevant to other eating disorders
Serotonin overall	✓ (after recovery)	✓ (during illness)		
SERT			✓	✓
5HT1A	✓			
5HT2A		✓		

References

Key information

Coppen, A. (1967). The biochemistry of affective disorders. *British Journal of Psychiatry, 113*(504), 1237–1264. doi:10.1192/bjp.113.504.1237

Hartline, J. T., Smith, A. N., & Kabelik, D. (2017). Serotonergic activation during courtship and aggression in the brown anole, Anolis sagrei. *PeerJ, 5*, e3331–e3331. doi:10.7717/peerj.3331

Silber, B. Y., & Schmitt, J. A. (2010). Effects of tryptophan loading on human cognition, mood, and sleep. *Neurosci Biobehav Rev, 34*(3), 387–407. doi:10.1016/j.neubiorev.2009.08.005

Finberg, J. P. M., & Rabey, J. M. (2016). Inhibitors of MAO-A and MAO-B in psychiatry and neurology. *Frontiers in Pharmacology, 7*, 340–340. doi:10.3389/fphar.2016.00340

Garthwaite, J. (2007). Neuronal nitric oxide synthase and the serotonin transporter get harmonious. *Proceedings of the National Academy of Sciences, 104*(19), 7739. doi:10.1073/pnas.0702508104

Joca, S. R. L., Sartim, A. G., Roncalho, A. L., Diniz, C. F. A., & Wegener, G. (2019). Nitric oxide signalling and antidepressant action revisited. *Cell and Tissue Research, 377*(1), 45–58. doi:10.1007/s00441-018-02987-4

Millan, M. J., Marin, P., Bockaert, J., & Mannoury la Cour, C. (2008). Signaling at G-protein-coupled serotonin receptors: recent advances and future research directions. *Trends Pharmacol Sci, 29*(9), 454–464. doi:10.1016/j.tips.2008.06.007

Charnay, Y., & Léger, L. (2010). Brain serotonergic circuitries. *Dialogues Clin Neurosci, 12*(4), 471–487. Retrieved from https://pubmed.ncbi.nlm.nih.gov/21319493 www.ncbi.nlm.nih.gov/pmc/articles/PMC3181988/

Serotonin and mood disorders

Mulinari, S. (2012). Monoamine theories of depression: Historical impact on biomedical research. *Journal of the History of the Neurosciences, 21*, 366–392. doi:10.1080/0964704X.2011.623917

Silber, B. Y., & Schmitt, J. A. (2010). Effects of tryptophan loading on human cognition, mood, and sleep. *Neurosci Biobehav Rev, 34*(3), 387–407. doi:10.1016/j.neubiorev.2009.08.005

Stockmeier, C. A. (2003). Involvement of serotonin in depression: evidence from postmortem and imaging studies of serotonin receptors and the serotonin transporter. *J Psychiatr Res, 37*(5), 357–373. doi:10.1016/s0022-3956(03)00050-5

Delgado, P. L. (2000). Depression: the case for a monoamine deficiency. *J Clin Psychiatry, 61* Suppl 6, 7–11. Retrieved from http://www.psychiatrist.com/JCP/article/Pages/depression-case-monoamine-deficiency.aspx

Delgado, P. L., Charney, D. S., Price, L. H., Aghajanian, G. K., Landis, H., & Heninger, G. R. (1990). Serotonin function and the mechanism of antidepressant action. Reversal of antidepressant-induced remission by rapid depletion of plasma tryptophan. *Arch Gen Psychiatry, 47*(5), 411–418. Retrieved from https://jamanetwork.com/journals/jamapsychiatry/article-abstract/495012

Murphy, S. E., Longhitano, C., Ayres, R. E., Cowen, P. J., & Harmer, C. J. (2006). Tryptophan supplementation induces a positive bias in the processing of emotional material in healthy female volunteers. *Psychopharmacology, 187*(1), 121–130. doi:10.1007/s00213-006-0401-8

Mann, J. J., Huang, Y. Y., Underwood, M. D., Kassir, S. A., Oppenheim, S., Kelly, T. M., . . . Arango, V. (2000). A serotonin transporter gene promoter polymorphism (5-HTTLPR) and prefrontal cortical binding in major depression and suicide. *Arch Gen Psychiatry, 57*(8), 729–738. doi:10.1001/archpsyc.57.8.729

Underwood, M. D., Kassir, S. A., Bakalian, M. J., Galfalvy, H., Dwork, A. J., Mann, J. J., & Arango, V. (2018). Serotonin receptors and suicide, major depression, alcohol use disorder and reported early life adversity. *Translational Psychiatry, 8*(1), 279. doi:10.1038/s41398-018-0309-1

Staley, J. K., Malison, R. T., & Innis, R. B. (1998). Imaging of the serotonergic system: interactions of neuroanatomical and functional abnormalities of depression. *Biol Psychiatry, 44*(7), 534–549. doi:10.1016/s0006-3223(98)00185-1

Kaufman, J., DeLorenzo, C., Choudhury, S., & Parsey, R. V. (2016). The 5-HT1A receptor in Major Depressive Disorder. *Eur Neuropsychopharmacol, 26*(3), 397–410. doi:10.1016/j.euroneuro.2015.12.039

Drevets, W. C., Thase, M. E., Moses-Kolko, E. L., Price, J., Frank, E., Kupfer, D. J., & Mathis, C. (2007). Serotonin-1A receptor imaging in recurrent depression: replication and literature review. *Nuclear Medicine and Biology, 34*(7), 865–877. doi:https://doi.org/10.1016/j.nucmedbio.2007.06.008

Richardson-Jones, J. W., Craige, C. P., Guiard, B. P., Stephen, A., Metzger, K. L., Kung, H. F., . . . Leonardo, E. D. (2010). 5-HT1A autoreceptor levels determine vulnerability to stress and response to antidepressants. *Neuron, 65*(1), 40–52. doi:10.1016/j.neuron.2009.12.003

Serotonin and sleep

Rancillac, A. (2016). Serotonin and sleep-promoting neurons. *Oncotarget, 7*(48), 78222–78223. doi:10.18632/oncotarget.13419

Denoyer, M., Sallanon, M., Kitahama, K., Aubert, C., & Jouvet, M. (1989). Reversibility of para-chlorophenylalanine-induced insomnia by intrahypothalamic microinjection of L-5-hydroxytryptophan. *Neuroscience, 28*(1), 83–94. doi:10.1016/0306-4522(89)90234-0

Silber, B. Y., & Schmitt, J. A. (2010). Effects of tryptophan loading on human cognition, mood, and sleep. *Neurosci Biobehav Rev, 34*(3), 387–407. doi:10.1016/j.neubiorev.2009.08.005

Ursin, R. (2002). Serotonin and sleep. *Sleep Medicine Reviews, 6*(1), 55–67. doi:https://doi.org/10.1053/smrv.2001.0174

Koella, W. P., & Czicman, J. (1966). Mechanism of the EEG-synchronizing action of serotonin. *American Journal of Physiology-Legacy Content, 211*(4), 926–934. doi:10.1152/ajplegacy.1966.211.4.926

Hipólide, D. C., Moreira, K. M., Barlow, K. B., Wilson, A. A., Nobrega, J. N., & Tufik, S. (2005). Distinct effects of sleep deprivation on binding to norepinephrine and serotonin transporters in rat brain. *Prog Neuropsychopharmacol Biol Psychiatry, 29*(2), 297–303. doi:10.1016/j.pnpbp.2004.11.015

Oikonomou, G., Altermatt, M., Zhang, R.-w., Coughlin, G. M., Montz, C., Gradinaru, V., & Prober, D. A. (2019). The serotonergic raphe promote sleep in zebrafish and mice. *Neuron, 103*(4), 686–701.e688. doi:https://doi.org/10.1016/j.neuron.2019.05.038

Monti, J. M. (2011). Serotonin control of sleep-wake behavior. *Sleep Med Rev, 15*(4), 269–281. doi:10.1016/j.smrv.2010.11.003

Cespuglio, R., Walker, E., Gomez, M.-E., & Musolino, R. (1976). Cooling of the nucleus raphe dorsalis induces sleep in the cat. *Neuroscience Letters, 3*(4), 221–227. doi:https://doi.org/10.1016/0304-3940(76)90077-X

Iwasaki, K., Komiya, H., Kakizaki, M., Miyoshi, C., Abe, M., Sakimura, K., . . . Yanagisawa, M. (2018). Ablation of central serotonergic neurons decreased REM sleep and attenuated arousal response. *Frontiers in Neuroscience, 12*, 535–535. doi:10.3389/fnins.2018.00535

Park, S.-H., & Weber, F. (2020). Neural and homeostatic regulation of REM sleep. *Frontiers in Psychology, 11*, 1662–1662. doi:10.3389/fpsyg.2020.01662

Sangare, A., Dubourget, R., Geoffroy, H., Gallopin, T., & Rancillac, A. (2016). Serotonin differentially modulates excitatory and inhibitory synaptic inputs to putative sleep-promoting neurons of the ventrolateral preoptic nucleus. *Neuropharmacology, 109*, 29–40. doi:10.1016/j.neuropharm.2016.05.015

Gallopin, T., Fort, P., Eggermann, E., Cauli, B., Luppi, P.-H., Rossier, J., . . . Serafin, M. (2000). Identification of sleep-promoting neurons in vitro. *Nature, 404*(6781), 992–995. doi:10.1038/35010109

Gallopin, T., Luppi, P. H., Cauli, B., Urade, Y., Rossier, J., Hayaishi, O., . . . Fort, P. (2005). The endogenous somnogen adenosine excites a subset of sleep-promoting neurons via A2A receptors in the ventrolateral preoptic nucleus. *Neuroscience, 134*(4), 1377–1390. doi:https://doi.org/10.1016/j.neuroscience.2005.05.045

Scammell, T. E., Arrigoni, E., & Lipton, J. O. (2017). Neural circuitry of wakefulness and sleep. *Neuron, 93*(4), 747–765. doi:https://doi.org/10.1016/j.neuron.2017.01.014

Murray, N. M., Buchanan, G. F., & Richerson, G. B. (2015). Insomnia caused by serotonin depletion is due to hypothermia. *Sleep, 38*(12), 1985–1993. doi:10.5665/sleep.5256

Serotonin and eating disorders

Kaye, W. H., Frank, G. K., Bailer, U. F., Henry, S. E., Meltzer, C. C., Price, J. C., . . . Wagner, A. (2005). Serotonin alterations in anorexia and bulimia nervosa: new insights from imaging studies. *Physiol Behav, 85*(1), 73–81. doi:10.1016/j.physbeh.2005.04.013

Kaye, W. H., Fudge, J. L., & Paulus, M. (2009). New insights into symptoms and neurocircuit function of anorexia nervosa. *Nature Reviews Neuroscience, 10*(8), 573–584. doi:10.1038/nrn2682

Riva, G. (2016). Neurobiology of anorexia nervosa: Serotonin dysfunctions link self-starvation with body image disturbances through an impaired body memory. *Front Hum Neurosci, 10*, 600. Retrieved from www.frontiersin.org/article/10.3389/fnhum.2016.00600 www.ncbi.nlm.nih.gov/pmc/articles/PMC5121233/pdf/fnhum-10-00600.pdf

Kaye, W. H., Wierenga, C. E., Bailer, U. F., Simmons, A. N., & Bischoff-Grethe, A. (2013). Nothing tastes as good as skinny feels: the neurobiology of anorexia nervosa. *Trends Neurosci, 36*(2), 110–120. doi:10.1016/j.tins.2013.01.003

Bulik, C. M., Sullivan, P. F., Tozzi, F., Furberg, H., Lichtenstein, P., & Pedersen, N. L. (2006). Prevalence, heritability, and prospective risk factors for anorexia nervosa. *Arch Gen Psychiatry, 63*(3), 305–312. doi:10.1001/archpsyc.63.3.305

Hoek, H. W., & van Hoeken, D. (2003). Review of the prevalence and incidence of eating disorders. *Int J Eat Disord, 34*(4), 383–396. doi:10.1002/eat.10222

Schalla, M. A., & Stengel, A. (2019). Activity based anorexia as an animal model for anorexia nervosa–A systematic review. *Frontiers in Nutrition, 6*(69). doi:10.3389/fnut.2019.00069

Haleem, D. J. (2012). Serotonin neurotransmission in anorexia nervosa. *Behav Pharmacol, 23*(5-6), 478–495. doi:10.1097/FBP.0b013e328357440d

van Gestel, M. A., Kostrzewa, E., Adan, R. A. H., & Janhunen, S. K. (2014). Pharmacological manipulations in animal models of anorexia and binge eating in relation to humans. *British Journal of Pharmacology, 171*(20), 4767–4784. doi:10.1111/bph.12789

Atchley, D. P., & Eckel, L. A. (2006). Treatment with 8-OH-DPAT attenuates the weight loss associated with activity-based anorexia in female rats. *Pharmacol Biochem Behav, 83*(4), 547–553. doi:10.1016/j.pbb.2006.03.016

Yokokura, M., Terada, T., Bunai, T., Nakaizumi, K., Kato, Y., Yoshikawa, E., . . . Ouchi, Y. (2019). Alterations in serotonin transporter and body image-related cognition in anorexia nervosa. *NeuroImage. Clinical, 23*, 101928–101928. doi:10.1016/j.nicl.2019.101928

Kuikka, J. T., Tammela, L., Karhunen, L., Rissanen, A., Bergström, K. A., Naukkarinen, H., . . . Uusitupa, M. (2001). Reduced serotonin transporter binding in binge eating women. *Psychopharmacology (Berl)*, *155*(3), 310–314. doi:10.1007/s002130100716

Kaye, W. H., Gwirtsman, H. E., George, D. T., & Ebert, M. H. (1991). Altered Serotonin activity in anorexia nervosa after long-term weight restoration: Does elevated cerebrospinal fluid 5-hydroxyindoleacetic acid level correlate with rigid and obsessive behavior? *Arch Gen Psychiatry*, *48*(6), 556–562. doi:10.1001/archp syc.1991.01810300068010

7

ENDOGENOUS OPIOIDS

Introduction

As you have made your way through the preceding chapters you have hopefully started to make and see connections between the various neurotransmitters discussed. You should have now started to build the picture where, although there are lots of differences, there are also lots of similarities in the mechanisms of synthesis, packaging, reuptake and neurotransmission. A good example of this are receptors. The majority of neurotransmitters have their own distinct receptors, although the vast majority confer to being either ionotropic or metabotropic. In the following two chapters things are going to get quite different as both endogenous opioids (the focus of this chapter) and substance P (the focus of the following chapter) are neuropeptides. Neuropeptides are quite distinct in their machinery compared to the neurotransmitters we have discussed in previous chapters, especially regarding the mechanisms of their synthesis, packaging and reuptake. So, prepare yourself for some quite distinct difference as we now delve into the neuropeptides, commonly known as endogenous opioids.

Outline of endogenous opioids synthesis, packaging and reuptake

So, why endogenous opioids? Well, endogenous means 'from within'. This is an important point as it emphasises the fact that there are many exogenous[1] opioids. I suspect that many of you are aware of this fact as opioids are one of the most abused drugs worldwide. However, it is important to remember that exogenous opioids have been hugely beneficial to humanity in the form of analgesics.[2] Their main issue lies in the fact that they are highly addictive. We will return to this point later in the chapter.

There are generally considered to be four core endogenous opioids, namely: β-endorphin, enkephalin, dynorphin and nociceptin.[3] Some of these have various subtypes, which we will come back to later. These endogenous opioids, and the receptors they have affinity with, are distributed widely in the central nervous system (CNS) and peripheral nervous system (PNS). This distribution is relatively specific

---

[1]External

[2]Something that reduces the feeling of pain; commonly referred to as painkillers.

[3]Nociceptin has a couple of different names, including orphanin FQ. Keep this in mind if you are doing any wider reading.

and they are especially prevalent in circuitry related to pain, stress and reward. We will return to some of these later in the chapter when we discuss endogenous opioids and their machinery's involvement in emotional states/atypical behaviour.

Test Yourself 7.1

The mechanisms of neuropeptides synthesis, packaging and reuptake are somewhat different from the other neurotransmitters covered in chapters up to now. Before we look at the difference, take a few seconds to reminder yourself of the common mechanisms for synthesis, packaging and reup-take of the neurotransmitters outlined in previous chapters, by completing the following table.

Table 7.1

	GABA	Glutamate	Acetylcholine	Dopamine	Serotonin
Precursors					
Synthesising enzyme(s)					
Vesicular transporter					
Which is the most important for vesicular transport? PH and/or membrane potential					
Reuptake transport (name and where expressed)					

Answers to all Test Yourself questions are at the end of the chapter.

Endogenous opioids synthesis

If you have read any of the other chapters in this book, at this point you are going to be thinking of precursor molecules, which are typically derived from metabolism or as dietary precursors. You are then likely to think of specific key enzymes that catalyse the synthesis of the neurotransmitter from these precursors. However, now is time for something completely different. Endogenous opioids, like other peptides/proteins are synthesised through the processes of DNA transcription and mRNA translation (see Figure 7.1).

You may first be thinking, what triggers the transcription of DNA? This is an excellent question and one that is not easily answered with reference to one biological mechanism. What is important to remember is that this is controlled by the cell, which carefully modulates the transcription of DNA, so that only one gene at a time is transcribed, based on need at that moment. This is a good point on which to build some connections between transcription and metabotropic receptors. If you recall from previous sections, secondary messengers synthesised as a result of metabotropic receptor activation, are key players in triggering transcription. So, what does this connection mean? Well, it means that activation of metabotropic receptors plays a key part in triggering the transcription of genes, which ultimately results in the synthesis of proteins/neuropeptides, such as the endogenous opioids. We will return to this later in the chapter, but for the moment this logically means that the synthesis of opioids is inherently intertwined with other neurotransmitters, neurotransmission and metabotropic receptors.

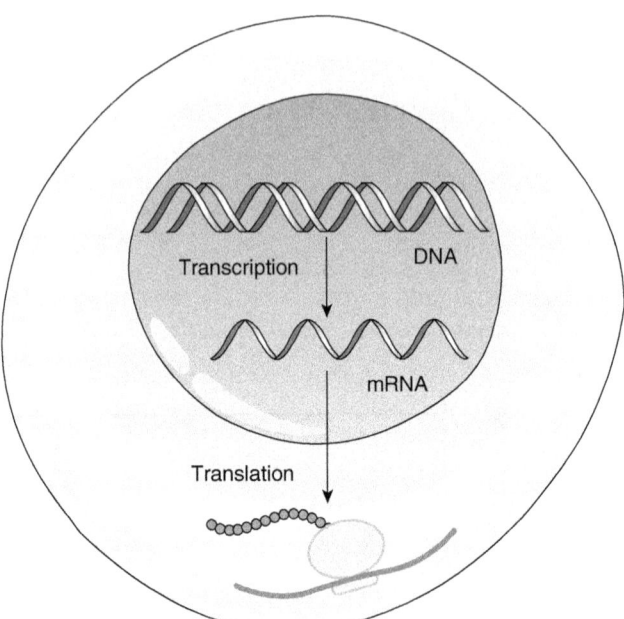

Figure 7.1   The synthesis of neuropeptides through transcription of DNA and translation of mRNA

As neuropeptides, the synthesis of endogenous opioids is not catalysed by enzymes in the same way as many of the other neurotransmitters covered in this book. The basic process involves the transcription of DNA into mRNA and the translation of this mRNA into the final peptide.

Source: Betts, G. J. et al. (2013) *Anatomy and physiology*. [Online]. Houston, TX: OpenStax. Available at: https://openstax.org/books/anatomy-and-physiology/pages/3-4-protein-synthesis (accessed 16 February 2022). Licensed under CC BY 4.0.

Once the relevant mRNA is transcribed from DNA, the mRNA migrates to the endoplasmic reticulum (still up in the cell body) where the process of translation[4] happens, specifically on the ribosomes. At this point, a large protein precursor molecule is typically produced. These large precursors are generally multiple copies of the same peptide/protein and have a high molecular weight. An enzyme is now needed to break this large molecule down into the final neuropeptide, in this case an endogenous opioid. We will discuss this in the next section as the packaging and final synthesis of endogenous opioids is inherently intertwined.

Packaging

The process of breaking down this large precursor protein into the final neuropeptide is called proteolytic cleavage. The process typically takes place in the vesicles that the peptide is packaged into, for transport down the axon to the synaptic terminal. The 'factory' that does this packaging of peptides is called the Golgi apparatus. The vesicles filled with the large precursor peptide bud off the Golgi apparatus. These vesicles are called large dense core vesicles (LDCVs). When they initially bud off the Golgi apparatus they are referred to as immature LDCVs. At this point they generally only contain one large precursor molecule; they then go through a process of homotypic fusion whereby several immature LDCVs fuse together. Then it is no longer one vesicle for one molecule and the final LDCVs typically contain a number of different large precursor peptides and/or proteins. This can include enzymes, such as the enzyme that performs proteolytic cleavage, and other proteins such as receptors and membrane transporters. These LDCVs now make their way down the axon to the terminal end attached to microfilaments – a process called anterograde transport. On their way the enzymes contained in the LDCVs are performing their functions, including proteolysis.[5] This means that by the time the LDCVs reach the terminal end the final neuropeptide molecule, in this case an endogenous opioid, is present within the LDCV.

---

[4]Essential translation is the decoding of the mRNA to produce the protein/peptide or, typically, a large precursor molecule.

[5]Also referred to as proteolytic cleavage.

So, what are the names of these large precursor molecules for endogenous opioids? Well, as there are four common endogenous opioids, there are four distinct large precursor molecules. These are called: POMC (preopiomelanocortin), PENK (pre-proenkephalin), PDYN (preprodynorphin)[6] and PNOC (prepronociceptin). By the time they reach the terminal end, POMC has been converted into β-endorphin; PENK into enkephalins, of which there are two common types, leu-enkephalin[7] and met-enkephalin; PDYN into dynorphins, of which there are two common types, dynorphin A and B; finally, PNOC will have been converted into nociceptin. β-endorphin, the enkephalins and dynorphin share a very similar structure and a common N terminal configuration. Nociceptin is distinct from these three and shares little sequence homology, although there is some cross-over with dynorphin.

It is worth spending a moment considering the distribution of neurons that express these precursors and consequently the final endogenous opioids. POMC-expressing neurons are restricted to two primary locations. These are within the hypothalamus[8] and in the dorsal medulla.[9] PENK- and PDYN-expressing neurons are expressed in a much more diffuse manner but the expression of each is still relatively heteroge-neous, with some brain regions expressing much higher numbers than others. For example, the hippocampus and nucleus accumbens contain high levels of PDYN-expressing neurons, while the thalamus expresses high levels of PENK neurons. If we consider the different role that each of these structures may play, it tells us some-thing about their relative contribution to behaviour. We will come back and consider this more later in the chapter, especially when we consider the role of opioids in emotional states/atypical behaviour.

The end point of these vesicles is fusion with the plasma membrane of the termi-nal end and release of their fully synthesised endogenous opioids into the synapse. Release at terminals occurs by cavicapture. This is a process by which the vesicle partially fuses with the cellular membrane. It is accompanied by depletion of cargo and reduction of size. This is a good time to highlight one final point, before we move on. I previously mentioned that these vesicles commonly include a number of

---

[6]Of course, as is a common theme in this book, this is not the whole story and there are a couple more that have yet to be fully elucidated.

[7]A bit of complexity is added here as leu-enkephalin is also a product of PDYN cleavage. In fact, the majority of leu-enkephalin is a product of POMC cleavage.

[8]Specifically, the arcuate nucleus.

[9]Specifically, the region known as the nucleus tractus solitarius.

other molecules as well as the endogenous opioids. Typically, with peptides such as endogenous opioids, these packages also include other neurotransmitters/neuropeptides. In fact, most peptides are rarely released alone. What does this mean? Well, simply that the effects of opioids released from the neuron are rarely in isolation and will contribute to shaping the target neuron's response to other neurotransmitters that are co-released. This is why it is always best to think of the role of neuropeptides as neuromodulators.

Reuptake/degradation

It should now be pretty clear that the processes of synthesis and packaging for neuropeptides is quite distinct from the synthesis and packaging of more traditional neurotransmitters covered in other chapters. Another aspect of distinction between classic neurotransmitters and neuropeptides is reuptake. In fact, how they differ is that neuropeptides, such as endogenous opioids, are not taken back up by the presynaptic neuron or other cells, such as astrocytes.[10] Once released from the presynaptic neuron, opioids are capable of diffusing relatively large distances and are certainly capable of making it out of the synaptic cleft and stimulating extra-synaptic receptors. This is known as volume transmission. We will return to this in a moment when we discuss opioid receptors. It is important to think about why endogenous opioids are not taken back up by the presynaptic neuron. Put simply, they do not need to be recycled as the presynaptic cell can simply produce more through the processes of transcription and translation. So, without reuptake, the degradation of opioid peptides has to happen either in the synapse or in the extra-synaptic space. There are a number of 'peptidases' that do this job. Some of these peptidases are thought to be relatively 'general' and can catabolise several peptides, resulting in inactive fragments. Two of these are neprilysin (NEP) and aminopeptidase N (APN), which as well as degrading enkephalin,[11] dynorphin and β-endorphin are also known to degrade substance P and neurotensin, among other peptides. This means that the findings of any research that focuses on modulating these, in order to modulate the levels of

---

[10]Which is common for some other neurotransmitters, such as glutamate.

[11]NEPs and APNs are thought to be particularly good at cleaving and therefore degrading enkephalin.

endogenous opioids, must be approached with caution as modulation will result in altered levels of a number of other neuropeptides. It is worth keeping all this in mind as we will return to substance P in the next chapter.

NEPs and APNs are known as zinc metallopeptidases. This is because in order for them to function correctly their catalytic element requires the binding of a metal, in this case zinc. These peptidases are typically found bound to the outside of the plasma membrane (see Figure 7.2). It is important to remember that although they are found bound to the plasma membrane, it is not always the case that they are bound to the membrane of the endogenous opioid-releasing neuron, as seen in Figure 7.2. This is important to remember due to endogenous opioids' role in volume transmission, modulating neurons far from the site of release.

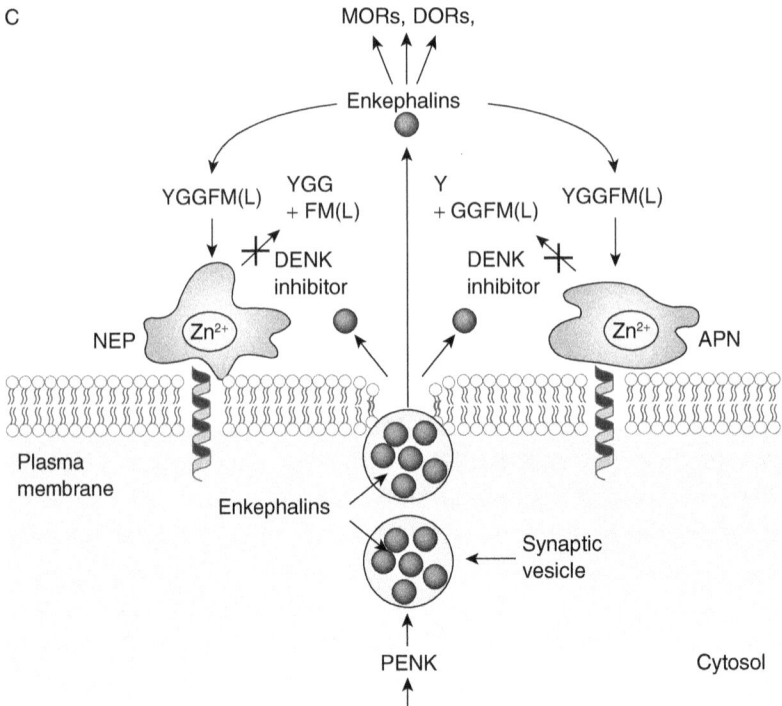

Figure 7.2 The involvement of zinc metallopeptidases in the degradation of endogenous opioids

The zinc metallopeptidases NEP and APN are instrumental in the breakdown of endogenous opioids and produce inactive metabolites. In this figure, they are found bound to the presynaptic membrane.

Source: Reprinted by permission from Springer Nature: Roques, B. P. et al. (2012). Inhibiting the breakdown of endogenous opioids and cannabinoids to alleviate pain. *Nature Reviews Drug Discovery, 11*(4), 292–310. ,Copyright (2012).

Although NEPs and APNs are known to degrade all endogenous opioids, dynorphin and β-endorphin also have their own specific peptidases. This suggests the potential for multiple targets if you wished to modulate the breakdown of each of the specific endogenous opioids. Over the last 20 years, this has been a particular focus of research concerned with the role of opioids as pain relief/analgesia.

Outline of opioid receptors

As referred to in the reuptake section, opioid neuropeptides can often travel far into the extra-synaptic space as they are involved in volume transmission. The consequence of this is that opioids can have effects on distant neurons and influence/modulate whole networks of neurons rather than simply targeting the presynaptic neuron in closest proximity at the synapse. This is one of the reasons why neuropeptides, such as endogenous opioids, are so interesting to researchers.

There are four general categories of endogenous opioids, as outlined in the introduction of this chapter and four general receptor types.[12] The receptors are commonly referred[13] to as MOR (µ-opioid receptors), DOR (delta opioid receptors), KOR (kappa opioid receptors) and NOR (nociception receptors). They are encoded for by the following genes: OPRM1, OPRD1, OPRK1 and OPRL1.

It would seem reasonable to assume that each endogenous opioid has its own receptor. However, this is not the case. A key point is that each receptor is not 100% specific for one endogenous opioid. If you think back to the packaging section of this chapter for a moment, we discussed how β-endorphins, the enkephalins and dynorphins share a very similar structure and a common N terminal configuration. This means that their binding domain is very similar and, as such, each endogenous opioid has at least some affinity for either all, or at least more than one, of the opioid receptors. The exception is nociceptin, which is only believed to have an affinity with NOP receptors. One interesting point about nociceptin and its receptors is revealed by their historical names. Nociceptin was historically

---

[12]Different receptors are continuously being identified and, recently, a zeta opioid receptor has been identified.

[13]There is a somewhat confusing nomenclature attached to opioid receptors. For example, MORs can also be called MOPs or OP2. Keep an eye on this in your wider reading.

referred to as orphanin FQ and its receptors 'ORL1', which was an acronym for opioid-like receptor. This was because initially they were believed to be a distant relative to endogenous opioid and not actually opioids. However, it was found that they actually shared connections with dynorphin and have subsequently become an accepted endogenous opioid, although both nociceptin and activation of NOP receptors have quite distinctly different behavioural effects than the other opioids. We will return to this later in the chapter.

So, what is the order of affinity for the endogenous opioids and the receptors? Well, all the endogenous opioids, bar nociception of course, have some affinity at MORs, with the highest affinity being for β-endorphin. β-endorphin and dynorphin A/B have high affinity with KORs. β-endorphin, leu- and met-enkephalin have high affinity with DORs, and, finally, nociceptin has the highest affinity with NORs, although dynorphins show some, low affinity.

Now to the receptor mechanisms, once the endogenous opioid has bound. As a general rule, the binding of endogenous opioids to all opioid receptors is thought to have an inhibitory effect or a reduction in excitability on the target neuron. Here, it is important to note that all these opioid receptors are GPCRs (G-protein-coupled receptors). This means that the binding of opioids and the activation of the receptor results in a complex chain of intracellular events and modulation of both secondary messenger and secondary messenger independent cascades. In the context of opioid receptors, the general effects of opioid binding are for the inhibitory G_i/G_o protein complex, which is attached to the receptor at its intracellular domain, to separate. The G protein separates into two: the α and the βγ complex. The result is inhibition of the enzyme adenylate cyclase and reduction in the synthesis of the secondary messenger cAMP. In addition, Ca^{2+} channels are inhibited[14] and GIRK (inwardly rectifying potassium) channels are activated. It is the α that travels across and inhibits adenylate cyclase. The βγ complex travels, presynaptically, to the membrane-embedded calcium channels, which it inhibits, and postsynaptically to the GIRK channels, which it activates. This all results in presynaptic inhibition of neurotransmitter/neuropeptide release and postsynaptic hyperpolarisation of the target neuron.

The mechanism discussed in the previous paragraph should be reasonably familiar to you, if you have read other chapters in this book, as these G protein effects are common to a number of neurotransmitters and their respective receptors.

---

[14]Especially P/Q- channels, which are typically located at presynaptic locations.

A new emerging concept in the world of opioid receptors is referred to as biased signalling. This is related to the fact that all the opioid receptors have at least some affinity for more than one endogenous opioid. What biased signalling means is that, dependent on the ligand that binds to the receptor, one of two (or more) different intracellular pathways can be activated. This means that activation of the receptor can have different effects on the cell and the behavioural state based on which ligand/molecule binds to it. For endogenous opioids, this is commonly thought to involve the activation of G protein-related mechanisms or the activation of β-arrestin-related mechanisms. So, certain molecules binding on opioid receptors will result in the alterations to cAMP level, calcium channels and GIRKs, as outlined in the previous paragraph, while binding of other molecules/agonists will result in β-arrestin-mediated responses. When β-arrestin is triggered, it results in a number of phosphorylation processes, which can result in desensitisation and/or endocytosis of the opioid receptor. It can also trigger a number of other intracellular signalling cascades, such as the mitogen-activated protein kinase cascade (MAPK). MAPK has a number of cellular effects, including regulating transcription factors,[15] cellular proliferation and protein scaffolding. This biased signalling also seems to have differential effects on behaviour, as outline in Figure 7.3. Keep this in mind when we discuss endogenous opioids and their mechanism's involvement in emotional states/behaviour later in this chapter.

A further point worth consideration is the number of binding sites found on opioid receptors that are not targets of the endogenous opioids. These sites are typically distinct from the binding site for the main, highest affinity endogenous opioid. Binding of ligands/molecules at these alternative sites results in what is referred to as allosteric modulation. This modulation can either be negative (NAMs) or positive (PAMs). Typically, what most people think is that occupation of the allosteric modulation sites results in alteration of the affinity or efficacy of the endogenous opioid at the main binding site. These allosteric binding sites have been of much interest to researchers over recent years as they theoretically present a mechanism by which you can modulate the efficacy, and ultimately the behavioural effect, of both endogenous and exogenous opioids. For example, some think that these allosteric sites, and the allosteric molecules that bind to

---

[15]Hopefully you can make the connection here to how this might modulate the synthesis of other neuropeptides.

them, can be used to reduce the amount of morphine needed for analgesia and therefore reduce the potential addictive elements of the analgesic. Further detail on this is beyond the scope of this book, but for a thorough discussion I strongly recommend the paper by Livingston and Traynor (2018).

Finally, a further layer of complexity is added to the functions of opioid receptors as they are known to form heterodimers. Heterodimisation is the interaction between different types of receptors that, when stimulated together,

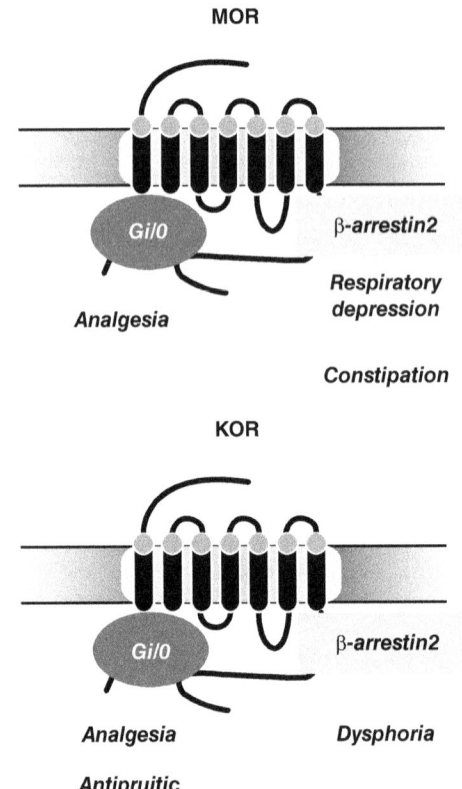

A. Activation of both MORs and KORs modulates β arrestin, although the effects of this modulation on the organism's behaviour are seen to be relatively distinct, with respiratory depression being a consequence of MOR activation and dysphoria a consequence of KOR activation.

Source: Figure A is reproduced from Valentino, R. J., & Volkow, N. D. (2018). Untangling the complexity of opioid receptor function. *Neuropsychopharmacology, 43*(13), 2514–2520. Licensed under CC BY 4.0

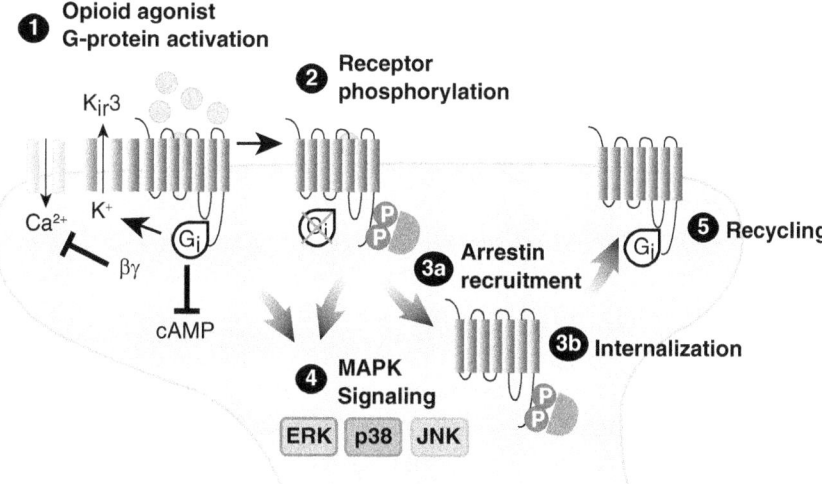

B. As metabotropic receptors, activation of endogenous opioid receptors produces a myriad of intracellular effects, including the modulation of cAMP levels, MAPK signalling and modulation of a number of membrane-bound channel proteins.

Source: Al-Hasani, R., & Bruchas, M, R. (2011). Molecular mechanisms of opioid receptor-dependent signaling and behavior. *Anesthesiology, 115*(6), 1363–1381 doi: https://doi.org/10.1097/ALN.0b013e318238bba6. Reproduced by permission from Wolters Kluwer Health, Inc.

Figure 7.3 The secondary messenger cascades activated in response to endogenous opioid binding at KORs and MORs

produce responses that are different from the response that would be elicited by stimulating the receptors independently. These heterodimers can be both opioid–opioid heterodimers (with other opioid receptor types) or with receptors that are targeted by distinctly different molecules, such as the known opioid–cannabinoid (CB1) receptor heterodimers and the MOR–NK1 receptor heterodimer. This MOR–NK1 receptor heterodimer is interesting on a number of levels. First, activation of this heterodimer results in distinct alterations in the sensitisation and endocytosis of receptors. Second, there is an interaction between opioid receptors and a class of receptors that are key targets for the other neuropeptide covered in this book: substance P. These heterodimers are a very current area of molecular research and one that many researchers are hopeful will produce novel therapeutic benefits.

Classic topics in opioid research

Both pleasure and pain can be conceived as evolutionary conserved mechanisms that aid our chances of survival. Pleasure represents the subjective[16] hedonic aspects of rewards and helps drive us towards repeating actions that benefit us. Pleasure is inherently intertwined with the concept of reward and reinforcement learning. Pain is best conceived as the subjective experience of suffering. It is inherently intertwined with the concepts of punishment and the feeling of pain motivates us to avoid stimuli or actions. Again, this is another key component of reinforcement learning, which conveys a survival advantage onto the organism.

Human history and experiences are strewn with examples of the competing demands to achieve pleasure but avoid pain. If you think to your personal history, I am sure you can think of multiple examples where you have been motivated by these two key drivers. This can be best thought of as the pain–pleasure dilemma and is explained well by the motivation-decision model proposed by Fields (2007). What this suggests is that when pain and pleasure are in direct competition, if the pleasurable target is more important for the organism's survival than the pain, then the biological system underpinning this should inhibit the pain. This should mean pleasure has anti-nocieceptive properties. So, imagine a situation where you can achieve something pleasurable, but it requires going through pain first. A good example might be a cyclist in the Tour de France going through the pain of climbing the Col de Tourmalet or Mont Ventoux to achieve the pleasure of winning. This would suggest the pursuit of pleasure should be able to inhibit, to some degree, the sensation of pain the cyclist feels. This is indeed the case, with many elite athletes reporting this phenomenon.

─**Focus on concepts** 💭─────────────────────

We are homeostatic beings

Homeostasis is best thought of as keeping something within a narrow range, neither too low nor too high; in other words, 'just right'. I like to think of this as the Goldilocks paradigm. A good biological example is the body's control of core temperature at 37°C, as seen in Figure 7.4.

---

[16]Although there are objective aspects to pleasure that can be measured in animals. See the work of Kent Berridge and colleagues for further information on this (Berridge, 2003; Berridge & Kringelbach, 2015).

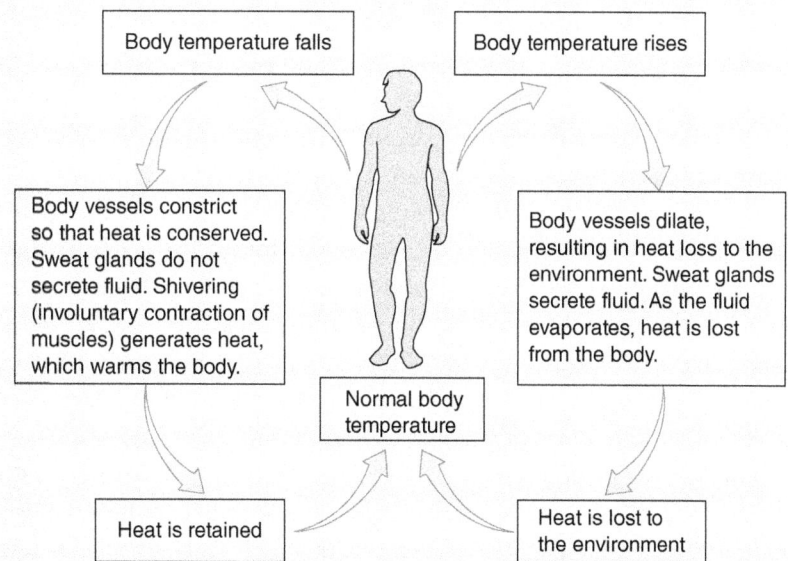

Figure 7.4   Homeostatic control of body temperature: the mechanisms by which the body regulates its temperature to keep core temperature within a very narrow set point

Source: Rye, C. et al. (2016). *Biology*. [Online]. Houston, TX: OpenStax. Available at: https://openstax.org/books/biology/pages/33-3-homeostasis (accessed 16 February 2022). Licensed under CC BY 4.0.

If you think closely about it, then pain and pleasure can be thought of as homeostatic mechanisms that keep us in the 'sweet spot'. Much like when body temperature increases or falls the biological system activates to return to this set point: this 'sweet spot'.

It makes theoretical sense that pleasure and pain may be conceived as end points for the same biological systems, modulated by a common currency. So, what is this common currency? Well, many believe it could be the endogenous opioid system. In this section, we will discuss the involvement of the endogenous opioid system in the context of both pleasure and pain, with a particular focus on atypical behavioural states associated with alterations in both pleasure and pain.

Endogenous opioids: pain-pleasure interaction

The interaction between pleasure and pain can be clearly evidenced by the fact that many people who experience chronic pain also develop depression, which is typified by a reduction in pleasure. Therefore, overactivation of the body's pain controlling systems (such as the endogenous opioids system) can perhaps be seen to contribute to reduced pleasure and the clinical symptomology of depression. Other examples of the interaction between pain and pleasure can be seen if we consider how normal reward seeking can be reduced if pain is too great. For example, researchers have shown that animals exposed to sustained pain have a significantly reduced drive to seek a pleasurable reward.

As well as pain modulating pleasure, there is also much evidence that suggests pleasure can produce a state of pain relief, generally termed 'affective analgesia'. This pleasure-related analgesia has been reported in both animal studies and research with humans. A fair body of research during the 1980s used sexual behaviour as the pleasant stimuli in rodent models. They typically found that sexual behaviour both decreased the animal's sensitivity to pain and modulated the release of endogenous opioids in various parts of the animal's midbrain. In humans, music, which is a stimulus well known to be highly effective in inducing both positive and negative emotional states, has been used to explore pleasure-induced analgesia. Research has shown that positively rated music can significantly reduce both self-reports of pain intensity and feelings of unpleasantness. Again, this all serves to suggest a common biological currency underpinning both pleasure and pain.

Therefore, what we can start to see here is that pain and pleasure are inextricably linked. A common underpinning biological system, namely the endogenous opioid system, for both is a logical assumption. We will now consider the role of endogenous opioids and their mechanisms in pleasure and pain responses separately.

Endogenous opioids and pleasure

So, let us now focus on the involvement of endogenous opioids in pleasure. When we discuss pleasure/hedonic responses we often conflate this with the drive towards achieving something positive; in other words, motivation. However, these have been shown to be two distinct aspects of reward and are both modulated by different neurotransmitter systems. The hedonic 'liking' aspect of reward is strongly believed to involve the endogenous opioids system, while the motivation aspect of reward is believed to be directly modulated by the dopamine system. A good example of research which clarifies this is the study of Peciña et al. (2003), which employed hyperdopaminergic mutant mice. These animals were shown to have an increased drive towards wanting a pleasurable substance but, crucially, they did not display an increased liking/hedonic response for that pleasurable substance. This seemingly confirms the distinction between the dopamine and the endogenous opioid system, with the endogenous opioid system being the key player in 'liking'/pleasure/hedonics.

For further information on this, I strongly recommend looking back to Chapter 5 on dopamine. For further reading, probably the most influential researcher on this topic is Kent Berridge, so I strongly recommend a good read of some of his publications.

Berridge and colleagues (Peciña et al, 2006) have established that there are small, localised regions in the brain (referred to as hedonic hotspots [HH]) that seem to mediate 'liking', therefore hedonics, independently of 'wanting'. One of these regions can be found in a part of the striatum known as the nucleus accumbens shell. You may remember this region from previous chapters. The nucleus accumbens has been heavily implicated as one of the key regions hijacked by drugs of abuse. It should therefore come as little surprise that within this region there can be found a subregion that modulates 'liking' responses. This HH, in the nucleus accumbens shell, is a 1 mm^3 region. Berridge and colleagues have shown that when this HH is stimulated with opioid receptor agonists,[17] animals exhibit up to 300 times increases in 'liking' responses to sucrose. This seems to be the case for mu, delta and kappa opioid receptors when targeted within this hedonic hotspot. Keep this in mind for when we next discuss the role of endogenous opioids and their mechanism in food consumption. What is also interesting is that agonists for all these receptors have also been shown to supress 'liking' within a named 'coldspot' located in the nucleus accumbens shell.

It is not just endogenous opioids that modulate 'liking' via these hotspots. Ho and Berridge (2013) have also shown that orexin can induce 'liking', via activation of these hedonic hotspots. Also, Mahler et al. (2007) have shown that endocannabinoid stimulation of this region can also increase 'liking' responses to a pleasurable substance. As ever, it is important to keep in mind that the brain's systems and neurotransmitters rarely operate in isolation; it is especially worth considering this in the context of the opioid–cannabinoid receptor heterodimers, discussed earlier in this chapter.

Although we have discussed the HH, specifically that in the nucleus accumbens shell,[18] and the work of Berridge and colleagues, there is also a rich seam of research that suggests opioids and their receptors modulate subjective feelings of pleasure and displeasure outside the HH region. For example, the PET scanning studies of Zubieta et al. (2003) involved the induction of a low mood using a sadness induction paradigm. Individuals were then scanned using PET with radioactive ligands targeting MORs. They found that reduced mood resulted in reduction of opioids

---

[17]Especially μ-opioid receptor agonists.

[18]There are a number of other hotspots that have been identified, including one in the ventral pallidum, which is a structure heavily connected to the nucleus accumbens.

binding to MORs. Therefore, there was an increased number of MORs that were not occupied by the endogenous opioids, compared to control participants. They also found that there was a strong correlation with the amount of binding reducing as mood reduced. This suggests that positive hedonics are mediated by endogenous opioids binding at MORs and negative hedonics by a corresponding reduction in binding at MORs. Another interesting study, this time employing music as the subjective hedonic stimulus, was conducted by Mallik, Chanda and Levitin (2017). This study involved participants listening to a piece of music they had identified as either positive or negative in terms of pleasure. Participants then listened to the music after administration of a broad-spectrum opioid receptor antagonist, namely naltrexone. This study found that both positive and negative hedonics were reduced by the blocking of opioid receptors. This suggests that the involvement of endogenous opioids in hedonics is perhaps more complicated than simply increased opioid binding resulting in increased positive hedonics. However, there is one caveat to this study and it lies with the opioid antagonist used. Naltrexone is a broad-spectrum antagonist, which means it blocks MORs, DORs, and to some extent KORs. Other studies have suggested that these receptors may play unique and specific roles in mediating hedonics, with binding at some increasing positive hedonics, while binding at others reduces positive hedonics. This means that, although the study is interesting, it really required more specific antagonists to clarify this relationship between endogenous opioids, their receptors and hedonics. A strand of research that holds the answer to this employed animal models: for example, studies such as that by Filliol et al. (2000), which used mouse models that are deficient in either DORs, KORs or MORs. This study suggests interesting opposing effects on hedonic responses, as the mice deficient in DORS displayed increased negative affect[19] while those deficient in MORs displayed reduced negative affect. This suggests that 'normally' binding at DORs may contribute to positive affective states and binding at MORs may contribute to negative affective states. Although very interesting, this is by no means a consistent finding for the involvement of DORs and MORs in hedonics. We will return to this shortly when we discuss endogenous opioids and depression.

Endogenous opioids and pleasure from food

Let us now focus on endogenous opioids in the context of pleasure from food. Much research has been conducted investigating this relationship in animal models.

---

[19]Actually, one of the big issues with this study is that the method they employ to measure emotional responses, namely the forced swim test, is more a measure of motivation than it is a measure of hedonic response.

Typically, rodents are given access to various liquid foods, including sucrose solution, which is typically used as the 'preferred/pleasurable' food. Various pharmacological manipulation then ensues and the intake levels for the different liquid foods are measured during these manipulations. Classic research on this topic employed the high affinity MOR antagonist naloxone. This was shown to significantly reduce the rodent's preference for sucrose. When a MOR-specific agonist is then given, it not only re-establishes the rodent's preference for sucrose, but increases it above baseline. So, what conclusion can we draw from this? Well, the general conclusion is that it suggests activation of MORs increases the palatability of food. For a good outline of this research field, see the references in the Focus on research box on Ann Kelley below. However, the method used by these studies to measure palatability of food, namely 'food intake', may not be a good measure of pleasure. Cast your mind back to what we discussed earlier in this chapter about 'liking' and 'wanting' and what you should conclude here is that 'food intake' is a measure of 'wanting', but not necessarily 'liking'. This seems quite damning, but it is worth remembering that these studies typically give the animals the free choice to select sucrose (the typically preferred food) or other solutions, including water. It means that there must be something more salient about sucrose. This could be about palatability/pleasure. Certainly, research such as that of Berridge and colleagues, using the taste reactivity paradigm, seems to strongly suggest that this salient quality is indeed increased pleasure/positive hedonics. So, what is the effect of exposure to lots of palatable food? Well, research has shown that this can dramatically alter endogenous opioids, specifically enkephalin gene expression. One study found that exposure to highly palatable food resulted in a decrease in the synthesis of the precursor molecule PPE. If endogenous opioids do modulate pleasure and palatability, then this suggests highly 'pleasurable' diets may reduce the value of the pleasurable stimulus by reducing the endogenous opioid response. This would produce a situation where more of the pleasurable stimulus is needed to get the same level of 'hit'. This is a common phenomenon with drugs of abuse.

Test Yourself 7.2

Which of the following methods do you think measures 'liking' and which measures 'wanting'?

1. Measure how many times an animal goes to eat a liquid meal.
2. Measure the amount an animal eats of a liquid meal.
3. Give the animal two bottles, one with a neutral substance (such as water) and the other with a substance that the animal is known to enjoy (such as sucrose). Measure the amount of each they drink; in other words, measure their preference.

(Continued)

4. Expose an animal to a preferred substance and measure their facial responses to the substance.
5. Place preference paradigm: measure how much time an animal spends in each of two rooms. One room has been paired with an 'addictive substance'.

Despite lots of evidence from animal research, there is surprisingly little linking opioids and their mechanisms to hedonic responses associated with food in humans. Much of the research certainly implicates endogenous opioids and their receptors in the consumption of food. For example, Burghardt et al. (2015) found that MOR availability decreased dramatically following ingestion of a liquid meal, suggesting that there was an increase in endogenous opioid release, binding to the receptors. While interesting, this study is indicative of the lack of connection made between endogenous opioids, hedonics and food consumption with human participants. One interesting study that does provide some important information in this regard was conducted by Tuulari and colleagues (2017). In this study, participants were asked to fast for 12 hours before a PET scan, or to ingest a meal with a high hedonic value, or to ingest a meal with a low hedonic value. The PET scan involved the participants being injected with the radioligand [^{11}C] carfentanil, which binds specifically to MOR. The findings from this study were interesting as they suggested that both the high and low hedonic food resulted in an increase in opioid release and increased binding to MORs. However, the pattern of release and binding was distinct in these two conditions, with opioid release being high across the brain in those who ingested the low hedonic food and specific to certain regions in those who had ingested the high hedonic food. We may interpret this as suggesting that foods with high hedonic value produce distinct alterations in endogenous opioid release and neurotransmission within specific circuits. If we think back in this chapter, then it is likely that these are the 'reward circuits' and include regions such as the nucleus accumbens.

Next, we are going to discuss the role of endogenous opioids and their mechanisms in addiction. Interestingly, many believe that addiction to controlled substances[20] is underpinned by the same mechanism that controls food hedonics and consumption. A finding that illustrates this link, from research with opioid addicts, is that addicts perceive sweet substances as more pleasant than drug-naïve controls. This suggests that the hedonic aspects of the abused substance can, to some degree, transfer over to other substances ingested. Indeed, we may view this, as many do, as an example of drugs of abuse[21] 'hijacking' the brain's pleasure and/or reward system.

---

[20]Such as heroin, cocaine, amphetamine.
[21]Which, in many instances, are exogenous opioids.

┏━ Focus on research 🔍 ━┓

Ann Kelley

Ann Kelley was a pioneering neuroscientist. She was among the first 13 women to be admitted to Trinity College, Cambridge, in 1976 to pursue her PhD, and developed some of the key theories relating to the endogenous opioid system's involvement in reward, specifically to its role in food consumption dependent on palatability.

For an excellent outline of some of her work, I strongly recommend reading the following introduction:

Richard, J. M., Castro, D. C., Difeliceantonio, A. G., Robinson, M. J., & Berridge, K. C. (2013). Mapping brain circuits of reward and motivation: in the footsteps of Ann Kelley. *Neuroscience and Biobehavioral Reviews, 37*(9 Pt A), 1919–1931. doi:10.1016/j.neubiorev.2012.12.008

For a great example of her own writing in the context of endogenous opioids and food, see:

Kelley, A. E., Bakshi, V. P., Haber, S. N., Steininger, T. L., Will, M. J., & Zhang, M. (2002). Opioid modulation of taste hedonics within the ventral striatum. *Physiology & Behavior, 76*(3), 365–377. doi:10.1016/s0031-9384(02)00751-510.1016/s0031-9384(02)00751-5

Endogenous opioids and drug addiction

As previously mentioned, exogenous opioids are one of the most heavily used and abused controlled substances worldwide. They are highly addictive, producing positive hedonic responses when taken. But abstinence/withdrawal from them produces pronounced negative affect and physiological pain. This serves to highlight a common theme in this section, namely the involvement of endogenous opioids and their mechanisms in pleasure and pain. This further suggests that the common biological substrate for both is likely to be found in the endogenous opioid system of the brain and central nervous system.

As discussed in previous chapters, 'addiction' is believed to consist of three core stages.[22] Much research suggests that each stage engages a different collection of opioid receptors. It is important to keep in mind that what we are now about to discuss, using research that employs specific genetically-modified animals, is likely to reveal only part of the story when it comes to endogenous opioid receptors' involvement in the whole experience of addiction.

Pre-clinical gene targeting studies have revealed a lot about the contribution of endogenous opioids and their mechanisms of neurotransmission to substance addiction. There has been much research focused on multiple abused substances.

---

[22]See references to Wise and Koob (2014) in Chapter 5 on dopamine for further information.

Research with MOR knockout (KO) animals has found increased motivation for heroin, while other researchers have found a reduced self-administration of alcohol. Other researchers using KORs knockout animals have found increased locomotor sensitisation to cocaine and alterations in tolerance levels to abused substances. Although there are obvious exceptions, the common theme is that opioid receptor KO seems to modulate addictive behaviour. Research using animals that overexpress the MOR or KOR receptor gene have also generally found an effect on addictive behaviour. For example, studies with these 'overexpressing' animals have found reduced locomotor sensitisation and reduced place-dependent relapse to substances of abuse.

There is also a relatively strong line of evidence from human gene polymorphism, specifically the A118G polymorphism in the OPR1 gene.[23] This polymorphism is thought to result in a gain of function, although this is still somewhat debated. These studies have generally shown an increased propensity towards substance addiction, such as increased pleasure ratings for nicotine and increased risk of alcohol or opioid addiction.

An important question to return to is: 'Is it really pleasure ('liking') that these receptors modulate to perpetuate addiction or is it motivation ('wanting')? Much research seems to suggest opioid receptors, especially MORs, appear to modulate motivation more than they do the pleasure aspects of addictive substances. In fact, the vast majority of research on addiction using animals does little to untangle the contribution of endogenous opioids to 'liking'. Most of the studies employ methods such as the conditioned place preference and self-administration paradigms, which tell us more about the drive/motivation ('wanting') than they do the 'liking' of the addictive substance. This is also the case with human studies, where they are typically focused on the 'wanting' of the addictive substance, rather than the 'liking'. One good study that does reveal something about the connection between opioid receptors, pleasure and addiction was conducted using MOR knockout animals by Ben Hamida and colleagues (2019). In this study the animals had specific loss of MORs in the striatum, while they remained intact in midbrain and hindbrain regions. These animals displayed significantly less alcohol consumption in a two-bottle choice task and significantly reduced alcohol-related place preference. This suggests that MORs specifically in the striatum territories are key to elements of addiction. Crucially, because this study measured the preference levels (for alcohol) of the animals, which were not altered in the MOR knockout animals, it suggests

---

[23]The gene that codes for MORs.

that MORs do not modulate hedonics to addictive substances but are indeed more involved with the motivational elements.

So, what can we conclude about endogenous opioids, pleasure and addiction? Well, although this area of research is complicated and it is difficult to untangle the contribution of opioids to hedonics and/or motivation changes associated with addiction, what seems likely is that opioids contribute more to the motivational 'wanting' elements than they do to altered hedonics. As mentioned previously in the chapter, this is likely due to their expression on dopaminergic neurons and the modulation of dopamine release by the binding of endogenous opioids.

Endogenous opioids and anhedonia/depression

Finally, we are going to have a brief look at the contribution of alterations in endogenous opioids mechanisms to depression. One line of research that you would think for sure implicates opioids and opioid receptors in pleasure responses is found in the literature focused on major depressive disorder (MDD), colloquially referred to as depression. The two core features of a clinical diagnosis of MDD are low mood and diminished capacity to find things pleasurable.

There is a vast literature involving measurement of various endogenous opioids in blood plasma from humans. I will not cover it here, but instead I will recommend a good starting point, in the review of Hegadoren et al. (2009), which specifically focuses on β-endorphins. The blood plasma studies are somewhat inconclusive, with some showing decreases of endogenous opioids in the plasma of those with a clinical diagnosis of depression, while others find the contrary, and yet others find no difference in the blood plasma opioid levels compared to controls. These studies are likely to be heavily confounded by two key factors. First, participants in the studies typically have comorbid diagnoses, for example, MDD and chronic pain. However, the degree of comorbidity and the types of comorbidity are likely to be different between studies, making the findings difficult to compare. Second, a large number of these studies include participants who are medicated. We cannot therefore conclusively say that the alterations in opioids found in the plasma are a consequence of the clinical condition; it is just as likely to be a biproduct of the medication. Again, this also means that the studies are somewhat difficult to compare as the number of medicated versus unmedicated participants is likely to vary greatly between studies.

So, what about studies that have higher levels of control, namely animal models of depression? Well, there are a number of pharmacological studies that elucidate the role of endogenous opioids and their receptors in animals' depressive-like behaviour.

One good example, which is typical of the literature, was conducted by Zomkowski, Santos and Rodriques (2005). They employed the forced swim test as a measure of depressive-like behaviour. When animals were given morphine, they showed a significant reduction in the depressive-like behaviour, whereas when MORs were blocked by naloxone these animals showed a significant increase in depressive-like behaviour. Research using a different method, known as the tail suspension test, has shown similar findings, with animals displaying reduced depressive-like behaviour when opioid compounds were administered. Typically, these studies have employed MOR agonists and antagonist. So, what about the other opioid receptors? Well, here is where it gets interesting as other studies have shown an increase in depressive-like behaviour[24] when specific KOR receptor agonists are administered. This was alluded to much earlier in this chapter. Indeed, many researchers believe that the different types of opioid receptors may antagonise each other's behavioural effects, being somewhat like an on-off for hedonics. Much of this animal research, especially in the context of opioid receptors, must be taken with a degree of caution as there are known significant differences between humans and rodents in the expression of these receptors, especially DORs and KORs, where there is seen to be much less DOR binding in humans and more KOR. Another interesting side point, although important, is that the KOR agonists used were also shown to depress dopamine release in the nucleus accumbens. This is an important mechanism of action for endogenous opioids and much research shows that opioids have their effect by their expression on dopaminergic neurons.

Again, we come to a common issue, namely, the methods used and behaviours measured reveal more about altered motivation ('wanting') than they do about altered pleasure ('liking'). The forced swim test measures how much the animal struggles and tries to escape the water. The less they struggle the more 'depressive-like' behaviour they are said to exhibit. You can probably tell from this description, this is much more a measure of the animal's 'motivation/drive' to escape, rather than their hedonic state. Again, what we see here is that the methods reveal more about the involvement of endogenous opioids and their mechanisms in motivation than hedonic states. This is indeed a major caveat of much of the research and something to really consider when reading wider research reporting to reveal information about pleasure/hedonics.

---

[24]Again, this was measured using the forced swim test.

Endogenous opioids and pain modulation

For many years exogenous opioids have been used for their pain relieving (analgesic) properties. During the 1970s and 1980s major advancements were made in our knowledge of the endogenous opioid system of the body, resulting in advancement of more specific and efficacious opioid-based analgesics. Today, we still rely heavily on exogenous opioids to help manage and alleviate pain. Unfortunately, opioids have several negative side effects, such as respiratory depression, nausea and vomiting. Probably the biggest issues, however, especially for their use in chronic pain conditions, is the development of tolerance and the fact that they are highly addictive. It should be pretty clear to you from reading the previous section on endogenous opioids and pleasure that this is likely to be the case. This has led to much research aimed at understanding the endogenous opioid system further and trying to identify ways to access their analgesic properties while mitigating their addictive aspects. As will shortly become apparent, much of this research has focused on understanding opioid receptors.

So, what do we know about the brain's endogenous opioid system as it relates to pain control? Well, opioid receptors are expressed diffusely throughout the brain and spinal cord. One particular area where all four opioid receptors are known to be expressed is the dorsal root ganglia (DRN). This is a key structure in the transmission of pain-related information and acts like a relay through which signals from the body's organs and somatosensory information travel to innervate the spinal cord. Somatosensory neurons within the DRN express the four key opioid receptors. It is well known that activation of these receptors by intrathecal[25] injection of opioid agonists can produce antinociception effects, blocking the sensation/awareness of the individual to painful stimuli. Recent methodological advances, such as single cell RNA sequencing, have revealed that the different types of opioid receptors seem to be expressed in separate and distinct subpopulations of somatosensory neurons in the DRN. This has led to the current belief that different neurons, expressing different opioid receptors, are uniquely involved in specific types of pain and somatosensory experiences.

---

[25]This is an injection into the fluid filled space between tissue that covers the brain and spinal cord.

MORS, DORs, KORs and NOPRs in the DRN

A number of studies have shown that MOR agonists produce analgesia when injected into DRN. Recent research by Sun et al. (2019) employed cross-breeding of two genetically-modified mouse strains. This resulted in mice (we shall refer to them as MOR 'knockout') with a complete loss of MORs in the DRN and significantly reduced levels in the spinal cord. When compared to control animals, morphine had significantly reduced antinociceptive effects on these MOR knockout animals. This suggests that MORs, expressed on peripheral sensory neurons in the DRN and the spinal cord, are key players in opioid-mediated analgesia. Although the weight of research evidence does seem to concur with these findings, it is important to remember two things. First, as previously discussed, different neuron types in the DRN are believed to be involved in different types of pain. Sun et al. (2019) focused on acute thermal and mechanical pain, so it is conceivable that MORs in the DRN specifically modulate this type of pain. Indeed, DORs and KORs are known to control other specific types of pain. For example, research has suggested that KORs are likely to be involved in the modulation of visceral pain.[26] Second, morphine has a relatively high affinity with other opioid receptors, such as KORs. We know that these are expressed in the DRN and therefore the reduction in antinociception response to morphine, seen in the MOR knockout mice, might be a result of the activation of these and a lack of corresponding activation of MORs. Therefore, it is likely that the picture is somewhat more complicated.

We also must not forget about NOPRs. They are known to be expressed in the DRN but little else is known about the involvement in pain gating. This is an emerging and exciting area for research on pain, with recent research suggesting their antagonism centrally (brain and spinal cord) and in the periphery (such as the DRN) can produce analgesic effects.

Into the spinal cord

As well as neurons in the DRN, several different populations of neurons within the dorsal horn of the spinal cord also express opioid receptors. These neurons range from interneurons to afferent and efferent projection neurons.

A significant amount of evidence using morphine has implicated the involvement of MORs within the spinal cord in pain modulation, such as the previously outlined

---

[26]This is generally considered to be pain caused by your internal organs

study by Sun et al. (2019). However, there is also evidence to suggest the involvement of DORs within the spinal cord in pain modulation. For example, DOR expression in somatostatin positive dorsal horn[27] interneurons contribute to the pain-relieving qualities of DOR agonists. But DORs are also expressed by other dorsal horn neurons, so it is likely that this is only part of the story. As well as the focus on opioid receptors, there has been some evidence to implicate regulation of endogenous opioid synthesis in the spinal cord in responses to painful stimuli. Dynorphin and enkephalin synthesising interneurons in the dorsal horn have been shown to increase in number directly after exposure to peripheral injury. This suggests that these two endogenous opioids and their respective receptors are perhaps involved in 'closing the gate' to pain signals – at least, perhaps, in response to acute pain.

Focus on research 🔍

Melzack and Wall's gate control theory

Although not directly linked to opioids, Melzack and Wall's gate control theory was and is one of the most influential theories in pain research. It proposed an elegant theory for how pain signals were controlled in the spinal cord. The theory is so influential because it was, and is, eminently testable. This is a key quality of a good theory. To discover more about the theory, I recommend reading the original article or the succinct anniversary article:

Melzack, R., & Wall, P. D. (1965). Pain mechanisms: a new theory. *Science, 150*(3699), 971-979. doi:10.1126/science.150.3699.971

Katz, J., & Rosenbloom, B. N. (2015). The golden anniversary of Melzack and Wall's gate control theory of pain: celebrating 50 years of pain research and management. *Pain Research & Management, 20*(6), 285-286. doi:10.1155/2015/865487

All this might make you think that the role endogenous opioids, especially their receptors, play in pain control is very much outside the brain, in the spinal cord and DRN. This is not the case and several studies have shown that pain can be controlled higher up the chain, in the brain.

---

[27]This is the location where all peripheral primary sensory afferents terminate in the spinal cord.

Back to the brain

If you take a minute to think back to the earlier section on pleasure and the pain–pleasure interaction, then you should be thinking that it can't all be about the DRN and spinal cord! Indeed, it is not and there is a significant body of research which has shown that the brain can modulate the affective elements of pain signals. For a great review of human studies that implicate specific brain regions common to pain and pleasure modulation, see the paper by Leknes and Tracey (2008). It is also worth taking a moment to think about how subjective pain can be and reflecting on this in terms of the relative contribution of the brain and peripheral nervous system to our experience of pain. It is perhaps best to think of the peripheral system as a simple open and closed gate, as per Melzack and Wall's (1965) gate control theory, while the brain ultimately decides how painful the stimulus is. Researchers have conceptualised and operationalised this as a difference between pain affect and sensation. These two elements to pain seem to be modulated independently. For example, analgesics such as morphine seem to reduce the affective elements of pain before they reduce the somatosensory 'sensation' of pain, with patients typically reporting a reduced affective response before they report the reduced pain sensation. This seems to be dose-dependent, with lower doses of analgesics reducing affective elements of pain and higher doses reducing the sensation of pain.

So, what brain regions are involved? If we work our way up from the spinal cord, then the first regions are the brain stem and midbrain. These include structures such as the medulla (part of the brain stem) and periaquatal grey (part of the midbrain). Several studies have implicated these regions in opioid modulation of pain signals. For example, research using microinjection of MOR agonists and antagonists directly into the PAG has shown that agonists have antinociception effects on animal models and antagonists have hyperalgesia effects. These studies have also found that MOR antagonists directly injected into the PAG also block the analgesic effects of morphine. The PAG is also heavily connected with the medulla, which then sends projections down towards the dorsal horn of the spinal cord. These connections are known to release endogenous opioids and different ones target different types of neurons within the dorsal horn. Many think this is the key descending pathway modulating the pain signal. It should not be surprising to find 'deep brain regions'[28] involved in pain modulation if you consider pain as an evolutionary conserved mechanism and the fact that the brains of primates evolved from the inside out.

---

[28]Such as midbrain and brain stem regions.

So, what about higher brain regions? Well, logically, it would be the higher brain regions that are more involved with the affective elements of pain. For once this appears to be what the research literature has found. Numerous studies in animal models of pain have found that there is an increased release of endogenous opioids in regions such as the amygdala and thalamus directly after exposure to a painful stimulus. One interesting study by Zubieta et al. (2003) using PET scans also found this to be the case in humans. Interestingly, the stronger the affective report of pain from the participants the higher the endogenous opioid release in these regions was found to be. A crucial factor here is that Zubieta and colleagues measured the participant's affective responses to the painful stimuli, suggesting these higher regions add the emotional response/interpretation to the painful stimuli. A number of other studies have implicated yet higher regions in endogenous opioid-mediated analgesia, especially the affective component. Two key regions are the anterior cingulate cortex (ACC) and the nucleus accumbens. Injections of MOR antagonists into the ACC have been shown to inhibit the positive affect associated with the administration of analgesics. In the nucleus accumbens the relations are more complex, but MORs activation has been shown to modulate dopamine release, which is associated with pain relief. This all links nicely with the known role these areas have in reward-related behaviour and affective disorders, as outlined previously in the pleasure section of this chapter. Therefore, it seems that endogenous opioids bring together and modulate multiple components of pain, including affective, motivation, reward-related and somatosensory components. Of course, they do not do this on their own, and it is important to remember that the endogenous opioids exert their effects like other neurotransmitters/neuropeptides by binding to receptors and modulating target neurons. In the case of certain aspects of pain related to affect and motivation, this is likely to be via the expression of opioid receptors on dopaminergic neurons.

Pain modulation: it's not just about opioids

Ultimately, as ever, it is not just about endogenous opioids. Of course, the classic analgesics are exogenous opioids and these drugs have greatly helped manage all kinds of pain. However, as discussed, they have some rather problematic side effects. With this in mind, researchers have heavily investigated other neuroactive molecules and receptor systems to identify novel analgesics that do not have the issues of tolerance and addiction. One such system targeted by researchers is the endocannabinoid system. This area of research has had rather inconsistent findings, with some studies finding positive analgesic effects of cannabinoid administration, while others find little analgesic effects of administered cannabinoids. One interesting study by

Haller, Stevens, and Welch (2008) found that administration of anandamide[29] at the same time as administering a substance that inhibits FAAHs (fatty acid amide hydrolase)[30] produced significant anticonception effects in a mouse model. This might explain the variability in research with cannabinoids, as the enzymes that degrade cannabinoids (FAAH) are highly effective; therefore, cannabinoids have a very short window of effect. Increasing this window seems to increase the analgesic properties. The major limitation of cannabinoids is that they typically only have analgesic effects after multiple administrations, whereas those drugs that target the endogenous opioid system are efficacious after a single dose. What can be said is that more research is needed in this field, but it perhaps has potential.

Conclusion

So, what can we conclude about the endogenous opioid system? Well, the key point is that it is an essential, evolutionary conserved system that contributes to fundamental behavioural processes that aid our survival. This has hopefully become clear from our discussion of pleasure, pain and their common biological substrate, namely endogenous opioids. Another important point to take from this chapter is the interaction between opioids, especially their receptors, and other core neurotransmitter systems in the nervous system. For example, we cannot underestimate how important the interaction between opioid receptors and their modulation of dopamine release is in integrating aspects of 'liking' with 'wanting' to produce the experience of reward. In summary, the endogenous opioid system and its molecular mechanisms are complex and essential for the function of an integrative system that maintains the organism.

─────(Test Yourself Answers)──────────────────────────────

Test Yourself 7.1

Table 7.2

	GABA	Glutamate	Acetylcholine	Dopamine	Serotonin
Precursors	Glutamate	Glutamine α-ketoglutarate	Acetyl-CoA and choline	Phenylalanine Tyrosine	Tryptophan 5-hydroxytryptophan

─────────

[29]This is a naturally-occurring endocannabinoid.

[30]FAAH is an enzyme that degrades endocannabinoids.

	GABA	Glutamate	Acetylcholine	Dopamine	Serotonin
Synthesising enzyme(s)	GAD	Glutaminase	ChAT	TH DDC	Tryptophan hydroxylase and AAADC (aromatic L-amino acid decarboxylase)
Vesicular transporter	VGATs	VGLUT1, 2 and 3	VAChT	VMATs (1 and 2)	VMATs (1 and 2)
Which is the most important for vesicular transport? PH and/or membrane potential	Zwitterions so both PH and membrane potential	Membrane potential	PH	PH	PH
Reuptake transport (name and where expressed)	The main two expressed in the brain and spinal cord are GAT1 (A1) and GAT3 (A13). These are both expressed on axon terminals and on astrocytes	EAAT1, 2, 3, 4 and 5 EAAT1-4 are all expressed in the brain, although there is a distinct difference in the type of cells that express them, with EAAT 1 and 2 predominantly being expressed in glia, while EAAT 3 and 4 are predominantly expressed in neurons	ChT (choline reuptake transporter) Presynaptic neuron	DAT Presynaptic neuron	SERT Presynaptic neuron

(Continued)

Test Yourself 7.2

1.   Measuring how many times an animal goes to eat a liquid meal is an example of 'wanting'.
2.   Measuring the amount an animal eats of a liquid meal is an example of 'wanting'. This seems a bit trickier as we naturally assume that the more we eat of something the more we 'like' it, but this is a subjective interpretation.
3.   Giving the animal two bottles, one with a neutral substance (such as water) and the other with a substance that the animal is known to enjoy (such as sucrose), and measuring the amount of each they drink is an example of 'liking'. Because the animal makes a choice this tends to suggest that the one they take the more of is the 'liked' substance.
4.   Exposing an animal to a preferred substance and measuring their facial responses to the substance is an instance of 'liking'. The taste reactivity paradigm using orofacial responses is believed by some to be the best objective measure of 'liking'.
5.   Measuring how much time an animal spends in each of two rooms when one room has been paired with an 'addictive substance' is an example of 'wanting'. This is difficult as the word 'preference' tends to suggest 'liking', but here the 'liking' is for the place, not the substance paired with it. Therefore, you may think of it as they like the room because they want the substance.

References

Key information

Benarroch, E. E. (2012). Endogenous opioid systems: current concepts and clinical correlations. *Neurology, 79*(8), 807–814. doi:10.1212/WNL.0b013e3182662098

Roques, B. P., Noble, F., Daugé, V., Fournié-Zaluski, M. C., & Beaumont, A. (1993). Neutral endopeptidase 24.11: structure, inhibition, and experimental and clinical pharmacology. *Pharmacol Rev, 45*(1), 87–146.

Reed, B., Bidlack, J. M., Chait, B. T., & Kreek, M. J. (2008). Extracellular biotransformation of beta-endorphin in rat striatum and cerebrospinal fluid. *J Neuroendocrinol, 20*(5), 606–616. doi:10.1111/j.1365-2826.2008.01705.x

Reed, B., Zhang, Y., Chait, B. T., & Kreek, M. J. (2003). Dynorphin A(1-17) biotransformation in striatum of freely moving rats using microdialysis and matrix-assisted laser desorption/ionization mass spectrometry. *J Neurochem, 86*(4), 815–823. doi:10.1046/j.1471-4159.2003.01859.x

Turner, A. J. (2013). Aminopeptidase N. *Handbook of Proteolytic Enzymes*, 397–403. doi:10.1016/B978-0-12-382219-2.00079-X

Valentino, R. J., & Volkow, N. D. (2018). Untangling the complexity of opioid receptor function. *Neuropsychopharmacology, 43*(13), 2514–2520. doi:10.1038/s41386-018-0225-3

Snyder, S. H., & Pasternak, G. W. (2003). Historical review: Opioid receptors. *Trends Pharmacol Sci*, *24*(4), 198–205. doi:10.1016/S0165-6147(03)00066-X

McDonald, J., & Lambert, D. G. (2015). Opioid receptors. *BJA Education*, *15*(5), 219–224. doi:10.1093/bjaceaccp/mku041

Bodnar, R. J. (2020). Endogenous opiates and behavior: 2017. *Peptides*, *124*, 170223. doi:https://doi.org/10.1016/j.peptides.2019.170223

Corder, G., Castro, D. C., Bruchas, M. R., & Scherrer, G. (2018). Endogenous and exogenous opioids in pain. *Annual Review of Neuroscience*, *41*, 453–473. doi:10.1146/annurev-neuro-080317-061522

Al-Hasani, R., & Bruchas, M. R. (2011). Molecular mechanisms of opioid receptor-dependent signaling and behavior. *Anesthesiology*, *115*(6), 1363–1381. doi:10.1097/ALN.0b013e318238bba6

Livingston, K. E., & Traynor, J. R. (2018). Allostery at opioid receptors: modulation with small molecule ligands. *British Journal of Pharmacology*, *175*(14), 2846–2856. doi:10.1111/bph.13823

Endogenous opioids and pleasure

Leknes, S., & Tracey, I. (2008). A common neurobiology for pain and pleasure. *Nature Reviews Neuroscience*, *9*(4), 314–320. doi:10.1038/nrn2333

Fields, H. L. (2007). Understanding how opioids contribute to reward and analgesia. *Reg Anesth Pain Med*, *32*(3), 242–246. doi:10.1016/j.rapm.2007.01.001

Nam, M.-H., Han, K.-S., Lee, J., Won, W., Koh, W., Bae, J. Y., . . . Lee, C. J. (2019). Activation of Astrocytic μ-Opioid Receptor Causes Conditioned Place Preference. *Cell Reports*, *28*(5), 1154–1166.e1155. doi:https://doi.org/10.1016/j.celrep.2019.06.071

Richard, J. M., Castro, D. C., Difeliceantonio, A. G., Robinson, M. J., & Berridge, K. C. (2013). Mapping brain circuits of reward and motivation: in the footsteps of Ann Kelley. *Neurosci Biobehav Rev*, *37*(9 Pt A), 1919–1931. doi:10.1016/j.neubiorev.2012.12.008. Epub 2012 Dec 19

Kelley, A. E., Bakshi, V. P., Haber, S. N., Steininger, T. L., Will, M. J., & Zhang, M. (2002). Opioid modulation of taste hedonics within the ventral striatum. *Physiol Behav*, *76*(3), 365–377. doi:10.1016/s0031-9384(02)00751-5 10.1016/s0031-9384(02)00751-5

Berridge, Kent C., & Kringelbach, Morten L. (2015). Pleasure systems in the brain. *Neuron*, *86*(3), 646–664. doi:https://doi.org/10.1016/j.neuron.2015.02.018

Berridge, K. C. (2007b). The debate over dopamine's role in reward: the case for incentive salience. *Psychopharmacology (Berl)*, *191*(3), 391–431. doi:10.1007/s00213-006-0578-x

Castro, D. C., & Berridge, K. C. (2014). Opioid hedonic hotspot in nucleus accumbens shell: Mu, delta, and kappa maps for enhancement of sweetness "liking" and "wanting". *The Journal of Neuroscience, 34*(12), 4239–4250. doi:10.1523/JNEUROSCI.4458-13.2014

Ho, C. Y., & Berridge, K. C. (2013). An orexin hotspot in ventral pallidum amplifies hedonic "liking" for sweetness. *Neuropsychopharmacology, 38*(9), 1655–1664. doi:10.1038/npp.2013.62

Pecina, S., & Berridge, K. C. (2005). Hedonic hot spot in nucleus accumbens shell: where do mu-opioids cause increased hedonic impact of sweetness? *J Neurosci, 25*(50), 11777–11786. doi:10.1523/jneurosci.2329-05.2005 10.1523/JNEUROSCI.2329-05.2005

Peciña, M., Karp, J. F., Mathew, S., Todtenkopf, M. S., Ehrich, E. W., & Zubieta, J. K. (2019). Endogenous opioid system dysregulation in depression: implications for new therapeutic approaches. *Mol Psychiatry, 24*(4), 576–587. doi:10.1038/s41380-018-0117-2

Peciña, S., Smith, K. S., & Berridge, K. C. (2006). Hedonic hot spots in the brain. *Neuroscientist, 12*(6), 500–511. doi:10.1177/1073858406293154 10.1177/1073858406293154.

Peciña, S., Cagniard, B., Berridge, K. C., Aldridge, J. W., & Zhuang, X. (2003). Hyperdopaminergic mutant mice have higher "wanting" but not "liking" for sweet rewards. *J Neurosci, 23*(28), 9395–9402. doi:10.1523/jneurosci.23-28-09395.2003

Tindell, A. J., Smith, K. S., Peciña, S., Berridge, K. C., & Aldridge, J. W. (2006). Ventral pallidum firing codes hedonic reward: when a bad taste turns good. *Journal of Neurophysiology, 96*(5), 2399–2409.

Berridge, K. C. (2003). Pleasures of the brain. *Brain Cogn, 52*(1), 106–128. doi:10.1016/s0278-2626(03)00014-9

Zubieta, J. K., Ketter, T. A., Bueller, J. A., Xu, Y., Kilbourn, M. R., Young, E. A., & Koeppe, R. A. (2003). Regulation of human affective responses by anterior cingulate and limbic mu-opioid neurotransmission. *Arch Gen Psychiatry, 60*(11), 1145–1153. doi:10.1001/archpsyc.60.11.1145 10.1001/archpsyc.60.11.1145

Mallik, A., Chanda, M. L., & Levitin, D. J. (2017). Anhedonia to music and mu-opioids: Evidence from the administration of naltrexone. *Scientific Reports, 7*(1), 41952. doi:10.1038/srep41952

Filliol, D., Ghozland, S., Chluba, J., Martin, M., Matthes, H. W., Simonin, F., . . . Kieffer, B. L. (2000). Mice deficient for delta- and mu-opioid receptors exhibit opposing alterations of emotional responses. *Nat Genet, 25*(2), 195–200. doi:10.1038/76061

Tindell, A. J., Smith, K. S., Berridge, K. C., & Aldridge, J. W. (2009). Dynamic computation of incentive salience: "Wanting" what was never "liked". *J. Neurosci.,* *29*(39), 12220–12228. doi:10.1523/JNEUROSCI.2499-09.2009

Burghardt, P. R., Rothberg, A. E., Dykhuis, K. E., Burant, C. F., & Zubieta, J. K. (2015). Endogenous opioid mechanisms are implicated in obesity and weight loss in humans. *J Clin Endocrinol Metab, 100*(8), 3193–3201. doi:10.1210/jc.2015-1783

Tuulari, J. J., Tuominen, L., de Boer, F. E., Hirvonen, J., Helin, S., Nuutila, P., & Nummenmaa, L. (2017). Feeding releases endogenous opioids in humans. *The Journal of Neuroscience, 37*(34), 8284. doi:10.1523/JNEUROSCI.0976-17.2017

Kelley, A. E., Bakshi, V. P., Haber, S. N., Steininger, T. L., Will, M. J., & Zhang, M. (2002). Opioid modulation of taste hedonics within the ventral striatum. *Physiol Behav, 76*(3), 365–377. doi:10.1016/s0031-9384(02)00751-5 10.1016/s0031-9384(02)00751-5

Roy, M., Peretz, I., & Rainville, P. (2008). Emotional valence contributes to music-induced analgesia. *Pain, 134*(1–2), 140–147. doi:10.1016/j.pain.2007.04.003

Szechtman, H., Hershkowitz, M., & Simantov, R. (1981). Sexual behavior decreases pain sensitivity and stimulated endogenous opioids in male rats. *Eur J Pharmacol, 70*(3), 279–285. doi:10.1016/0014-2999(81)90161-8

Wise, R. A., & Koob, G. F. (2014b). The development and maintenance of drug addiction. *Neuropsychopharmacology : official publication of the American College of Neuropsychopharmacology, 39*(2), 254–262. doi:10.1038/npp.2013.261

Ben Hamida, S., Boulos, L. J., McNicholas, M., Charbogne, P., & Kieffer, B. L. (2019). Mu opioid receptors in GABAergic neurons of the forebrain promote alcohol reward and drinking. *Addict Biol, 24*(1), 28–39. doi:10.1111/adb.12576 10.1111/adb.12576. Epub 2017 Nov 2.

Ziauddeen, H., Nestor, L. J., Subramaniam, N., Dodds, C., Nathan, P. J., Miller, S. R., . . . Bullmore, E. T. (2016). Opioid antagonists and the A118G polymorphism in the μ-opioid receptor gene: Effects of GSK1521498 and Naltrexone in healthy drinkers stratified by OPRM1 genotype. *Neuropsychopharmacology, 41*(11), 2647–2657. doi:10.1038/npp.2016.60

Charbogne, P., Gardon, O., Martín-García, E., Keyworth, H. L., Matsui, A., Mechling, A. E., . . . Kieffer, B. L. (2017). Mu opioid receptors in gamma-aminobutyric acidergic forebrain neurons moderate motivation for heroin and palatable food. *Biol Psychiatry, 81*(9), 778–788. doi:10.1016/j.biopsych.2016.12.022

Hegadoren, K. M., O'Donnell, T., Lanius, R., Coupland, N. J., & Lacaze-Masmonteil, N. (2009). The role of beta-endorphin in the pathophysiology of major depression. *Neuropeptides, 43*(5), 341–353. doi:10.1016/j.npep.2009.06.004 10.1016/j.npep.2009.06.004. Epub 2009 Aug 3.

Berrocoso, E., Ikeda, K., Sora, I., Uhl, G. R., Sánchez-Blázquez, P., & Mico, J. A. (2013). Active behaviours produced by antidepressants and opioids in the mouse tail suspension test. *Int J Neuropsychopharmacol, 16*(1), 151–162. doi:10.1017/s1461145711001842

Zomkowski, A. D. E., Santos, A. R. S., & Rodrigues, A. L. S. (2005). Evidence for the involvement of the opioid system in the agmatine antidepressant-like effect in the forced swimming test. *Neuroscience Letters, 381*(3), 279–283. doi:10.1016/j.neulet.2005.02.026

Carlezon, W. A., Jr., Béguin, C., DiNieri, J. A., Baumann, M. H., Richards, M. R., Todtenkopf, M. S., . . . Cohen, B. M. (2006). Depressive-like effects of the kappa-opioid receptor agonist salvinorin A on behavior and neurochemistry in rats. *J Pharmacol Exp Ther, 316*(1), 440–447. doi:10.1124/jpet.105.092304

Darcq, E., & Kieffer, B. L. (2018). Opioid receptors: drivers to addiction? *Nature Reviews Neuroscience, 19*(8), 499–514. doi:10.1038/s41583-018-0028-x

Mahler, S. V., Smith, K. S., & Berridge, K. C. (2007). Endocannabinoid hedonic hotspot for sensory pleasure: anandamide in nucleus accumbens shell enhances 'liking'of a sweet reward. *Neuropsychopharmacology, 32*(11), 2267–2278.

Endogenous opioids and pain

Fields, H. L. (2007). Understanding how opioids contribute to reward and analgesia. *Reg Anesth Pain Med, 32*(3), 242–246. doi:10.1016/j.rapm.2007.01.001

Roques, B. P., Fournié-Zaluski, M. C., & Wurm, M. (2012). Inhibiting the breakdown of endogenous opioids and cannabinoids to alleviate pain. *Nat Rev Drug Discov, 11*(4), 292–310. doi:10.1038/nrd3673

Southerland, W. A., Gillis, J., Kuppalli, S., Fonseca, A., Mendelson, A., Horine, S. V., . . . Gulati, A. (2021). Dual enkephalinase inhibitors and their role in chronic pain management. *Curr Pain Headache Rep, 25*(5), 29. doi:10.1007/s11916-021-00949-0

Bodnar, R. J. (2020). Endogenous opiates and behavior: 2017. *Peptides, 124*, 170223. doi:https://doi.org/10.1016/j.peptides.2019.170223

Corder, G., Castro, D. C., Bruchas, M. R., & Scherrer, G. (2018). Endogenous and exogenous opioids in pain. *Annual review of neuroscience, 41*, 453–473. doi:10.1146/annurev-neuro-080317-061522

Peciña, M., Love, T., Stohler, C. S., Goldman, D., & Zubieta, J. K. (2015). Effects of the Mu opioid receptor polymorphism (OPRM1 A118G) on pain regulation, placebo effects and associated personality trait measures.

Neuropsychopharmacology, 40(4), 957–965. doi:10.1038/npp.2014.272 10.1038/npp.2014.272. Epub 2014 Oct 13.

Zubieta, J. K., Ketter, T. A., Bueller, J. A., Xu, Y., Kilbourn, M. R., Young, E. A., & Koeppe, R. A. (2003). Regulation of human affective responses by anterior cingulate and limbic mu-opioid neurotransmission. *Arch Gen Psychiatry, 60*(11), 1145–1153. doi:10.1001/archpsyc.60.11.1145 10.1001/archpsyc.60.11.1145

Zubieta, J. K., Smith, Y. R., Bueller, J. A., Xu, Y., Kilbourn, M. R., Jewett, D. M., . . . Stohler, C. S. (2001). Regional mu opioid receptor regulation of sensory and affective dimensions of pain. *Science, 293*(5528), 311–315. doi:10.1126/science.1060952

Katz, J., & Rosenbloom, B. N. (2015). The golden anniversary of Melzack and Wall's gate control theory of pain: Celebrating 50 years of pain research and management. *Pain Research & Management, 20*(6), 285–286. doi:10.1155/2015/865487

Melzack, R., & Wall, P. D. (1965). Pain mechanisms: a new theory. *Science, 150*(3699), 971–979. doi:10.1126/science.150.3699.971

Podvin, S., Yaksh, T., & Hook, V. (2016). The emerging role of spinal dynorphin in chronic pain: A Therapeutic Perspective. *Annu Rev Pharmacol Toxicol, 56*, 511–533. doi:10.1146/annurev-pharmtox-010715-103042

Leknes, S., & Tracey, I. (2008). A common neurobiology for pain and pleasure. *Nature Reviews Neuroscience, 9*(4), 314–320. doi:10.1038/nrn2333

Sun, J., Chen, S. R., Chen, H., & Pan, H. L. (2019). μ-Opioid receptors in primary sensory neurons are essential for opioid analgesic effect on acute and inflammatory pain and opioid-induced hyperalgesia. *J Physiol, 597*(6), 1661–1675. doi:10.1113/jp277428

Vanderah, T. W. (2010). Delta and kappa opioid receptors as suitable drug targets for pain. *Clin J Pain, 26* Suppl 10, S10–15. doi:10.1097/AJP.0b013e3181c49e3a

Dagnino, A. P. A., da Silva, R. B. M., Chagastelles, P. C., Pereira, T. C. B., Venturin, G. T., Greggio, S., . . . Campos, M. M. (2019). Nociceptin/orphanin FQ receptor modulates painful and fatigue symptoms in a mouse model of fibromyalgia. *Pain, 160*(6), 1383–1401. doi:10.1097/j.pain.0000000000001513

Zhang, Y., Du, L. N., Wu, G. C., & Cao, X. D. (1998). Modulation of intrathecal morphine-induced immunosuppression by microinjection of naloxone into periaqueductal gray. *Zhongguo Yao Li Xue Bao, 19*(6), 519–522.

Navratilova, E., Xie, J. Y., Okun, A., Qu, C., Eyde, N., Ci, S., . . . Porreca, F. (2012). Pain relief produces negative reinforcement through activation of mesolimbic reward-valuation circuitry. *Proc Natl Acad Sci U S A, 109*(50), 20709–20713. doi:10.1073/pnas.1214605109

Haller, V. L., Stevens, D. L., & Welch, S. P. (2008). Modulation of opioids via protection of anandamide degradation by fatty acid amide hydrolase. *European Journal of Pharmacology, 600*(1), 50–58. doi:https://doi.org/10.1016/j.ejphar.2008.08.005

8

SUBSTANCE P (AN EXAMPLE TACHYKININ)

Chapter outline

In this chapter we will cover:

- Outline of substance P and other tachykinins' mechanisms of:
    - synthesis
    - packaging
    - reuptake
- Outline of the neurokinin receptors
- Classic topics in substance P research:
    - Substance P and pain
    - Substance P and anxiety
- Novel topics in substance P research:
    - Substance P and addiction

Introduction

Substance P (SP) was one of the first neuropeptides to be identified, back in the early 1930s. It was initially identified by von Euler and Gaddum, who were investigating the distribution of acetylcholine in various tissue. They identified that when the fluid they had extracted was applied to intestine tissue, it produced a contraction, even when you blocked the receptors for acetylcholine with atropine. At this time it was referred to as Preparation P. It took 30 more years for it to be successfully isolated from mammalian (horse) tissue. So, why is it called substance P? Well, this is a name that derives from the fact it was initially an unidentified component of a powder that had been extracted from horse brain and intestine tissue. The name therefore tells us little about what the molecule actually does and what biological systems and behaviour it is involved in. The answer to this is complex. It will be unravelled to some degree in this chapter, although it is vital to keep in mind that SP is known to be involved in many diverse roles, of which many are non-neuronal and therefore beyond the scope of this book.

SP as a molecule is what is known as an undecapeptide. This means it has 11-amino-acid residues/units. As mentioned above, it has many diverse roles in the body but most important for us, like other neuropeptides, it modulates the activity of neurons. It is known to modulate neurons in both the central nervous system (CNS) and the peripheral nervous system (PNS). Its general effects on target neurons are believed to be excitatory. This is somewhat of an oversimplification and it is important to keep in mind that its effects are mediated by binding to G-protein-coupled receptors (GPCRs), therefore its intracellular effects are varied and complex. Historically, SP has commonly been grouped with a number of other neuropeptides under the heading gut-brain peptides. As the name implies, it is known to be heavily expressed in the gut.[1] In fact, it was initially identified from samples of intestinal tissue. We will return to substance P's role in the intestines/gastrointestinal (GI) tract in Chapter 10. Obviously, the group name also implies it is expressed in the brain. It is expressed heterogeneously in the brain, meaning that it is found in some regions but not others. This tells us that unlike neurotransmitters, such as GABA and glutamate, that are homogeneously distributed, substance P's involvement in behaviour is likely to be more specific and involved in the modulation of distinct circuits. Even within the regions where it is expressed the levels vary, with some regions expressing higher

---

[1]The gut is a generally used colloquial term for our lower gastrointestinal tract, which includes parts of the small and large intestine.

levels than others. For humans, high levels are found in brain regions such as the hippocampus, amygdala, caudate putamen, nucleus accumbens and neocortex. A moment's thought about these regions should enable you to start identifying the likely behaviour SP is known to be involved in modulating. It may help in your thoughts to know that SP is typically co-expressed in these brain regions with other neurotransmitters we have covered in other chapters, such as dopamine, serotonin and GABA.

The grouping of SP as a gut-brain peptide is somewhat of a historical misnomer as it is not just expressed in the gut and brain; in fact, it is found in a number of regions of the body, including the liver, kidney and salivary glands. Further to this wide regional distribution, SP is also synthesised and expressed by a number of different types of cells, not just neurons. For example, SP is known to be synthesised by immune cells, such as lymphocytes and macrophages, and by enterochromaffin cells in the gut. This perhaps helps to explain the vast array of biological processes SP is involved in: from wound healing to inflammatory response, from gastrointestinal processes to vasodilation.

As well as SP traditionally belonging to a group of neuropeptides based on their regions of expression (gut-brain peptides), SP also belongs to a group of neuropeptides based on their structural similarity. This is a much more important grouping in many ways as their structural similarity means they interact. The name of this grouping is tachykinins[2] and it includes a number of other neuropeptides, which will be outlined in the synthesis section of this chapter. SP belongs to a subfamily of this group known as neurokinins, of which there are neurokinin A (NKA) and neurokinin B (NKB), among others. So, what is common between these neuropeptides that groups them as tachykinins? Well, the key factor is their common carboxyl terminal end sequence (C terminal consensus -Phe-X-Gly-Leu-Met-NH2). This is the section of their structure that binds with receptors. Keep this in mind when we consider neurokinin receptors (NKRs) as the fact that they share a similar binding motif tells us something essential about their likely affinity for the different NKRs.

Earliest research on SP typically focuses on its role in pain. We will look at this later in the chapter as well as considering SP and its mechanism's involvement in other states and atypical behaviour. But first let's consider the synthesis, packaging and degradation of substance P.

---

[2]This name is derived from their function in the gut where they stimulate rapid contraction of gut muscle.

Outline of Substance P: synthesis, packaging and degradation

As mentioned above SP is a neuropeptide. This means that the mechanisms of its synthesis, packaging and degradation are very similar to the endogenous opioids discussed in Chapter 7. I am not going to cover these mechanisms extensively again here. Instead, for an overview of the general mechanisms of synthesis, packaging and degradation of neuropeptides I recommend going back over these sections in Chapter 7.

To see if you need to do this or if you can continue to read on in this chapter, why not try completing the following self-assessment activities?

Test Yourself 8.1

Synthesis

Which of the following terms relate to neuropeptide synthesis and what do they mean?

Table 8.1

Translation	Transcription	Ribosomes	Secondary messengers	Cavicapture
Proteolysis	LDCVs	Homotypic fusion	NEPs	Volume transmission

Answers to all Test Yourself questions are at the end of the chapter.

Test Yourself 8.2

Packaging

Put the following boxes in order from the packaging process that happen in the cell body to the vesicle's eventual release of its content (hint: not all boxes are relevant and more than one may fit at each level).

Table 8.2

Immature vesicles bud off the trans Golgi network	Immature vesicles fuse to form LDCVs	This process is called homotypic fusion and results in mature LDCVs that typically contain more than one peptide	The mature LDCVs are transported down the axon via microfilaments	This process is called anterograde transport and, via a process of condensation, the vesicle reduces in size

During this stage, the large precursor molecule typically goes through the process of proteolysis	Once the LDCV arrives at this point, the final neuropeptide has been synthesised	Cavicapture occurs	This is a process of partial fusion with the cell membrane and release of the vesicle's content	Translation of mRNA occurs at the ribosome producing a large precursor molecule

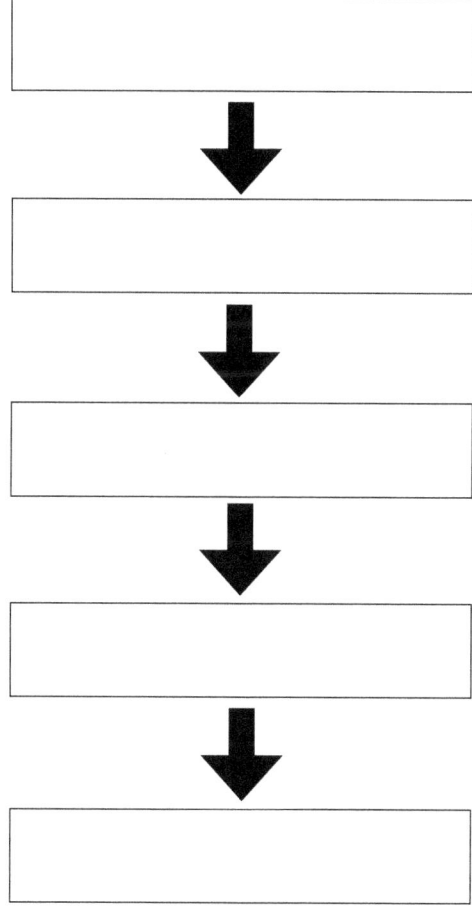

Test Yourself 8.3

Reuptake/degradation

Have a look at Figure 8.2 and then answer the questions that follow.

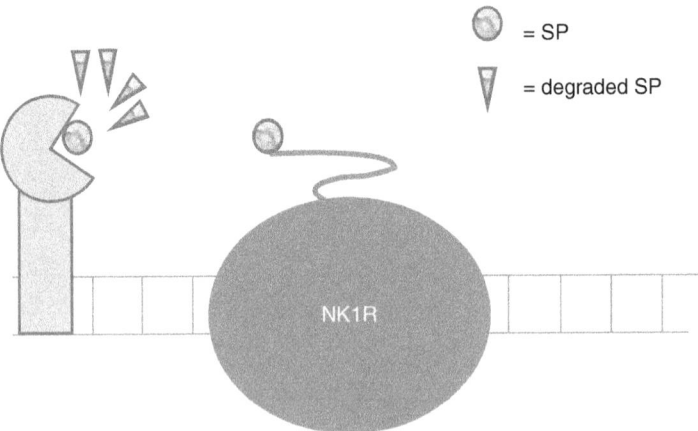

Figure 8.1   The degradation of substance P

1.   What is the common name for the enzyme (on the left of the figure) that degrades substance P?
2.   What metal is a common co-factor in the effective functioning of these enzymes?
3.   Based on Figure 8.1, where would you suggest these enzymes are commonly expressed?
4.   Based on what you know about neuropeptides, are these enzymes likely to be present on the postsynaptic terminal within the synapse? Try to explain your answer.

So, hopefully you now feel confident with the general mechanisms of SP synthesis. Next, there are a few specific bits of information that should round off your understanding.

Transcription of the preprotachykinin A gene (known as PPTA and TAC1), on chromosome 7 in humans, is responsible for the synthesis of SP. However, there are four splice variants[3] of this gene, resulting in its involvement in the synthesis

---

[3]This means it can be cut at four points, producing four distinct mRNA strands.

of other tachykinins. α and δ variants produce SP alone, although the β and y variants also produce NKA, neuropeptide K and neuropeptide y[4] (see Figure 8.2). Two other genes are also involved in tachykinin synthesis. These are PPTB (TAC3) and PPTC (TAC4). You may be thinking at this point what has happened to TAC2? Well, TAC2 was initially considered to be the gene for NKA synthesis, but further research identified that it was simply a splice variant of TAC1. PPTB (TAC3) is responsible solely for NKB, while PPTC (TAC4) is responsible for hemokinin-1[5] and the endokinins: EKA, EKB, EKC and EKD. While the majority of other tachykinin receptors are conserved across species, hemokinin-1 is found to be quite different in humans compared to rodents. Therefore, anything you come across in your wider reading regarding hemokinin-1 in animal models is worth treating with caution. Similarly, EKC and EKD do not share the common C terminal motif[6] that other tachykinins do. Thus, they have minimal interaction with tachykinin receptors such as the high-affinity SP NK1 receptor. These are of less interest to us in the context of neuromodulation.

Substance P packaging

Substance P is very much packaged in the same manner as other neuropeptides, such as the endogenous opioids, but the enzymes that catalyse proteolytic cleavage of the large precursor molecules are distinct. There are six of these enzymes and they are collectively referred to as convertases. As can be seen in Figure 8.2, the posttranslational cleavage processes that occur as the large precursor peptide is being transported down to the terminal end in the LDCVs, play a key role in the final synthesis of the different tachykinins. These cleavage processes and enzymes are relatively understudied and would make an exciting and potentially fruitful target for future neuropeptide-related research.

---

[4]Neuropeptide K and y are slightly altered forms of NKA.

[5]These are tachykinins largely expressed in the hematopoietic system. This is the system that controls the production of various cellular constituents of blood.

[6]This means that many people do not view these as tachykinins and they are commonly referred to as tachykinin gene-related peptides.

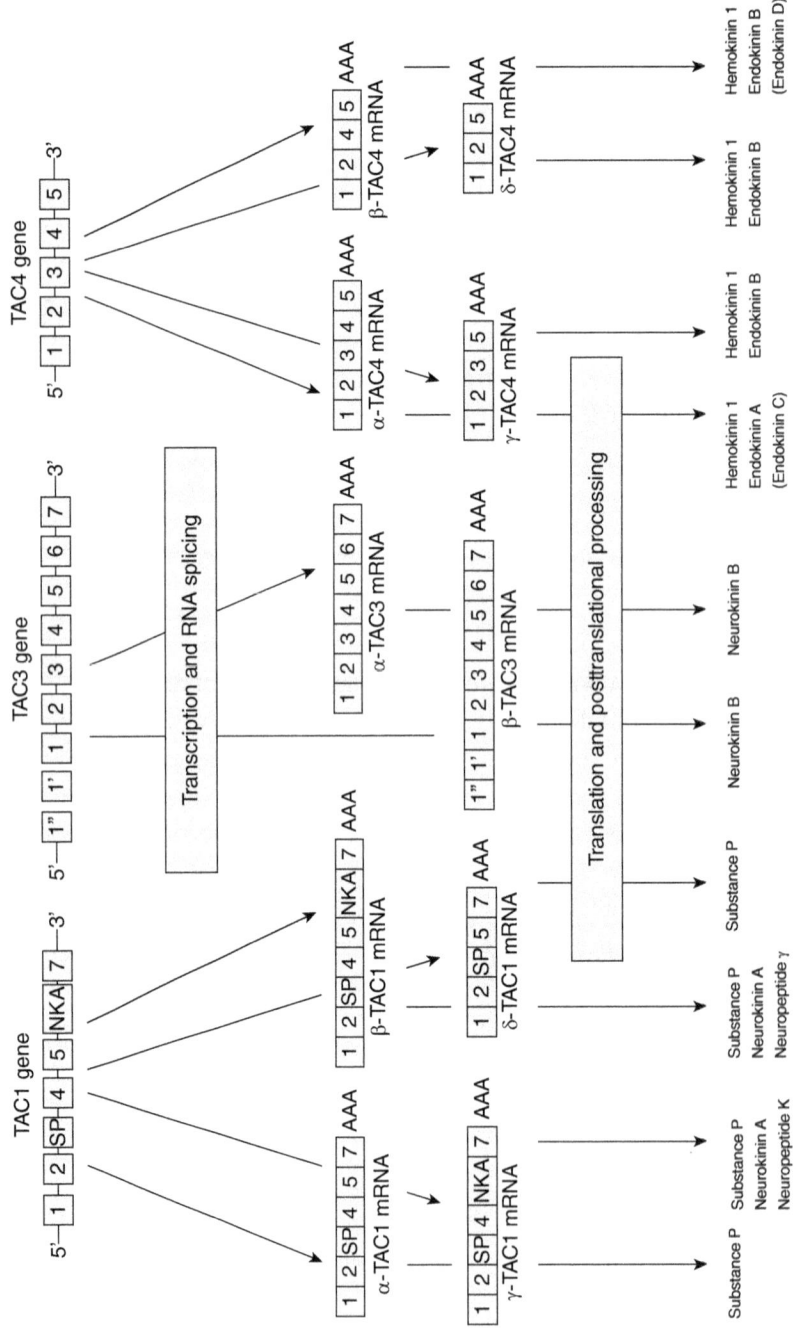

Figure 8.2 The involvement of transcription and translational processes in the formation of tachykinins from three key genes

The many tachykinin peptides are transcribed from TAC1, TAC3 and TAC4 genes, with substance P being transcribed from TAC1. Notice that a number of post-transcription processes impact the translation of the mRNA into the final neuropeptides.

Source: Reprinted from: Joos, G. F., & De Swert, K. (2006). Kinins and neuropeptides. In *Encyclopedia of respiratory medicine* (pp. 509–517). Oxford: Elsevier. , Copyright (2006), with permission from Elsevier.

Substance P degradation

Again, the degradation of substance P is very similar to that outlined for the endogenous opioids in the previous chapter. However, one key difference to note is that, while other peptidases degrade endogenous opioids, neprilysins (NEPs) and angiotensin-converting enzymes (ACEs) are very much the key players for the degradation of SP. These different degrading enzymes splice SP at various locations, resulting in a number of different SP metabolites. The consequence of this is under-explored in the literature. However, the main area they splice is just before the C terminal end. This deactivates the peptides as this is the core region, as previously discussed, that binds to the neurokinin receptors.

Outline of neurokinin receptors

The receptors for substance P are named neurokinin receptors (NKRs). They are derived from the TACR genes.[7] There are three key receptors that have an affinity for SP: NK1, NK2 and NK3. SP has the highest affinity at NK1 receptors. It also has affinity at NK2 and NK3 receptors, but this is much lower. The other commonly expressed mammalian tachykinins, namely neurokinin A (NKA) and neurokinin B (NKB), also have an affinity for all the NKRs, although the ones they have highest affinity for are different from substance P. It is generally considered that NKA has the highest affinity at NK2 receptors and NKB has the highest affinity at NK3 receptors with NKA having a 100 times lower affinity than SP at the NK1R and NKB having a 500 times lower affinity than SP at the NK3R. So, what is it that means these three tachykinins (SP, NKA and NKB) have at least some affinity for all the NKRs (1, 2 and 3)? Well, the simple answer, as discussed earlier, is that all these tachykinins share the same C terminal motif (-Phe-X-Gly-Leu-Met-NH2). This is the region, which binds to the active zone on the NKR; therefore, the same motif means that they all bind. The fact that all three tachykinins mentioned here have at least some affinity at all of the receptors is important to keep in mind. This is because research using specific receptor subtype agonists and antagonists is unlikely to reveal the whole picture when it comes to both the receptor's and the specific tachykinin's involvement in behaviour of the cell and the organism as a whole.

---

[7]NK1R being coded for by TACR1, NK2R by TACR2 and NK3R, TACR3.

So, where are these NK receptors found? Well, like SP, NKRs are expressed in abundance in lots of different tissue throughout the body, from the liver and kidney to the gut and peripheral nervous system. But what about in the mammalian brain? Well, NK1R and NK3Rs are believed to be widely distributed in the brain. However, most research suggests that the pattern of NK2R expression is much more specific, with expression in the thalamus cortex and hippocampus. Although NK1 and NK3Rs are widely distributed in the brain, they are known to be in particularly high expression levels in specific regions and circuits. For example, NK1Rs are expressed in high amounts in regions associated with stress responses, especially the amygdala. They are also found in relatively high numbers in regions such as the striatum, the raphe nucleus and a range of regions throughout the cortex. This should get you thinking again about the potential behavioural role of these receptors, especially NK1Rs, as we have come across these regions in other chapters of the book, when discussing topics such as addiction, pleasure and pain. If you are starting to make some connections, you may notice that, although similar in regional distribution, these receptors can be found in regions that are not known to synthesise SP. How can this be the case? The key thing to think about here is that neuropeptides are involved in volume transmission, so it is worth remembering that they can act on targets far from their release site. In your wider reading, especially in older papers, you may come across quite a lot of contradiction as to the distribution of NKRs. This is partly due to the initial use of non-specific ligands and antibodies used to detect their expression, typically in post-mortem tissue. However, there are also known species differences in terms of receptor expression, affinity for agonists/antagonists and distribution, which are important to keep in mind. This will become apparent when we discuss NKRs in the context of pre-clinical and clinical pain management research, later in the chapter.

So, what are the common features of NKRs? A key thing to keep in mind is that they are all GPCRs, so the binding of tachykinins (such as SP, NKA and NKB) at these receptors results in complex intracellular responses mediated by G proteins. However, it gets complicated as the type of G proteins bound to these receptors varies depending on what type of cell the receptor is expressed on. For example, NK receptors expressed in tumours activate distinct G proteins which interact with the Rho (Rhodopsin) rock pathway and are involved in the synthesis of cytokines and other inflammatory responses. However, when NK1Rs expressed on astrocytes are activated, different G proteins are involved, which results in activation of the enzyme PLC (phospholipase C) and increases in the secondary messengers IP3 and DAG. Yet still, when NK1Rs bound to smooth muscle cells are activated, it results in the enzyme adenylate cyclase producing increased levels of cAMP. This again is mediated through a different G protein. This is extremely interesting, but as we are focused on neuronal activity, going any further is beyond the scope of this book.

For further information on the NK receptors and the secondary messengers involved beyond neurons, I strongly recommend the review by Steinhoff and colleagues (2014), especially the section entitled 'Initiation of NKR signalling'.

Because this section is mainly focused on SP, we are now going to go a little deeper into the receptor that shows by far the greatest affinity for SP: NK1 receptors (NK1R). As mentioned above, the secondary messenger system activated by the binding of SP at NK1 receptors very much depends on what type of cell the NK1R is expressed on. In the brain, and when expressed on neurons, common secondary messenger pathways activated by binding of SP at NK1 receptors are cAMP, IP3 and DAG. This suggests that on neurons the most common G proteins coupled to NK1Rs are $G\alpha q$ and $G\alpha s$-proteins. This is because $G\alpha s$-protein is commonly associated with upregulation of cAMP by activating adenylate cyclase and $G\alpha q$ protein is commonly associated with activation of PLC and production of IP3 and DAG. The complexity does not stop here as there are also known splice variants of the gene that codes for NK1Rs, which results in a long and short form of the NK1R. These two variants recruit different secondary messenger systems. There is a great amount of similarity across species in terms of the structure of NK1Rs, although there is enough difference that the results from pharmacological studies using mouse models and various NK1R agonists should be treated with some caution. This is perhaps evidenced best later in the chapter when we discuss the use of NK1 receptor targeting for pain modulation. The final layer of complexity is revealed when we return to the fact that NKRs, such as NK1Rs, have at least some affinity for SP, NKA and NKB. It is well known that intracellular response to binding at these receptors varies depending on which neurokinin binds. As you may remember from Chapter 7, different messenger cascades can be activated depending on the molecule that binds. This is worth thinking about, especially in the context of research which modulates the level of one of the tachykinins, as this is likely to modulate the ability of other tachykinins to bind at the receptor, influencing the secondary cascade and intracellular signalling, as can be seen in Figure 8.3.

Classic topics in substance P research

Substance P (tachykinin receptors) and pain

Substance P and NKRs, like endogenous opioids, have been associated with modulating pain responses for many years. Two regions implicated in the modulation of pain sensation, as discussed in Chapter 7 on endogenous opioids, are the dorsal horn of the spinal cord and the dorsal root ganglion (DRG). Indeed, a number of radioligand

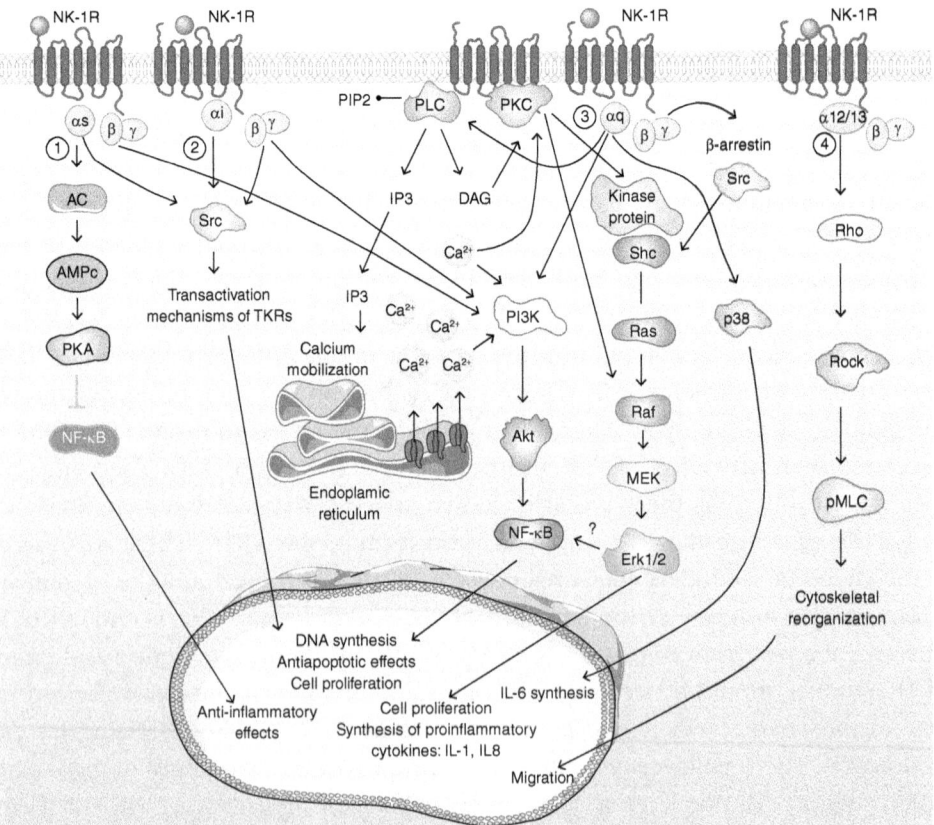

Figure 8.3   The myriad of intracellular responses to binding at the NK1 receptor

Neuropeptide binding at extracellular site of the metabotropic NK1 receptor causes dissociation of the intracellular-bound G protein. The G-protein subunits effect a number of intracellular processes, including: activity of adenylyl cyclase (AC) and protein kinase C (PKC). These consequently impact cell proliferation and DNA transcription amongst other processes. The intracellular effect is believed to be determined by the neuropeptide which binds at the NK1 receptor.

Source: Garcia-Recio, S., & P. Gascón (2015). Biological and pharmacological aspects of the NK1-receptor. *BioMed Research International*, 2015: 495704. Copyright © 2015 Susana Garcia-Recio and Pedro Gascón, licensed under CC BY 3.0.

binding studies in both animal and human tissue have established that SP and NKRs[8] are heavily expressed within the dorsal horn and DRG circuitry. Consistent with this is the fact that SP is known to be specifically expressed in C fibres found in the DRG. These are well known to convey nociceptive signals. This provides the first circumstantial evidence for the involvement of SP and NKRs in nociception.

---

[8]Especially NK1R and NK2Rs.

A further line of research focused on the use of locally administered SP/NKR agonists and antagonists in animal models of pain have provided much stronger evidence for the role of SP and NKR receptors in the DRG in the sensation of pain. A number of highly influential studies in the late 1980s and through the 1990s convincingly showed that when NK1R antagonists were administered to the DRG, pain-related responses were dramatically reduced in animal models. Conversely, when agonists for NK1R or SP were locally administered in the DRG, then pain-related responses in these animals were increased, as you might expect.

─ Focus on methods   ─────────────────────────────

Pain induction in animal models

Obviously, pain research with humans is difficult due to the understandable ethical issues with causing pain. This has resulted in many animal models of pain, such as the tail flick test (see Figure 8.4), formalin paw, carrageenan paw edema, nerve injury (neuropathic pain) and nociceptive reflex responses. Reading further on these methods might raise many questions for you regarding the ethics of animal research, but it is worth keeping in mind that all research of this nature is heavily controlled and monitored by the Home Office in the UK. Researchers also ascribe to the principles of the NC3Rs (National Centre for replacement, refinement and reduction in animal research), which you can find more information on via this link: https://nc3rs.org.uk/the-3rs.

It is important to consider the differences between these methods. For example, the tail flick test is generally considered a model of acute pain, while formalin paw in some respects is a model of both acute and chronic pain. There are also further differences in the type of pain. For example, carrageenan paw is an example of acute inflammatory pain, while nerve injury is a form of neuropathic pain and the tail flick typically involves either pressure or heat related pain. So why is it important to consider this? Well, simply, SP and its mechanisms may be involved to different degrees and via different mechanisms depending on the types of pain. This is therefore an important methodological issue to consider when doing wider reading on pain and the involvement of tachykinins. Indeed, it may well be a key point to consider if you notice any inconsistencies in the published literature.

(Continued)

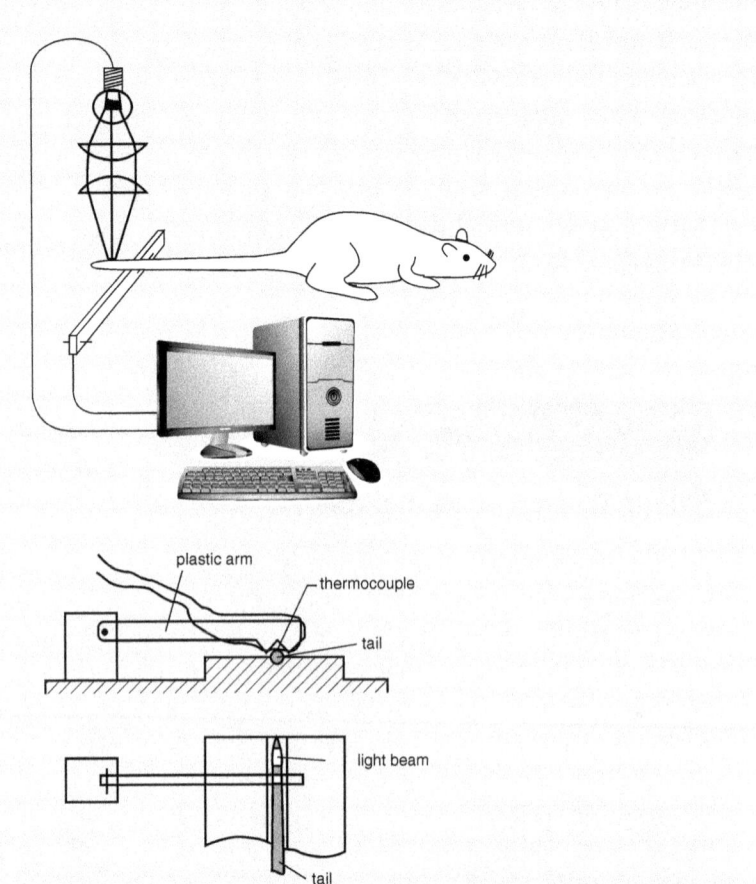

Figure 8.4   A schematic of the typical equipment set-up for studies employing the tail flick test

The tail is generally held in place with a plastic arm and a heat source either attached to the tail or directly above. A computer records the movement of the tail when a light beam is broken by the movement.

Source: Hole, K., & Tjølsen, A. (2007). Tail flick test. In R. F. Schmidt & W. D. Willis (Eds.), *Encyclopedia of pain* (pp. 2392–2395). Berlin, Heidelberg: Springer. Reprinted by permission from Springer Nature.

There is also much research that suggests that certain neurons in the DRG respond to pain/noxious stimulation by increasing their expression of NKRs. This research typically involves the application of a known noxious, pain-inducing substance, then immunohistochemical and/or electrophysiological techniques are used to visualise/ measure the expression levels of receptors. One example study of this type was carried out by Shanley and colleagues (2011). They applied the noxious stimuli capsaicin[9] and found that this induced increased synthesis/transcription of SP from the TAC1 gene in various sensory neurons innervating the DRG, but not all. In addition, the study also found an increase in NK1R expression within the DRG to this noxious capsaicin stimulation. Further research shows that this increase of NK1Rs in the DRG to noxious stimuli can be blocked if NK1R antagonists are also administered. It suggests that the increased expression of NK1Rs is likely to be a mechanism controlling the chronicity of pain sensation. Early on in this section I identified the dorsal horn of the spine as a key player in the sensation of pain. To support this point, a number of studies have also found similar results as outlined above in the DRG, namely an increase in NK1R receptors in this dorsal horn to noxious/painful stimuli and corresponding blocking of this increase with the application of NK1R antagonists. The weight of the research therefore seems to strongly suggest that SP release and NK1R receptor activation/upregulation convey pain-related signals and that increased levels of release, by specific populations in the DRG and dorsal horn, potentially contribute to the continued sensation of pain. Of course, on a mechanistic level it is also important to consider the intracellular effects of receptor activation. In the case of NK1Rs, a host of complex intracellular cascades and channel proteins, such as those identified in Figure 8.5, are modulated. There is a vast amount of research on the involvement of these mechanisms in SP-associated nociception. A good starting point for your wider reading is the review by Chang, Jiang, and Chen (2019).

─Focus on research 🔍─

Kenneth McCarson and colleagues

Many researchers and research groups spend their career focused on specific areas, so it is often a good idea when looking for published literature on a topic to search for these researchers. Kenneth E. McCarson, who has been Professor with tenure at the University of Kansas (US) since 2015, is a good name to search when looking for research generally on pain. If you wish to read further on this topic, I recommend using his name as a good search term in search engines such as Pubmed: https://pubmed.ncbi.nlm.nih.gov/26331888/

Further information on McCarson and his research can be accessed through the website of the University of Kansas's Medical Center: www.kumc.edu/school-of-medicine/pharmacology-toxicology-and-therapeutics/faculty/tenured-and-tenure-track-faculty/kenneth-e-mccarson-phd.html

---

[9]This is the active ingredient in chillis that give them their 'heat', producing a burning sensation on the pallet when ingested.

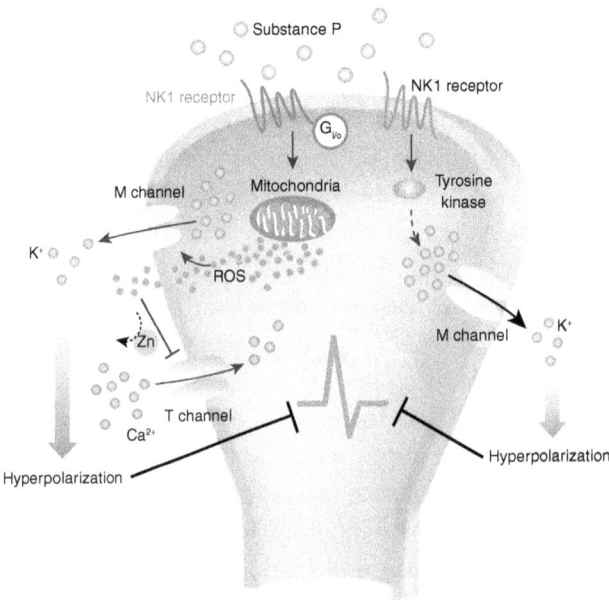

Figure 8.5   The intracellular modulation of ROS by substance P activation of NK1 receptors

The intracellular effect of NK1 receptor activation by substance P are myriad. Many of these identified in the figure are believed to be involved in the modulation of pain by substance P, such as the downstream release of reactive oxygen species (ROS), which have dramatic and toxic effects on neurons.

Source: Chang, C.-T., Jiang, B.-Y., & Chen, C.-C. (2019). Ion channels involved in substance P-mediated nociception and antinociception. *International Journal of Molecular Sciences, 20*(7), 1596, licensed under CC BY 4.0.

One thing worth spending a moment to consider here is that pain can typically be split into two types: acute and chronic. Acute pain is short-term and transient; chronic pain is long-term and sustained. Many of the animal models of pain, as discussed above in the Focus on methods section, are acute pain models. It is therefore conceivable that the mechanisms that modulate acute and chronic pain signals may be different. However, there is significant research evidence which still implicates NK1R in chronic pain. For example, Henry and colleagues (1999) find that there is marked continuous tonic activity at NK1 receptors in the formalin paw rodent pain model long after the initial administration of formalin to the paw. Also, other research, such as that by Khasabov and collegaues (2017), implicates NK1R activation higher up in brain stem regions in hyperalgesia.[10]

---

[10]This is an increased sensitivity to pain.

There was much excitement surrounding NKR antagonists as potentially effica-cious analgesics based upon pre-clinical work with animal models. Unfortunately, when this moved into clinical trials there seemed to be little efficacy. Indeed, over the last 20 years there has only been one licensed NK1R antagonist, which is used to reduce post-operative and cancer treatment related vomiting – what is referred to as an antiemetic effect. So why the lack of efficacy in humans? Well, simply put, there are many differences in the expression and affinity of NKRs and tachykinins between species, with known differences between humans and rodents. I would generally say that rodents are an excellent model of behaviours/responses that have been evolutionarily conserved, although in this instance, it may not be the case.

Beyond the issue related to animal research and substance P, it is also impor-tant to remember that SP and NK1Rs are expressed beyond the central nerv-ous system and by many different types of cells. Therefore, when you read about SP's involvement in pain-related biological processes, such as inflamma-tory responses, it is likely that as well as relating to sensory neuronal responses, such as the involvement of A and B fibres, this research may be focused on non-neuronal cells. In this book we are interested in the neuron and SP's role as a neuromodulator, but it is worth keeping in mind that SP and NKRs may also be involved in modulating pain by modulating processes, such as inflammation, via non-neuronal mechanism and 'reducing' the sensation of pain at the source, rather than modulating the pain signal in the DRG or high up in the CNS via modulation of neuronal circuits.

One final point of consideration is the interaction between SP, NKRs and endogenous opioids. As previously mentioned in this section and extensively covered in Chapter 7, endogenous opioids act as analgesics. If we consider for a moment that SP and NK1Rs seem to do the opposite, namely to act to increase the sensation of pain, then we can conceptualise these neuropeptides as a puta-tive switching mechanism. Indeed, research shows that endogenous opioids can modulate the release of SP from neurons in regions such as the spinal cord. Several studies also show that the ability of opioid receptors[11] to modulate SP release in regions of the spinal cord is significantly reduced in animal models

---

[11]Generally, μ opioid receptors.

of chronic, neuropathic pain.[12] A good example of research on this topic has been conducted by Wan and colleagues (2016) using diabetic neuropathic pain rodents. This serves to illustrate a common theme in this book, namely that neurotransmitters rarely function in isolation and behaviour is typically modulated by the complex interplay between neurotransmitters and their mechanisms.

We will next consider the role of SP and its mechanisms in anxiety. It is worth knowing at this point that anxiety and depression are common outcomes for both animal models of chronic pain and humans experiencing chronic pain. This may therefore suggest some involvement of altered SP and its mechanisms in anxiety.

Substance P and anxiety

Like pain, anxiety is considered by many to be an evolutionarily conserved mechanism that conveys a survival advantage onto the organism. Indeed, the behaviour that typifies anxiety includes caution and avoidance in response to either a real or imagined stimulus. If you think about it in evolutionary terms, this makes a lot of sense as caution would have potentially helped us avoid harm and increased the chance of survival. However, in the modern world, anxiety can be debilitating and is often in response to stimuli that cause no real physical danger. A good way to think about anxiety is to consider phobias, which are a specific type of anxiety disorder. Phobias can exist for a range of different stimuli, many of which pose no real physical danger to the individual, for example, anthophobia, which is an intense and persistent fear of flowers, or agrizoophobia, which is an intense and persistent fear of teddy bears. Another good example of dysfunctional anxiety is the intense anxiety many feel in response to impending exams. In evolutionary terms, the exam cannot really cause any physical harm to the individual, but it can often provoke intense fear responses that are persistent and result in caution and avoidance behaviour.

So, what do we mean by clinical anxiety? Well, anxiety, as classified by the DSM-5 (American Psychiatric Association, 2013), is a group of different disorders, including GAD (generalised anxiety disorder) and phobias. It is important to keep this in mind as molecular neuroscience studies, typically employing animal models, often talk about increased or decreased anxiety responses in more general terms. Therefore, what we see in these models may not be a complete reflection of what we mean

---

[12]Although an interesting point, this is not true for all types of pain. See Chen, McRoberts, and Marvizón (2014) for a good example study.

when we discuss clinical anxiety disorders experienced by humans. Remember this as we discuss animal research on anxiety in this section. There are, however, key characteristics of altered anxiety that are evolutionarily conserved and can be measured objectively in animal models. Therefore, it is important not to 'write off' these studies as invalid for human anxiety. See the Focus on methods box on p. 235 for further information.

So, if we consider anxiety as an evolutionarily conserved mechanism, then it makes sense that the mechanisms that underpin it are also conserved. Much research suggests that these mechanisms may revolve around SP, especially the regulation of NK receptors and the consequences of their activation both intracellularly and on neural circuits.

One starting point, although somewhat circumstantial, is research focused on the expression level of SP in the mammalian brain. Much research has identified high levels in brain regions such as the amygdala and hypothalamus. These structures are heavily associated with mediating anxiety responses. It would therefore make sense that SP and its respective receptors would be involved in metering anxiety responses. Further evidence comes from studies employing pharmacological manipulation of SP and its receptors. There was much research using these methods in the early to mid-2000s. These studies typically involved intracerebral injection of either substance P, NK1R agonists or NK1R antagonists followed by measurement of anxiety-like behaviour using a number of different methods,[13] including the elevated plus maze, measurement of distress vocalisation, measurement of startle responses and measurement of reinforced place aversion, to name a few. The vast majority of these studies found that SP and NK1R agonists had anxiogenic[14] effects, while SP and NK1R antagonists produce anxiolytic effects. However, there were a significant number of studies that found the contrary, namely that administration of SP and NK1R agonists produced anxiolytic effects.[15] So how was this resolved? Well, the difference in SP and NK1R agonist effects seemed to be largely due to the brain region to which they were administered, with certain regions, such as parts of the amygdala, producing anxiogenic effects and other regions, such as the ventral pallidum, producing anxiolytic effects. Of course, it is also worth keeping in mind when doing wider reading on this subject that these studies also used different methods to measure anxiety, which may therefore explain at least some of the difference. So, what does

---

[13]See the Focus on research methods box (p. 235) on measuring anxiety responses in animals for a further outline of these methods.

[14]The generation of anxiety-like behaviour.

[15]Reduction in anxiety-like behaviour.

this tell us? Well, it seems to suggest that SP and its receptors can play different roles in metering anxiety dependent upon the region administered and the circuitry and neurons that are being modulated.

This gets more complicated as some really interesting research, carried out along the same lines as that described above, instead of injecting SP, injected fragments of the N and C terminal end[16] of SP. Depending on the fragment, there were various behaviour responses in the animals from anxiolytic to anxiogenic. For example, the research by De Araújo, Huston, and Brandão (2001) found that SP fragment 6-11 (C terminal fragment) had anxiogenic effects when injected into the PAG (periaqueductal grey) and SP fragment 1-7 (N terminal fragment) had anxiolytic effects when injected into this same region. What this perhaps suggests is that there is a role for the other tachykinins and their receptors in anxiety. Remember that all NKRs have at least some affinity with SP, NKA and NKB. If you recall from earlier in the chapter, this is because they share a similar C terminal end. Therefore, what we see here is that SP may modulate circuitry involved in anxiogenic effects through binding at one NKR, while it modulates circuitry involved in anxiolytic effects by binding at a different NKR.

Research that can be seen to concur with these findings focuses on a brain region known as the ventral pallidum.[17] This region is an area of great convergence and some have recently conceptualised it as a final common region, where information is processed before behavioural output. It is also believed to be heavily involved in reward-related and reward-motivated behaviour. Therefore, if you think of anxiety in evolutionary terms as a mechanism to motivate the organism, it makes sense that if SP is involved in anxiety responses, then application to this region should modulate anxiety. Indeed, this seems to be the case as the research of Nikolaus et al. (1999, 2000) suggests SP affects anxiety-like behaviour when injected into the ventral pallidum (VP). What is perhaps most interesting in Nikolaus' studies is that different SP fragments[18] seem to all produce anxiolytic responses when injected into the VP. This seems to fit with the fact that the VP is a region of circuitry convergence and we can perhaps see SP and NKRs in this region as being involved in the selection of behaviour – in this case, reduced anxiety.

---

[16]This being the section, as outlined earlier in the chapter, that binds to the receptor.

[17]For an excellent review of this region, see Root et al. (2015).

[18]Both C terminal fragments, which have a high affinity for NK1Rs and N terminal fragments, have little affinity for NK1Rs but higher affinity for other NKRs.

Measuring anxiety responses in animals

There are a number of different methods employed to measure anxiety-like behaviour in animals. We have mentioned above that there are objective measures of anxiety in animal models. These are cautious and avoidant behaviour in response to a 'feared' stimuli or the expectation of a feared stimuli. Rodents typically fear open spaces as they are predated upon. This means that in a novel environment, rodents will generally exhibit anxiety-like behaviour by avoiding open spaces, staying close to the walls of rooms and showing increased caution by spending less time in the open areas and moving rapidly through the open (feared) areas.

One particularly well-validated method that allows you to measure these responses is the elevated plus maze (see Figure 8.6 for an outline) and Kraeuter, Guest, and Sarnyai's (2019) protocol for information on how to use it to measure anxiety.

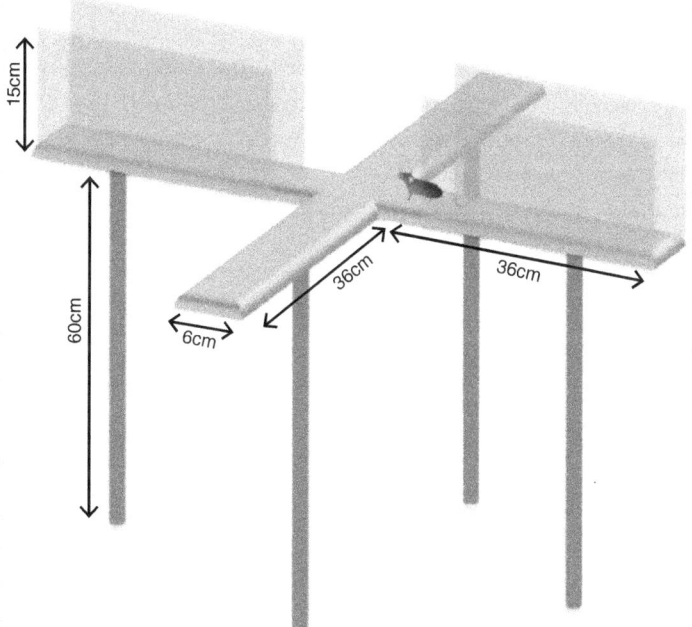

A. The typical structure and layout of an elevated plus maze, as typically used in studies of mouse behaviour

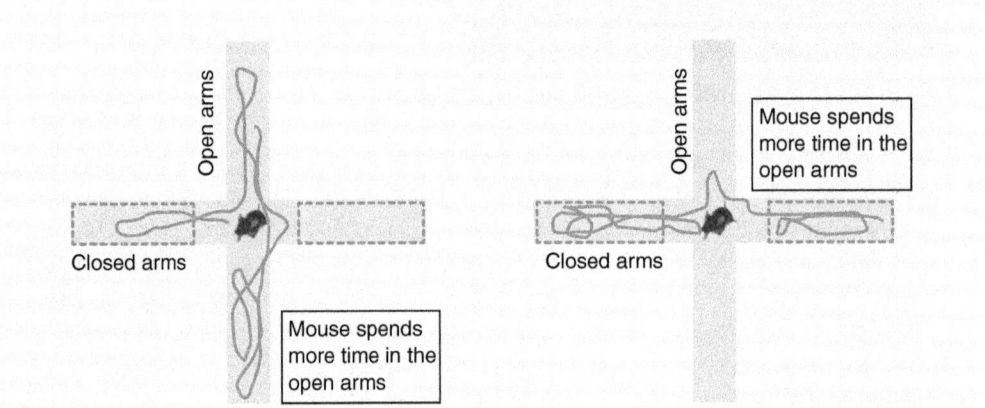

B. Animals with low levels of anxiety typically exhibit the behaviour shown on the left, while animals with high levels of anxiety typically exhibit the behaviour on the right

Figure 8.6   A schematic of the typical equipment set-up for studies employing the elevated plus maze to investigate anxiety-like behaviour in rodents

Source: Reprinted by permission from Springer Nature: Kraeuter, A. K., Guest, P. C., & Sarnyai, Z. (2019). The elevated plus maze test for measuring anxiety-like behavior in rodents. In P. Guest (Ed.), *Pre-clinical models: methods in molecular biology* (vol. 1916). New York: Humana Press. https://doi.org/10.1007/978-1-4939-8994-2_4. Copyright (2019).

It is important to keep in mind with animal studies using a range of different methods, that these can only really be said to be anxiety-like behaviours. Certainly, they do concur with the behaviour seen in humans reporting anxiety, but of course it is impossible to directly ask the animal if what they are experiencing is 'anxiety'.

Pharmacological studies, such as those outlined above, always have caveats related to the specificity of the compounds used. This is a particularly important issue with SP and NKRs as there is so much cross-affinity. One 'cleaner' method for studying the involvement of receptors in behaviour is by employing genetically-modified animal models. So, what have genetically-modified animal models told us about the involvement of NKRs in metering anxiety-like behaviour? Well, early studies, such as that by Santarelli and colleagues (2001), provided convincing evidence to implicate NK1Rs in anxiety-related behaviour. In this study, NK1R gene deletion mice were bred. These animals were then subjected to a number of procedures to measure anxiety-like behaviour. This included the elevated plus maze and the maternal separation paradigm. This study and that by Bilkei-Gorzo, Racz, Michel, and Zimmer (2002) are really good examples of top-quality research and I strongly recommend reading the original articles. What did they find? Well, they found that consistently across measures anxiety-like behaviour was reduced in the NK1R gene deletion mice. Within the study they also applied NK1R antagonists and found that in 'typical control animals' these antagonists reduced anxiety-like behaviours but had no effect on the NK1R gene deletion animals. This further supports the role of these NK1Rs in anxiety-like behaviour and their agonism in anxiogenic behaviour. Although this seems convincing, it is important to consider, as discussed earlier in the chapter, that posttranslational processes can result in different isoforms of the NK1R. This means that studies such as Santarrelli et al. (2001) are perhaps too broad and fail to fully acknowledge the role that specific receptor isoforms play in anxiety-like behaviour.

What about humans? Well, like the excitement that surrounded the potential efficacy of NK1 antagonists in pain modulation, there was also a considerable amount of research centred around SP and NK1R antagonists as potential anxiolytic agents. Unfortunately, as was the case with NK1R antagonists as analgesics, there is little evidence from randomised control trials that NK1R antagonists are efficacious for the treatment of anxiety disorders, including generalised anxiety disorder (GAD) and specific types of anxiety disorders, such as social anxiety disorder (SAD), as well as for anxiety-related disorders such as post-traumatic stress disorder (PTSD).

One thing to keep in mind when considering research on anxiety is the connection between anxiety and physiological stress responses. When someone experiences anxiety, this typically results in the activation of the body's stress response system, namely the fight versus flight response, which is partially controlled by the hypothalamic-pituitary adrenal axis. There is much evidence that shows SP is a key player in this response and the circuitry involved in its control, for example many regions of the hypothalamus are heavily innervated by SP expressing fibres. Also, many studies have shown that a number of brain regions respond to stressful stimuli by increasing the release of substance P. This therefore poses the question: is SP involved in anxiety responses or the ensuing stress response that is a consequence of exposure to the feared (anxiety-inducing) stimuli? Unfortunately, this discussion is beyond the scope of this book, but for a good overview of research investigating SP and stress, I strongly recommend the review by Ebner and Singewald (2006).

As a final point, as ever, we must remember that anxiety is not just about SP and NKRs. A large body of research also implicates other neurotransmitters in anxiety. A particular strand of research suggests that SP and NK1R may be involved in anxiety-like behaviour through their modulation of serotonin and serotonergic neuron populations. This gets more complicated as many believe that the modulation of serotonin levels by SP is via its modulation of GABA and/or glutamatergic neurons in the raphe nucleus, which then innervate and modulate serotonergic neurons. This is a complicated picture, which evidences the complex interaction between neurotransmitters and their impact on behaviour of the cell and the whole organism.

Novel topics in substance P research

Substance P and addiction

One key thing to first remind ourselves about addiction is the brain regions and circuits believed to be implicated. It is good to remember the phrase: 'the reward circuits are hijacked by drugs of abuse'. This means that the likely candidate regions where SP and NKRs might have some involvement are these same reward circuits. Indeed, there is much clear circumstantial evidence in the fact that SP is known to be synthesised and released by neurons in this circuitry, including cholinergic neurons and medium spiny neurons (MSNs). NKRs are also known to be expressed in this circuitry and modulate neuronal inputs to areas heavily implicated in addiction. The case for SP's involvement in addiction builds further when you consider the known role of these neurons in this circuitry, in action selection and goal-directed behaviour. These are two of the key behaviours that addictive substances subvert to produce habit-seeking and increased seeking behaviour – the hallmarks of addiction.

There is significant evidence to suggest that SP administration alone within these 'reward' circuits[19] can have reinforcing effects. For example, the research of Nikolaus, Huston, and Hasenöhrl, (1999) showed that a place preference could be established by intracerebral injections of SP and this same place preference could be blocked by coadministration of a NK1R antagonist. Another very interesting finding from this research is that fragments of SP that are known to have an affinity with NK3R receptors can still produce the reinforced place preference even when administered with the NK1R antagonist. What this suggests is that the reinforcing effects of SP may be mediated by different types of NKRs, therefore it is conceivable that these mechanisms may be hijacked by addictive substances to produce behaviour associated with addiction.

---

[19]Specifically, when administered in the ventral pallidum.

Indeed, this seems very plausible when you consider that pharmacological studies, such as that by Barbier and colleagues (2013), have demonstrated reduced heroin-related reinforcement when NKR antagonists, specifically NK1R antagonists, are administered to rodent models. This is further supported by studies with genetically-modified 'knockout'[20] mice that have consistently found a reduction in drug-related behaviours with the knockout of NK1Rs, especially reduced drug-associated place preferences and reduced self-administration of drugs, such as opioids. Although these findings are fairly consistent, an interesting caveat is that a number of these studies using knockout animals found no alteration in drug-related behaviour for non-opioid drugs, such as cocaine. This suggests that the role of SP and NK1Rs may be specific to metering the addictive qualities of opioids.

As mentioned above, it does not seem to be just about SP and NK1Rs. There is also a significant body of research that implicates NK3Rs in addiction, especially in alcohol and cocaine dependence.[21] Genetic studies with individuals that have a family history of alcoholism show mutations on chromosome 4q and linked regions. One of these genes on the linked region is TACR3, which codes for the NK3 receptor. This suggests that alteration in the coding and, ultimately, expression of NK3Rs may underpin the risk of substance abuse. Indeed, family cohort studies, such as Foroud et al (2008) have found a strong relationship between mutations in this NK3R gene and a family history of alcohol dependence. The relationship was found to be particularly strong for members of the families who showed the highest levels of alcohol dependence and increased co-morbid cocaine abuse. At this point it is important to think about what this means for SP's involvement in addiction. Remember, SP does have some affinity at the NK3R, although the main tachykinin with affinity for NK3R is NKB. Further evidence from pharmacological studies with animal models of addiction seem to solidify the involvement of NK3Rs in addiction. A number of studies have demonstrated that administration of NK3R agonists reduce alcohol preference and cocaine preference. (For a good review on this topic see Schank (2020).) For example, studies have shown that NK3R agonists are able to inhibit alcohol intake

---

[20]This involves either the deletion of the gene that codes for the receptor in these animals, or the insertion of a gene, which interferes with the transcription of DNA and typically results in a non-functioning version of the receptor.

[21]If you make the connection here to the previous paragraph, you might be thinking different tachykinins/ NKRs meter the addictive responses to different classes of drugs.

in rodent animal models without reducing the intake of other necessary substances, such as food and water.

What we seem to have here is a dichotomy with NK1 agonism increasing reinforcement and drug-related behaviour, while NK3R agonism seems to reduce reinforcement and drug-related behaviours. However, there is a caveat to this. As we have discussed elsewhere in this chapter, there are significant known species differences in the form and function of NKRs. At first, this would not seem to be an issue as the human genetic studies in the previous paragraph seem to concur with the animal research regarding NK3R's involvement. However, the human genetic studies do not tell us whether it is increased or decreased activity at NK3Rs. Rodent studies seem to suggest agonism of NK3Rs reduces addiction-related behaviours. However, research with primates seems to suggest agonism of NK3Rs, like agonism at NK1Rs, actually increases drug-related addictive behaviour. So, after all, it may not be a functional dichotomy in primates. All we can perhaps conclude is that NKRs are involved in the addictive responses to drugs of abuse.

─ Focus on research 🔍 ─

Making connections to improve treatment

As mentioned previously, there was a lot of excitement about the potential of efficacious analgesics targeting the NKRs. Unfortunately, there was little evidence of efficacy in clinical trials. This left opioid analgesics, such as morphine, still as the main form of clinically prescribed pain relief. However, as previously mentioned, opioid analgesics have the unfortunate problem of being highly addictive.

So how do we block the addictive/rewarding elements of opioids? Well, researchers think the answer might lie with antagonising SP with NK1R antagonists. As discussed in this chapter, NK1R antagonists are known to reduce the rewarding effects of addictive substances. They are also biologically known to reduce the release of dopamine. So, the answer may be to develop drugs that combine an opioid agonist with a NK1R receptor antagonist. I strongly recommend the following paper, which suggests the potential of this approach:

Sandweiss, A. J., McIntosh, M. I., Moutal, A., Davidson-Knapp, R., Hu, J., Giri, A. K., & Vanderah, T. W. (2018). Genetic and pharmacological antagonism of NK1 receptor prevents opiate abuse potential. *Molecular Psychiatry, 23*(8), 1745–1755. doi:10.1038/mp.2017.102

Of course, as ever there are things to keep in mind when considering research that investigates the involvement of SP and NKRs in addiction-related behaviour. Several have already been mentioned in this section. However, one other thing to consider, outlined in a number of chapters in this book, is the fact that addiction is not commonly thought of as a unitary concept. Rather, it is believed to be formed from several steps/components, as outlined by Wise and Koob (2014). This means that we need to be cautious when looking at research implicating SP and its mechanistics in addiction as, depending on the methods employed, it may reveal SP's and/or NKR's involvement in certain aspects of addiction rather than the whole experience.

One final point of interest is the known links between stress and drug taking. It is well known that stressful events and stimuli increase the chances that both animal models and humans will return to seeking drugs, even after significant periods of abstinence. Given how we previously discussed the links between SP, stress and anxiety, it seems likely that there is a convergent underpinning mechanism, which perpetuates a return to drug seeking after stressful events. A few studies have suggested that the biological mechanism underpinning this could revolve around NK1Rs. Indeed, NK1R antagonism has been shown in these studies to attenuate drug seeking (cocaine and opioid) after exposure to stressful stimuli. This is a potentially important and fruitful area for future research.

Conclusion

What can we conclude about SP and more broadly the neurokinins and their respective receptors? Well, their effects on behaviour are certainly complicated, with different receptors, and the type of tachykinin that binds with them, having a profound and unique effect on the behaviour exhibited by the organism. The last section on addiction is a good example of this, with some evidence showing that different receptors mediate dichotomous behaviour effects on addiction-related behaviour. Of course, their complex effects are not just at the level of the whole organism, and we have seen that the tachykinins and NKRs can trigger distinct and complex intracellular effects that profoundly affect the behaviour of the cell. Indeed, it is wise to remember tachykinins, like other neuropeptides, have complex and modulatory effects.

─Test Yourself Answers─

Test Yourself 8.1: Synthesis

Table 8.3

Translation - this is the process by which mRNA is decoded and proteins/peptides are synthesised at the ribosomes	Transcription - this is the process by which mRNA is produced from DNA	Ribosomes - these are the structures, located on the endoplasmic reticulum, where mRNA is translated into proteins/peptides	Secondary messengers - although more commonly associated with neurotransmission and receptors, secondary messengers can trigger cascades that ultimately result in transcription and translation	Cavicapture - this is the process by which LDCVs partially fuse with the cell membrane to release contents into the synapse
Proteolysis - this is the enzymic controlled process by which large precursor peptides are broken up into the final active neuropeptide. This happens in the LDCVs as they travel towards the terminal end	LDCVs - these are the large dense core vesicles that package the large peptide precursor molecules up in the cell body. They bud off the Golgi body as immature LDCVs	Homotypic fusion - this is the process by which immature LDCVs fuse together to produce the final LDCVs that contain multiples proteins and peptides	NEPs	~~Volume transmission~~

Test Yourself 8.2: Packaging

Immature vesicles bud off the trans Golgi network	Translation of mRNA occurs at the ribo-some producing a large precursor molecule

Immature vesicles fuse to form LDCVs

This process is called homotypic fusion and results in mature LDCVs that typically contain more than one peptide

The mature LDCVs are transported down the axon via microfilaments

This process is called anterograde trans-port and, via a process of condensation, the vesicle reduces in size

During this stage, the large precursor molecule typically goes through the pro-cess of proteolysis

Once the LDCV arrives at this point the final neuropeptide has been synthesised

Cavicapture occurs

This is a process of partial fusion with the cell membrane and release of the vesicle's content

(Continued)

c: Reuptake/Degradation

1.   NEP (neprilysin).

2.   Zinc

3.   Postsynaptic neuron in close proximity to NK receptors.

4.   No - they are more likely to be expressed on distant terminal ends, due to the role neuropeptides play in volume transmission. Although this is true, it is worth keeping in mind that they can still be found on the postsynaptic terminal end within the synapse.

References

Key information

Garcia-Recio, S., & Gascón, P. (2015). Biological and Pharmacological Aspects of the NK1-Receptor. *BioMed Research International, 2015*, 495704–495704. doi:10.1155/2015/495704

Steinhoff, M. S., von Mentzer, B., Geppetti, P., Pothoulakis, C., & Bunnett, N. W. (2014). Tachykinins and their receptors: contributions to physiological control and the mechanisms of disease. *Physiological Reviews, 94*(1), 265–301. doi:10.1152/physrev.00031.2013

US, V. E., & Gaddum, J. H. (1931). An unidentified depressor substance in certain tissue extracts. *J Physiol, 72*(1), 74–87. doi:10.1113/jphysiol.1931.sp002763

Satake, H., & Kawada, T. (2006). Overview of the primary structure, tissue-distribution, and functions of tachykinins and their receptors. *Curr Drug Targets, 7*(8), 963–974. doi:10.2174/138945006778019273

Ebner, K., & Singewald, N. (2006). The role of substance P in stress and anxiety responses. *Amino Acids, 31*(3), 251–272. doi:10.1007/s00726-006-0335-9

Rigby, M., O'Donnell, R., & Rupniak, N. M. (2005). Species differences in tachykinin receptor distribution: further evidence that the substance P (NK1) receptor predominates in human brain. *J Comp Neurol, 490*(4), 335–353. doi:10.1002/cne.20664

Lai, J. P., Douglas, S. D., & Ho, W. Z. (1998). Human lymphocytes express substance P and its receptor. *J Neuroimmunol, 86*(1), 80–86. doi:10.1016/s0165-5728(98)00025-3

Ho, W. Z., Lai, J. P., Zhu, X. H., Uvaydova, M., & Douglas, S. D. (1997). Human monocytes and macrophages express substance P and neurokinin-1 receptor. *J Immunol, 159*(11), 5654–5660.

Ribeiro-da-Silva, A., & Hökfelt, T. (2000). Neuroanatomical localisation of Substance P in the CNS and sensory neurons. *Neuropeptides, 34*(5), 256–271. doi:10.1054/npep.2000.0834

Severini, C., Improta, G., Falconieri-Erspamer, G., Salvadori, S., & Erspamer, V. (2002). The tachykinin peptide family. *Pharmacol Rev, 54*(2), 285–322. doi:10.1124/pr.54.2.285

Schank, J. R., & Heilig, M. (2017). Substance P and the neurokinin-1 receptor: The new CRF. *Int Rev Neurobiol, 136*, 151–175. doi:10.1016/bs.irn.2017.06.008

Lecci, A., & Maggi, C. A. (2009). Substance P/tachykinins and its/their receptors. In L. R. Squire (Ed.), *Encyclopedia of Neuroscience* (pp. 599–615). Oxford: Academic Press.

Nässel, D. R., Zandawala, M., Kawada, T., & Satake, H. (2019). Tachykinins: Neuropeptides that are ancient, diverse, widespread and functionally pleiotropic. *Frontiers in Neuroscience, 13*(1262). doi:10.3389/fnins.2019.01262

Skidgel, R. A., & Erdös, E. G. (2004). Angiotensin converting enzyme (ACE) and neprilysin hydrolyze neuropeptides: a brief history, the beginning and follow-ups to early studies. *Peptides, 25*(3), 521–525. doi:10.1016/j.peptides.2003.12.010

Varnäs, K., Finnema, S. J., Stepanov, V., Takano, A., Tóth, M., Svedberg, M., . . . Farde, L. (2016). Neurokinin-3 receptor binding in guinea pig, monkey, and human brain: In vitro and in vivo imaging using the novel radioligand, [18F]Lu AF10628. *Int J Neuropsychopharmacol, 19*(8). doi:10.1093/ijnp/pyw023

SP and pain

De Koninck, Y., & Henry, J. L. (1991). Substance P-mediated slow excitatory postsynaptic potential elicited in dorsal horn neurons in vivo by noxious stimulation. *Proc Natl Acad Sci U S A, 88*(24), 11344–11348. doi:10.1073/pnas.88.24.11344

Hole, K., & Tjølsen, A. (2007). Tail flick test. In R. F. Schmidt & W. D. Willis (Eds.), *Encyclopedia of Pain* (pp. 2392–2395). Berlin, Heidelberg: Springer Berlin Heidelberg.

Shanley, L., Lear, M., Davidson, S., Ross, R., & Mackenzie, A. (2011). Evidence for regulatory diversity and auto-regulation at the TAC1 locus in sensory neurones. *Journal of Neuroinflammation, 8*, 10. doi:10.1186/1742-2094-8-10

Henry, J. L., Yashpal, K., Pitcher, G. M., Chabot, J., & Coderre, T. J. (1999). Evidence for tonic activation of NK-1 receptors during the second phase of the formalin test in the rat. *J Neurosci, 19*(15), 6588–6598. doi:10.1523/JNEUROSCI.19-15-06588.1999

Khasabov, S. G., Malecha, P., Noack, J., Tabakov, J., Giesler, G. J., Jr., & Simone, D. A. (2017). Hyperalgesia and sensitization of dorsal horn neurons following activation of NK-1 receptors in the rostral ventromedial medulla. *J Neurophysiol, 118*(5), 2727–2744. doi:10.1152/jn.00478.2017

McCarson, K. E., & Krause, J. E. (1996). The neurokinin-1 receptor antagonist LY306,740 blocks nociception-induced increases in dorsal horn neurokinin-1 receptor gene expression. *Molecular Pharmacology, 50*(5), 1189–1199.

Wan, F. P., Bai, Y., Kou, Z. Z., Zhang, T., Li, H., Wang, Y. Y., & Li, Y. Q. (2016). Endomorphin-2 inhibition of substance P signaling within lamina I of the spinal cord is impaired in diabetic neuropathic pain rats. *Front Mol Neurosci, 9*, 167. doi:10.3389/fnmol.2016.00167

Chen, W., McRoberts, J. A., & Marvizón, J. C. (2014). μ-Opioid receptor inhibition of substance P release from primary afferents disappears in neuropathic pain but not inflammatory pain. *Neuroscience, 267*, 67–82. doi:10.1016/j.neuroscience.2014.02.023

King, T. E., Cheng, J., Wang, S., & Barr, G. A. (2000). Maturation of NK1 receptor involvement in the nociceptive response to formalin. *Synapse, 36*(4), 254–266. doi:10.1002/(sici)1098-2396(20000615)36:4<254::Aid-syn2>3.0.Co;2-a

Mantyh, P. W. (2002a). Neurobiology of substance P and the NK1 receptor. *J Clin Psychiatry, 63* Suppl 11, 6–10.

Mantyh, P. W. (2002b). Neurobiology of substance P and the NK1 receptor. *Journal of Clinical Psychiatry, 63*(SUPPL. 11), 6–10. Retrieved from www.scopus.com/inward/record.uri?eid=2-s2.0-0036448830&partnerID=40&md5=d75ac6a25c50c9fe7c3b1e3b542e0e31

Zieglgänsberger, W. (2019). Substance P and pain chronicity. *Cell and Tissue Research, 375*(1), 227–241. doi:10.1007/s00441-018-2922-y

Parent, A. J., Beaudet, N., Beaudry, H., Bergeron, J., Bérubé, P., Drolet, G., . . . Gendron, L. (2012). Increased anxiety-like behaviors in rats experiencing chronic inflammatory pain. *Behavioural Brain Research, 229*(1), 160–167. doi:10.1016/j.bbr.2012.01.001

Hill, R. (2000). NK1 (substance P) receptor antagonists--why are they not analgesic in humans? *Trends Pharmacol Sci, 21*(7), 244–246. doi:10.1016/s0165-6147(00)01502-9

Picard, P., Boucher, S., Regoli, D., Gitter, B. D., Howbert, J. J., & Couture, R. (1993). Use of non-peptide tachykinin receptor antagonists to substantiate the involvement of NK1 and NK2 receptors in a spinal nociceptive reflex in the rat. *Eur J Pharmacol, 232*(2–3), 255–261. doi:10.1016/0014-2999(93)90782-d

Chang, C.-T., Jiang, B.-Y., & Chen, C.-C. (2019). Ion channels involved in substance P-mediated nociception and antinociception. *International Journal of Molecular Sciences, 20*(7). doi:10.3390/ijms20071596

SP and anxiety

American Psychiatric Association. (2013). Diagnostic and statistical manual of mental disorders (5th ed.) (DSM-5). Washington, DC: APA.

Ebner, K., & Singewald, N. (2006). The role of substance P in stress and anxiety responses. *Amino Acids, 31*(3), 251–272. doi:10.1007/s00726-006-0335-9

Ebner, K., Muigg, P., Singewald, G., & Singewald, N. (2008). Substance P in stress and anxiety: NK-1 receptor antagonism interacts with key brain areas of the stress circuitry. *Annals of the New York Academy of Sciences, 1144*, 61–73. doi:10.1196/annals.1418.018

Bilkei-Gorzo, A., Racz, I., Michel, K., & Zimmer, A. (2002). Diminished anxiety- and depression-related behaviors in mice with selective deletion of the Tac1 gene. *J Neurosci, 22*(22), 10046–10052. doi:10.1523/JNEUROSCI.22-22-10046.2002

Ribeiro, S. J., Teixeira, R. M., Calixto, J. B., & De Lima, T. C. (1999). Tachykinin NK(3)receptor involvement in anxiety. *Neuropeptides, 33*(2), 181–188. doi:10.1054/npep.1999.0021

Hasenöhrl, R. U., Jentjens, O., De Souza Silva, M. A., Tomaz, C., & Huston, J. P. (1998). Anxiolytic-like action of neurokinin substance P administered systemically or into the nucleus basalis magnocellularis region. *European Journal of Pharmacology, 354*(2–3), 123–133. doi:10.1016/s0014-2999(98)00441-5

Kraeuter, A.-K., Guest, P. C., & Sarnyai, Z. (2019). The Elevated Plus Maze Test for Measuring Anxiety-Like Behavior in Rodents. In P. C. Guest (Ed.), *Pre-Clinical Models: Techniques and Protocols* (pp. 69–74). New York, NY: Springer New York.

McCarson, K. E. (2015). Models of inflammation: Carrageenan- or complete Freund's adjuvant (CFA)-induced edema and hypersensitivity in the rat. *Curr Protoc Pharmacol, 70*, 5.4.1–5.4.9. doi:10.1002/0471141755.ph0504s70

Root, D. H., Melendez, R. I., Zaborszky, L., & Napier, T. C. (2015). The ventral pallidum: Subregion-specific functional anatomy and roles in motivated behaviors. *Progress in Neurobiology.* doi:10.1016/j.pneurobio.2015.03.005

De Araújo, J. E., Huston, J. P., & Brandão, M. L. (2001). Opposite effects of substance P fragments C (anxiogenic) and N (anxiolytic) injected into dorsal periaqueductal gray. *Eur J Pharmacol, 432*(1), 43–51. doi:10.1016/s0014-2999(01)01460-1

Nikolaus, S., Huston, J., & Hasenöhrl, R. (2000). Anxiolytic-like effects in rats produced by ventral pallidal injection of both N- and C-terminal fragments of substance P. *Neuroscience Letters, 283*, 37–40. doi:10.1016/S0304-3940(00)00902-2

Nikolaus, S., Huston, J. P., & Hasenöhrl, R. U. (1999). Reinforcing effects of neurokinin substance P in the ventral pallidum: mediation by the tachykinin NK1 receptor. *European Journal of Pharmacology, 370*(2), 93–99. doi:https://doi.org/10.1016/S0014-2999(99)00105-3

Gavioli, E. C., Canteras, N. S., & De Lima, T. C. (1999). Anxiogenic-like effect induced by substance P injected into the lateral septal nucleus. *NeuroReport, 10*(16), 3399–3403. doi:10.1097/00001756-199911080-00026

Ebner, K., Rupniak, N. M., Saria, A., & Singewald, N. (2004). Substance P in the medial amygdala: Emotional stress-sensitive release and modulation of anxiety-related behavior in rats. *Proceedings of the National Academy of Sciences, 101*(12), 4280. doi:10.1073/pnas.0400794101

Santarelli, L., Gobbi, G., Debs, P. C., Sibille, E. T., Blier, P., Hen, R., & Heath, M. J. (2001). Genetic and pharmacological disruption of neurokinin 1 receptor function decreases anxiety-related behaviors and increases serotonergic function. *Proc Natl Acad Sci U S A, 98*(4), 1912–1917. doi:10.1073/pnas.041596398

Lesch, K. P. (2001). Mouse anxiety: the power of knockout. *Pharmacogenomics J, 1*(3), 187–192. doi:10.1038/sj.tpj.6500016

Michelson, D., Hargreaves, R., Alexander, R., Ceesay, P., Hietala, J., Lines, C., & Reines, S. (2013). Lack of efficacy of L-759274, a novel neurokinin 1 (substance P) receptor antagonist, for the treatment of generalized anxiety disorder. *Int J Neuropsychopharmacol, 16*(1), 1–11. doi:10.1017/s1461145712000065

Tauscher, J., Kielbasa, W., Iyengar, S., Vandenhende, F., Peng, X., Mozley, D., . . . Marek, G. (2010). Development of the 2nd generation neurokinin-1 receptor antagonist LY686017 for social anxiety disorder. *Eur Neuropsychopharmacol, 20*(2), 80–87. doi:10.1016/j.euroneuro.2009.10.005

Mathew, S. J., Vythilingam, M., Murrough, J. W., Zarate, C. A., Jr., Feder, A., Luckenbaugh, D. A., . . . Charney, D. S. (2011). A selective neurokinin-1 receptor antagonist in chronic PTSD: a randomized, double-blind, placebo-controlled, proof-of-concept trial. *Eur Neuropsychopharmacol, 21*(3), 221–229. doi:10.1016/j.euroneuro.2010.11.012

Ribeiro, S. J., Teixeira, R. M., Calixto, J. B., & De Lima, T. C. (1999). Tachykinin NK(3)receptor involvement in anxiety. *Neuropeptides, 33*(2), 181–188. doi:10.1054/npep.1999.0021

SP and addiction

Schank, J. R. (2020). Neurokinin receptors in drug and alcohol addiction. *Brain Research, 1734*, 146729–146729. doi:10.1016/j.brainres.2020.146729

Schank, J. R. (2014). The neurokinin-1 receptor in addictive processes. *J Pharmacol Exp Ther, 351*(1), 2–8. doi:10.1124/jpet.113.210799

Foroud, T., Wetherill, L. F., Kramer, J., Tischfield, J. A., Nurnberger, J. I., Jr., Schuckit, M. A., . . . Edenberg, H. J. (2008). The tachykinin receptor 3 is associated with alcohol and cocaine dependence. *Alcohol Clin Exp Res, 32*(6), 1023–1030. doi:10.1111/j.1530-0277.2008.00663.x

Barbier, E., Vendruscolo, L. F., Schlosburg, J. E., Edwards, S., Juergens, N., Park, P. E., . . . Heilig, M. (2013). The NK1 receptor antagonist L822429 reduces heroin reinforcement. *Neuropsychopharmacology, 38*(6), 976–984. doi:10.1038/npp.2012.261

Nikolaus, S., Huston, J. P., & Hasenöhrl, R. U. (1999). Reinforcing effects of neurokinin substance P in the ventral pallidum: mediation by the tachykinin NK1 receptor. *European Journal of Pharmacology, 370*(2), 93–99. doi:https://doi.org/10.1016/S0014-2999(99)00105-3

DeLong, M. R. (1990). Primate models of movement disorders of basal ganglia origin. *Trends in Neurosciences, 13*(7), 281–285. doi:https://doi.org/10.1016/0166-2236(90)90110-V

Gerfen, C. R. (1991). Substance P (neurokinin-1) receptor mRNA is selectively expressed in cholinergic neurons in the striatum and basal forebrain. *Brain Research, 556*(1), 165–170. doi:10.1016/0006-8993(91)90563-B

Gerfen, C. R., & Scott Young, W. (1988). Distribution of striatonigral and striatopallidal peptidergic neurons in both patch and matrix compartments: an in situ hybridization histochemistry and fluorescent retrograde tracing study. *Brain Research, 460*(1), 161–167. doi:https://doi.org/10.1016/0006-8993(88)91217-6

Clark, M. (2020). Effects of electrical stimulation of NAc afferents on VP neurons' tonic firing. *Front Cell Neurosci, 14*(401). doi:10.3389/fncel.2020.599920

Sandweiss, A. J., McIntosh, M. I., Moutal, A., Davidson-Knapp, R., Hu, J., Giri, A. K., . . . Vanderah, T. W. (2018). Genetic and pharmacological antagonism of NK1 receptor prevents opiate abuse potential. *Mol Psychiatry, 23*(8), 1745–1755. doi:10.1038/mp.2017.102

Gadd, C. A., Murtra, P., De Felipe, C., & Hunt, S. P. (2003). Neurokinin-1 receptor-expressing neurons in the amygdala modulate morphine reward and anxiety behaviors in the mouse. *J Neurosci, 23*(23), 8271–8280. doi:10.1523/jneurosci.23-23-08271.2003

Sandweiss, A. J., & Vanderah, T. W. (2015). The pharmacology of neurokinin receptors in addiction: prospects for therapy. *Substance abuse and rehabilitation, 6*, 93–102. doi:10.2147/SAR.S70350

Ripley, T. L., Gadd, C. A., De Felipe, C., Hunt, S. P., & Stephens, D. N. (2002). Lack of self-administration and behavioural sensitisation to morphine, but not cocaine, in mice lacking NK1 receptors. *Neuropharmacology, 43*(8), 1258–1268. doi:10.1016/s0028-3908(02)00295-2

Murtra, P., Sheasby, A. M., Hunt, S. P., & De Felipe, C. (2000). Rewarding effects of opiates are absent in mice lacking the receptor for substance P. *Nature, 405*(6783), 180–183. doi:http://www.nature.com/nature/journal/v405/n6783/suppinfo/405180a0_S1.html

Ciccocioppo, R., Panocka, I., Polidori, C., Froldi, R., Angeletti, S., & Massi, M. (1998). Mechanism of action for reduction of ethanol intake in rats by the tachykinin NK-3 receptor agonist aminosenktide. *Pharmacology Biochemistry and Behavior, 61*(4), 459–464. doi:10.1016/S0091-3057(98)00090-2

Melamed, J., Silva, M., Tomaz, C., Müller, C., Huston, J., & Barros, M. (2012). Sensitization of hypervigilance effects of cocaine can be induced by NK3 receptor activation in marmoset monkeys. *Drug and Alcohol Dependence, 128*. doi:10.1016/j.drugalcdep.2012.08.020

Fulenwider, H. D., Nennig, S. E., Hafeez, H., Price, M. E., Baruffaldi, F., Pravetoni, M., . . . Schank, J. R. (2020). Sex differences in oral oxycodone self-administration and stress-primed reinstatement in rats. *Addict Biol, 25*(6), e12822. doi:10.1111/adb.12822

Schank, J. R. (2014). The neurokinin-1 receptor in addictive processes. *J Pharmacol Exp Ther, 351*(1), 2–8. doi:10.1124/jpet.113.210799

9

NITRIC OXIDE

Introduction

Despite the many differences between neurotransmitters covered in this book so far, there is one thing that binds all these neurotransmitters together and that is that they are solid molecules in their normal state within the body. Nitric oxide (NO) is unique as it is a gaseous neurotransmitter. This brings with it some unique qualities that allow it to diffuse readily both intra- and extracellularly. It also means that the molecule is very unstable[1] and therefore breaks down rapidly. Most estimates of NO's half-life suggest around 1–10 seconds. This is the main limiting factor in its ability to modulate the behaviour of cells distal from its site of release.

Nitric oxide was originally coined EDRF (endothelial derived relaxing factor) as it was initially discovered in samples of vascular endothelial cells.[2] It is now well known to be synthesised in other cells, of which, for us, the most important are neurons, where it is conceived as a neurotransmitter/neuromodulator. The molecule consists of a nitrogen and an oxygen atom bound by a double covalent bond. It is electrically neutral, which is one of its key properties that allows it to cross biological membranes with ease. This is an important point, which we will return to when we discuss the receptors for NO. As a point of interest, though, the diffusion rate of NO through membranes is believed to be approximately 800 square micrometres a second. It is worth considering this in the context of NO's known half-life, which is identified above. We may say that NO's ability to act on target cells is confined by its high reactivity.

Nitric oxide is best thought of as a neuromodulator as its effects on the neuron are complex and seldom result in a stereotyped excitatory or inhibitory response. It is typically released postsynaptically, normally as the consequence of other neurotransmitters, such as glutamate, binding at its target receptor on the postsynaptic membrane. This results in calcium influx, which is generally the main factor triggering the synthesis and consequent release of NO. This means that NO is generally referred to as a retrograde neurotransmitter[3] (see Figure 9.1). Although this is the most common mechanism of neurotransmission for NO, it is important to remember, as is so often the case, that NO can also be synthesised and released presynaptically, and therefore is also involved in anterograde neurotransmission. This is especially the case at NO terminals in the peripheral nervous system.

---

[1]Mainly due to the lack of one electron in the sp2 orbital of the nitrogen atom.

[2]These are cells that make up the inner layer of veins, arteries and capillaries.

[3]This means to work backwards. In this case, diffuse from the postsynaptic to the presynaptic neuron.

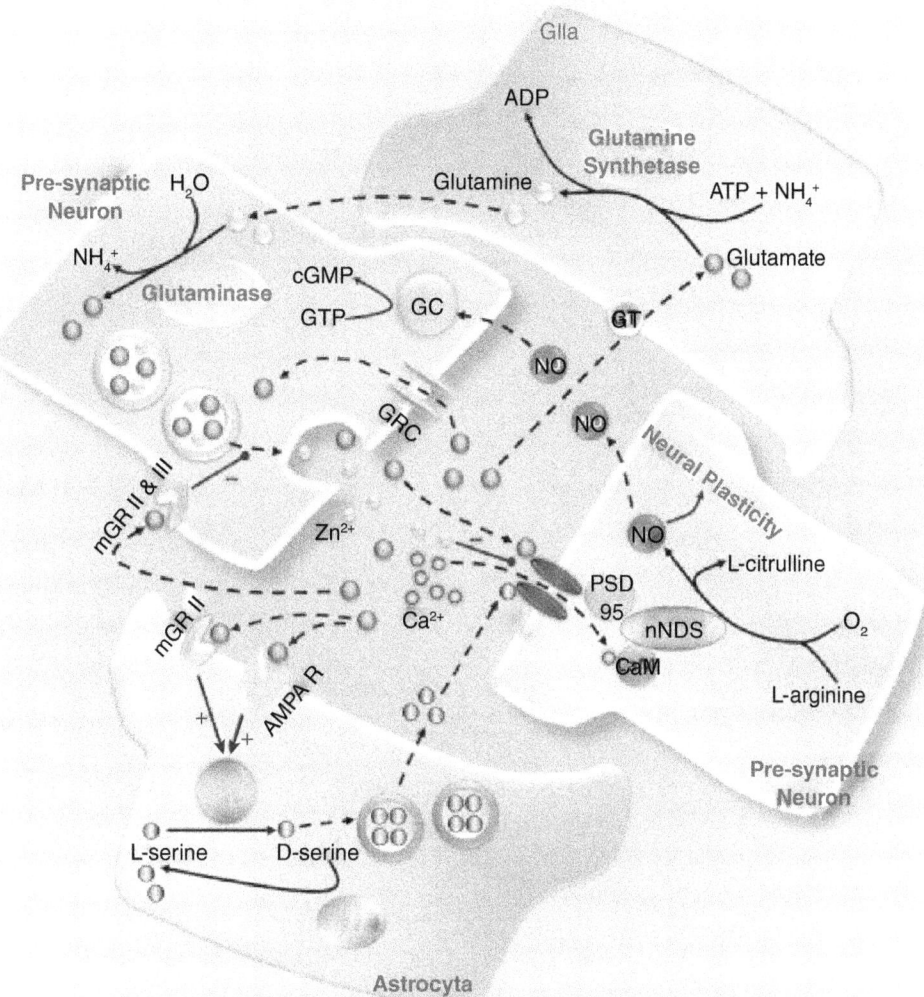

Figure 9.1 The complex retrograde effects of NO synthesis in the postsynaptic neuron

Notice that NO is synthesised here because of nNOS (neuronal nitric oxide synthase) bound to the glutamate NMDA (N-Methyl-D-aspartate) receptor. Therefore, it could be said that, among its many actions, glutamate triggers NO synthesis.

Source: Reprinted from: Ghasemi, M., Claunch, J., & Niu, K. (2019). Pathologic role of nitrergic neurotransmission in mood disorders. *Progress in Neurobiology, 173*, 54–87. Copyright (2019), with permission from Elsevier.

─ Focus on concepts 🔍 ─

Nitric oxide is a free radical

A free radical is a molecule that has one less electron in a free orbital. This is the case for nitric oxide, which lacks one electron in its sp² orbital of its nitrogen atom (see Figure 9.2). This makes the molecule highly reactive.

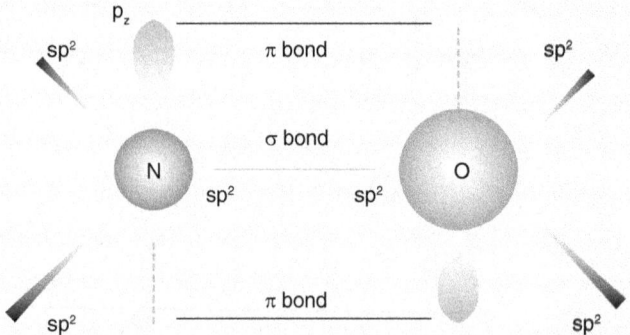

Figure 9.2   The molecular structure of NO typified by one nitrogen and one oxygen molecule

Source: Picón-Pagès, P., Garcia-Buendia, J., & Muñoz, F. J. (2019). Functions and dysfunctions of nitric oxide in brain. *Biochimica et Biophysica Acta (BBA) – Molecular Basis of Disease, 1865*(8), 1949–1967. doi:https://doi. org/10.1016/j.bbadis.2018.11.007. Licensed under CC BY 4.0.

Free radicals have been heavily associated with a number of processes, such as oxidative stress and ageing. In the case of NO, this reactivity means it forms a number of secondary/intermediate molecules that are considered to be highly cytotoxic. We will discuss this later in the chapter.

For now, if you are interested in a bit of light reading related to free radicals, then I recommend the following: Lobo, V., et al. (2010). Free radicals, antioxidants and functional foods: impact on human health. *Pharmacognosy Reviews, 4*(8), 118-126.

It is particularly interesting to keep in mind that research on free radicals, especially in the context of disease states, has not been as conclusive as you may expect.

Outline of nitric oxide's synthesis, packaging and reuptake

Nitric oxide synthesis

The vast majority of NO is synthesised in a one-step process from L-arginine by the enzyme's family nitric oxide synthases (NOS). L-arginine, in the presence of molecular oxygen, is oxidised by NOS and the products are citrulline and NO. It is a little bit more complex than this as a number of other molecules/co-factors are involved. These will be discussed later in this section. It is also important to note at this point, and to keep in mind for subsequent sections of this chapter, that the process does not just happen within neurons. Other cells, such as glia and vascular endothelial cells, also contain NOS and therefore produce and release NO.

L-arginine is a semi-essential amino acid and, as such, is not considered to be a limiting factor in the synthesis of NO. In fact, if levels of L-arginine are depleted, then more can be synthesised readily from citrulline and other α amino acids, such as glutamic acid, which are non-essential amino acids. As a point of interest, under some conditions, depletion of L-arginine can result in an increase in reactive oxygen species (ROS), such as superoxide from NOS actions. Build-up of these can have profound cytotoxic effects. We will return to this later in the chapter.

Notice that at the start of this section I said: 'the vast majority of NO' is synthesised by NOS. This obviously implies that there are other molecular mechanisms that can result in the synthesis of NO. Other enzymes, such as the reductase family[4] of enzymes can transform nitrates into nitrites, which will ultimately result in NO. These nitrates can come from our diet and research has shown that dietary nitrites do seem to modulate NO synthesis. Nitrites and nitrates can also be thought of as a way for the body to store NO, and the body therefore has a ready supply if other mechanisms and precursors are depleted. This is not the end of the story for NO synthesis. Indeed, NO can also be synthesised via non-enzymic processes, such as via the oxidation of L-arginine. Although all these other mechanisms exist for the synthesis of NO, it is still the case that the vast majority of NO is believed to be synthesised by the family of NOS enzymes. There are generally considered to be three types of NOSs: eNOS (endothelial), iNOS (inducible) and nNOS (neuronal), the last of which is the one of most importance to us. It is important to keep this in mind as you do wider reading, as some functions and behavioural effects on the organism are mediated by the distinct types. There is also the somewhat controversial suggestion that a fourth type exists: mtNOS (mitochondrial). The existence of mtNOS is somewhat controversial as several studies have shown that there is no apparent mtNOS activity in mitochondria extracted from heart tissue and that nNOS knockout animals also do not seem to express mtNOS. This has led many to suggest that nNOS and mtNOS are one and the same.

---

[4]This includes xanthine oxidase (XO).

Let's focus for a moment on the ones that we are confident exist, namely iNOS (NOS2), eNOS (NOS3) and nNOS (NOS1). iNOS gets its name as it is produced in response to a cellular challenge/insult. It is expressed most abundantly in immune and glia cells when they are challenged, although it is also known to be expressed in other cells, such as muscle cells and cells in the retina. More specifically, in the brain it is expressed in vascular cells, astrocytes and microglia. Generally, its role seems to be to promote apoptosis in cells that have been damaged/insulted. eNOS, as the name implies, is primarily expressed by vascular endothelial cells, although it has also been shown to be expressed in skeletal muscle, motor neurons and astroglia. It is activated by calcium/calmodulin. This is also true of nNOS. It is a key player in platelet aggregation and vascular tone. It is therefore essential to the health of the cardiovascular system. We will come back to this later in the chapter when we discuss NO's role in vascular diseases.

Finally, but by no means least for us, is nNOS. nNOS is known to be widely distributed in the brain and expressed by a number of different interneuron types. It is a key player, via NO production, in the control of the neuron's function. Although the name implies it is expressed solely in neurons, confusingly it is also known to be expressed in a number of other cells, such as smooth muscle cells and various cells of the vascular system. This caveat is key to keep in mind when we discuss nNOS later in the chapter in the context of atypical behaviour. nNOS is known to have four different isoforms: αnNOS, μnNOS, ɣnNOS and βnNOS. α and ɣ are attached to intracellular structures, such as the intracellular domain of certain receptor proteins, such as NMDA receptors and certain serotonin receptors. This is believed to be the main reason that NO plays a role in synaptic plasticity. μ and β are referred to as cytoplasmic as they are free in the cytoplasm and unbound to any intracellular structure.

For all the NOS family to function correctly there need to be co-factors. The co-factors bind at one of two domains, these being either the N-terminus domain or the C-terminus domain of NOS. These two domains facilitate different processes, with the N-terminus domain acting as an oxygenase[5] and the C-terminus domain acting as a reductase.[6] So, what factors and co-factors bind to which domain? Well, of primary importance is the main precursor to NO, namely L-arginine, which binds at the

---

[5]This is an enzyme that catalyses the addition of oxygen onto another atom. In the case of NO, oxygen is added to a nitrogen atom.

[6]Reductase enzymes catalyse reduction reactions. This generally involves the addition/gain of electrons.

N-terminus domain. Hopefully, you were able to work this out for yourself as earlier in the chapter it was stated that L-arginine is oxidised to produce NO. However, in addition, the co-factors BH4 and iron protoporphyrin IX also bind to the N-terminus domain. As for the C-terminus domain, there are a number of co-factors that bind, including the calcium-binding protein calmodulin and calcium itself. As well as calcium, FMN (flavin adenine mononucleotide), FAD (flavin adenine dinucleotide) and NADPH[7] (nicotinamide adenine dinucleotide phosphate) also bind to the C-terminus domain (see Figure 9.3).

So how is the catalytic action of NOS modulated? Well, calcium and calmodulin binding are essential for the correct functioning of NOS,[8] with many suggesting that calcium levels are the main regulator of NOS activity as calcium opens the main channel that allows electron flux. This is essentially the mechanism by which NOS is activated. Although calcium levels seem to be the key regulatory mechanism, there are others. For example, phosphorylation by cAMP (cyclic adenosine monophosphate), protein kinase or phosphorylation by PKC (protein kinase C).

Nitric oxide packaging

This is a relatively short section compared to other chapters in this book. Nitric oxide is not packaged into lipid-lined vesicles as is the case with the vast majority of other neurotransmitters. It is consequently not released via fusion of the vesicles with the lipid membrane and exocytosis. This is due to its nature as an electrically neutral, lipophilic molecule. So how is it stored? Well, the simple answer is that it is not; rather, it is produced on demand by the NOS family. As mentioned previously, because L-arginine is readily available, this 'on demand' production rarely causes any issues.

Nitric oxide transformation

The physical nature of NO prevents it from being degraded through hydrolytic cleavage by hydrolytic cleaving enzymes, as is the typical method of degradation for most other neurotransmitters. Also, the vast majority of other neurotransmitters,

---

[7]This is one of the most important biological electron donors.

[8]This is true for nNOS and eNOS, but not iNOS.

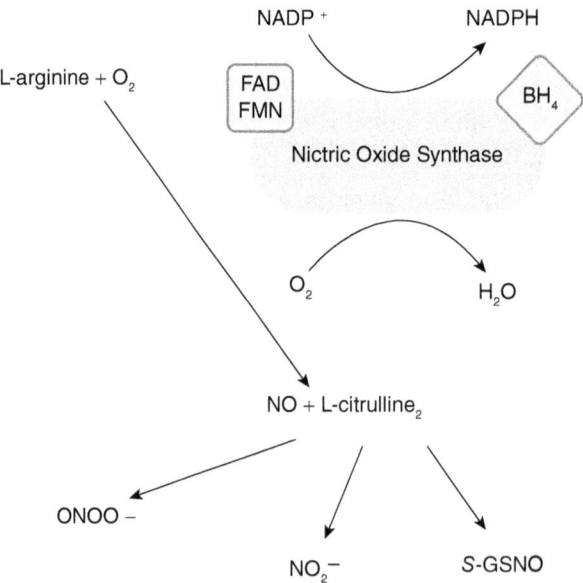

Figure 9.3   The multiple molecules needed for the synthesis of NO by NOS, including co-factors, such as FAD and FMN

Also note the multiple end points for NO, including the molecules formed because of interaction with reactive oxygen species, such as superoxide. Note: FAD = flavin adenine dinucleotide; FMN = flavin mononucleotide; BH_4 = tetrahydrobiopterin; H_2O = water; NADP+ = glutamate dehydrogenase; NADPH = reduced nicotinamide adenine dinucleotide phosphate; O_2 = oxygen; S-GSNO = S-nitrocysteine; ONOO- = peroxynitrite; NO_2- = nitrite.

Source: Król, M., & Kepinska, M. (2021). Human nitric oxide synthase – its functions, polymorphisms, and inhibitors in the context of inflammation, diabetes and cardiovascular diseases. *International Journal of Molecular Sciences, 22*(1). doi:10.3390/ijms22010056. Licensed under CC BY 4.0.

or the product of their degradation, are typically taken back into the cell. This again is not the case with NO. Fundamentally, the mechanism by which NO is mediated is by it reacting with another substrate. Once it has reacted there are a number of molecules that can be formed. Commonly NO, in aqueous solutions, as is the case within the cell, produce nitrites, but these, like NO, are highly unstable molecules and rapidly bind with other molecules to form nitrates. NO in the presence of O_2, or more precisely the free radical superoxide anion ($O_2.^-$), is oxidised and forms peroxynitrite ($OONO^-$), which is what is known as a reactive oxygen species. We will discuss this further in the following Focus on concepts box, but, in short, many believe peroxynitrite to be the key reason for the cyto-toxic effects of NO.

─Focus on concepts 💭─

Nitric oxide and reactive oxygen species

Reactive oxygen species (ROS) are a group of oxygen molecules ($O^{.-}$) that, as the name implies, are highly reactive and can cause significant damage to the cell. They are a product of metabolism in all aerobic organisms; therefore, cellular mechanisms are present to control their levels. The main positives of ROSs are that, although highly reactive, they are relatively short-lived intermediates and are not capable of easily permeating through the cell membrane. It means that the damage they can do is normally localised to one cell. They are also readily broken down by the catabolising enzyme referred to as SOD (superoxide dismutase). Thus, the damaging effect they may have on cells is kept to a minimum. This is not completely true, as a consequence of catabolising $O^{.-}$ is that SODs produce H_2O_2, which can also have damaging oxidative effects on the cell. This issue is resolved by the cell, but further discussion is beyond the scope of this book and I recommend Radi (2018) for more detail on this.

However, this is where NO comes into the picture as these reactive oxygen species, such as $O^{.-}$, can form bonds with NO and this produces compounds that are highly reactive, cytotoxic and far more capable of permeating out of the cell and having a subsequent damaging influence on many other cells. Further adding to the likelihood of these compounds having a negative impact on the cell is the fact that SODs cannot readily stop their formation, as $O^{.-}$ has a much greater affinity for NO than SODs. Therefore, these products of NO synthesis are considered to be one of the key reasons NO is implicated in mechanistic dysfunction at the cellular level and dysfunction/degeneration at the behavioural level of the organism.[9] As mentioned above, the compound most heavily implicated is peroxynitrite.

For further reading on reactive oxygen species and peroxynitrite, I recommend the following as an excellent starting point:

Radi, R. (2018). Oxygen radicals, nitric oxide, and peroxynitrite: Redox pathways in molecular medicine. *Proceedings of the National Academy of Sciences, 115*(23), 5839. doi:10.1073/pnas.1804932115

---

[9]As we will discuss later in the chapter, oxidate forms of NO are heavily implicated in changes associated with degenerative disorders, such as Alzheimer's disease.

Outline of nitric oxide receptors/signalling

Nitric oxide signalling is an extremely interesting area of NO's mechanisms, which relates to its state as a soluble gas, as an electrically neutral molecule and ultimately its lipophilic properties. This all means that NO can readily permeate through the plasma membrane of a neuron. Therefore, rather than the receptors being embedded in the plasma membrane,[10] the classic receptor/signalling pathway for NO is actually found free in the intracellular cytoplasm. In many ways it is perhaps best to think of NO as though it is one of the secondary messengers typically synthesised as a result of neurotransmitters binding to metabotropic receptors.

Once NO has permeated through the phospholipid cell membrane of a neuron, it can bind directly to amino acids in a process referred to as nitrosylation. This mechanism, unlike nitration,[11] is a reversible process. It is therefore more classically considered to be a method of cell signalling that involves NO. There are two distinct types of nitrosylation. The first we will look at is metal nitrosylation and the second is referred to as s-nitrosylation.

The main 'receptor' for NO is called soluble guanylyl cyclase (sGC). As the name implies, it is soluble and found free within the intracellular cytoplasm. NO binds to sGC in the process of metal nitrosylation. sGC consists of a N terminal domain with a bound iron[12] and a catalytic domain including an α and β subunit. These two domains are bound together by a PAS (Per-Arnt-Sim) domain and a coiled helix. See Figure 9.4 opposite for a schematic representation of sGC.

Binding of NO to the iron[13] at the N terminal of sGC results in nitrosylation of the heme iron. This results in a confirmational shape change in the enzyme and the synthesis of the secondary messenger cGMP (cyclic guanosine monophosphate) from GTP (guanosine triphosphate), as illustrated in Figure 9.4. This synthesis is dramatic, producing a several hundredfold increase in cGMP levels. The process cannot be directly switched off *per se*, but the product, namely cGMP, can be hydrolysed by phosphodiesterase (PDE), which results in reduced downstream effects. Also, when NO dissociates from the heme iron, this also returns the cGMP to basal levels. So, what are the downstream effects of cGMP? Well, the main effect is for cGMP to activate protein kinase G. But it also facilitates the gating of certain membrane-bound ion channels and modulates the function of a number of PDEs.

---

[10]This is typically the case discussed in other chapters of this book for other neurotransmitters.

[11]Nitration is the addition of NO_2, which is a non-reversible process that generally result in deactivation.

[12]This is the binding site for NO.

[13]This iron is commonly referred to as the heme co-factor.

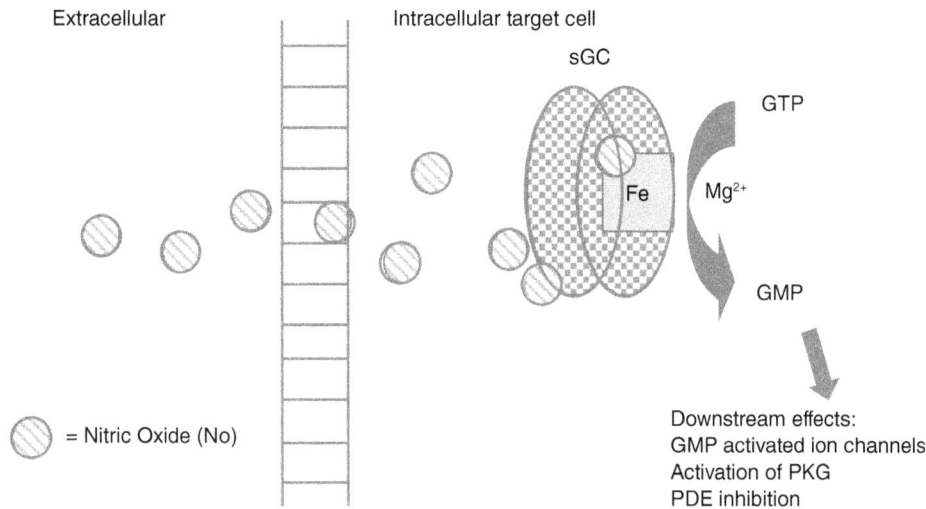

Figure 9.4 The diffusion of NO through the lipid membrane of a neuron followed by its subsequent binding to the heme iron attached to soluble guanylate cyclase (sGC)

It is important to bear in mind that it is not just sGC that NO targets. In fact, there are a number of other enzymes that contain heme irons and therefore are targets for metal nitrosylation by NO. For example, NO binding to the enzyme HRI (heme regulated inhibitor) results in phosphorylation, which triggers translation of mRNAs. This mRNA codes for cellular scaffolding proteins and ultimately promotes the development and growth of neuronal spines.

Through a similar process to metal nitrosylation, NO can bind and alter the function of other proteins via the process of s-nitrosylation. Instead of binding to a metal, in this instance NO binds to a cysteine residue. The binding of NO has regulator effects and can result in both gain of function and loss of function. For example, s-nitrosylation of glyceraldehyde-3-phosphate dehydrogenase (GAPDH), which is involved in DNA replication, is known to result in cellular apoptosis. S-nitrosylation also modulates a range of metabolic enzymes. This serves to modulate the redox state of the neuron and can contribute to processes like the increased catabolism of fatty acids chains. S-nitrosylation is also extensively involved in the direct regulation of a number of ion channels. One key player is the NMDA receptor, where s-nitrosylation results in inhibition and closure of the channel. This is believed to be a key process in the regulation of LTP (long term potential) by glutamate signalling and NMDA receptors. For more detail and links to further primary research related to cellular effects of metal nitrosylation and s-nitrosylation, I strongly recommend the excellent review by Picón-Pagès, Garcia-Buendia, and Muñoz (2019) or the paper of Mannick and Schonhoff (2004).

Metals and cysteine residues are typically found as part of enzymes and many other proteins. In fact, virtually all cellular proteins contain cysteine residues and a large number contain metals. However, despite this, nitrosylation by NO is specific to only certain targets. How can this be the case? Well, the general consensus is that there are two factors that make targets specific. One of these is the co-localisation of NOS with nitrosylation targets, a good example of this being the NMDA receptors mentioned above. The other factor that makes s-nitrosylation targeted is the existence of specific consensus motifs expressed on certain cysteine residues. The existence of unique consensus motifs is also the case for other kinases that NO targets, and ultimately means that NO can be very specific and have extremely targeted effects upon the mechanisms of the target neuron.

Test Yourself 9.1

Sort the following information under the correct heading

Table 9.1

Nitration	Nitrosylation	Reactive oxygen species

- A reversible process
- A non-reversible process that causes inactivation of many lipids, proteins and nucleic acids
- $O.^-$
- NO binds to a metal or a cysteine residue
- Binding of NO to soluble guanylyl cyclase
- NO binding to amino acids
- Addition of an NO_2 group
- Peroxynitrite is a key player in ...

Answers to all Test Yourself questions are at the end of the chapter.

Classic topics in nitric oxide research

Nitric oxide and affective disorders

The class of clinical disorders referred to collectively as affective disorders is one of the most frequently experienced worldwide. Colloquially, they are commonly referred to as mood disorders and this is how we discussed them in Chapter 6 on serotonin. The class includes a variety of disorders, including major depressive disorder (MDD) and bipolar disorder (BPD). What unifies them is that they all involve altered emotional responses and changes in motivation level.[14] In the case of MDD, this involves an extreme depressed/negative emotional state and a lack of drive/motivation to carry out even activities that were once considered pleasurable by the individual. In the case of BPD, the individual's emotional state cycles between periods of extreme mania[15] and extreme low/depressed mood. These two states are also typically accompanied by increased drive/motivation and reduced drive/motivation, respectively.

As mentioned above, the monoamine neurotransmitter serotonin has long been considered a key player in affective disorders and many of the main therapeutics used affect the brain's serotonergic system, especially in the case of MDD, where many drug therapies are SSRIs (selective serotonin reuptake inhibitors) or MAOIs (monoamine oxidase inhibitors). However, these drugs are not efficacious for all and can have some extremely negative side effects for certain groups. For example, Olfson, Marcus, and Shaffer (2006) found that children and teenagers (ages 8–18) being treated with SSRIs have an increased risk of suicidal intention compared to those that are not on SRRIs and adults on SRRIs. This kind of observation has led many researchers to consider other mechanisms and neurotransmitter systems that could be targeted for the treatment of affective disorders. One such target is the NO system. If you take a moment to reflect on what we have covered so far in this chapter, then this is a rational system to target. Why so? Well, as previously mentioned, there is known interplay between serotonin mechanisms and NO mechanisms. For example, bound isoforms of NOS are known to be linked to serotonin receptors. NOS also binds to the serotonin transporters (SERTS) and there is a known connection between NO expression and serotonin expression levels.

Early research in this area typically measured blood plasma levels of nitrates using these as a proxy for the relative levels of NO being synthesised. This research has typically found that nitrate levels are elevated in those with a diagnosis of MDD. What we might expect is that if NO/nitrate levels are elevated in those with a MDD diagnosis, then the opposite should be true for patients with a BPD diagnosis, who are in a state of mania. Unfortunately, it is not as clear cut as this and a number of studies, such as that by Savas and colleagues (2006), suggest that there is a positive correlation both between the number of depressive episodes and NO levels and between the number of manic episodes and NO levels.

---

[14]See DSM-5 (American Psychiatric Association, 2013) for more detailed outlines of the diagnostic criteria for the affective disorders.

[15]Mania is typically defined as a state of extreme euphoria and excitement.

There are some obvious caveats with much of this research, especially studies that use nitrate levels in the blood plasma as a proxy for NO levels. First, NO is synthesised by a variety of different cells in many different regions of the body. Therefore, altered blood plasma levels may be as a result of non-neuronal sources and non-CNS sources. For example, it could relate to NO produced by endothelial cells or perhaps by neurons, but in regions outside the central nervous system, such as the gut. Further evidence that adds to this argument comes from research looking at the relative expression and activity of the NOS family of enzymes in those with a diagnosis of an affective disorder. Research, such as that by Xing and colleagues (2002), found there to be no significant reduction in nNOS levels in post-mortem cortical tissue from those with a diagnosis of both unipolar and bipolar depression. However, these researchers and others have found alterations in other members of the NOS family. For example, researchers have shown higher iNOS levels in the pre-frontal cortex of post-mortem tissue from patients who had a diagnosis of BPD. Although not nNOS, iNOS is known to be expressed in neurons. Therefore, these findings do not preclude the involvement of altered NO signalling within neurons to be involved in the pathology of affective disorders. It is also worth keeping in mind that there is research that contradicts Xing and colleague's findings (2002), especially when nNOS levels are measured within specific brain regions, rather than in the brain as a whole. For example, one study found a significant elevation in nNOS expression in the hippocampus (specifically the CA1 region) of BPD patients. Therefore, it seems likely that neuronal NO and nNOS are elevated in affective disorders, although we can also say that affective disorders may also be contributed to by NO mechanisms beyond the neuron and beyond the molecule's role in neurotransmission.

Blood plasma studies have also found interesting differences in patients with a diagnosis of MDD. One interesting study by Kim et al. (2006) found that there were significantly higher NOx blood plasma levels in depressive patients who were suicidal compared to depressive patients who were non-suicidal. This suggests that rather than NO levels mediating all the symptoms of a patient diagnosed with MDD, it may be more specifically involved in mediation of suicidal thoughts/behaviour. This is an important point to keep in mind. Although simplistic explanations and biological mechanisms are attractive, it is unlikely that complex clinical conditions are solely mediated by the mechanisms of one neurotransmitter. Indeed, we can see already from the known involvement of serotonin and the interplay between serotonin and NO that complex disorders, such as depression, are likely to be mediated by complex biological processes.

Although there is a reasonable amount of evidence that NO elevations do contribute to affective disorders, other research points out that it is not just NO that seems to be elevated. This comes back to the fact that many believe the negative effects of NO are mediated by peroxynitrite. Research, such as that by Savas and colleagues (2006), has found that reactive oxygen species, such as superoxide ions, are also elevated in those with an affective disorder diagnosis. This means that peroxynitrite levels are also likely to be higher. Consequently, it may not be NO *per se* that causes biological alterations that result in affective behaviour change, but products of the reaction with these reactive oxygen species. This is certainly supported by research that suggests alterations in 'waste' products of NO degradation in those with a diagnosed affective disorder. It suggests that it is aspects of the neuronal 'waste management' system that may be altered, resulting in more of the toxic byproducts, such as peroxynitrite.

So, what happens if you knockout the enzymes that synthesise NO in neurons, namely nNOS? Well, much of the research using genetically-modified animal models suggests what we may expect based on the findings of most blood plasma studies. Namely, that NOS knockout animals seem to express less depressive-like behaviour, such as reduced immobility times in the forced swim test, and general hyperactivity compared to typically-developing animals that have been subjected to a depression induction paradigm. These nNOS knockout animals also seem to consistently display reduced depressive-like responses to stress induction paradigms. In typically-developing animals, these paradigms commonly produce behavioural despair. This all seems to present a very clear picture regarding NO's role in depression, suggesting that if you reduce it (by removing the main synthesising enzyme), you reduce 'depressive-like' behaviour. It would therefore intuitively suggest that increasing NO levels must result in more 'depressive-like' behaviour, as is suggested by the majority of human plasma studies. However, there are a few important considerations with this animal research. One is that the vast majority of the research employs measures that reflect the animal's 'motivation' levels. Therefore, it is conceivable that altered NO levels contribute largely to this aspect of affective disorders, rather than the whole experience. The second point of consideration is that many of these studies suggest that other neurotransmitters and their mechanisms are also involved. For example, the research of Tanda and colleagues (2009), which employed nNOS knockout animals, found that dopamine D1 receptor-mediated signalling was also altered in these animals. This highlights the fact that although nNOS is likely to be involved, it may largely exert its effect through other moderating mechanisms, which also contribute to the affective disorder.

Although there is much complexity attached to the involvement of NO signalling in affective disorders, many NO modulators have emerged as a potential novel antidepressant. Many of these compounds are inhibitors of either nNOS or the 'receptor' sGC and are quite consistently found to reduce the 'depressive-like' symptoms in rodent models. Three that have been extensively studied in pre-clinical animal research are Methylene blue, L-NAME and L-NA. These studies typically involve depression induction paradigms that involve an environmental stressor, such as crowding, or chronic stress from physiological pain, such as shock. Researchers then typically measure the depressive-like behaviour by employing methods, such as the forced swim test and the tail suspension test. For an excellent review of this litera-ture, I strongly recommend the paper by Ghasemi, Claunch, and Niu (2019) or that by Zhou, Zhu, Nemes, and Zhu, (2018). It is well worth taking a bit of time to think about these methods and returning to other chapters in this book where we have previously discussed them. There are certainly a number of reasons why we might not think that they fully capture the human experience of an affective disorder. It is perhaps best to say that this research seems to suggest that NO and its mechanisms contribute, to some extent, to the symptomology of affective disorders. Therefore, NO may, to some degree, present as a rationale for an efficacious therapeutic.

Nitric oxide and Parkinson's disease

NO and its mechanisms have been implicated in a number of disorders. One group is the degenerative disorders, which includes Alzheimer's disease,[16] and a range of other disorders which are typified by alterations in the ability to produce controlled movement. This group is collectively referred to as the motor degenerative disorders. One specific type that has garnered a lot of research focus is Parkinson's disease. We have discussed Parkinson's disease in other chapters of this book, such as Chapter 5 on the neurotransmitter dopamine. To see what you remember about its characteristics/symptomology, have a go at the following Test Yourself section.

—Test Yourself 9.2—

From the following list, select the correct motor and non-motor symptoms of Parkinson's disease.

- Brady dyskinesia
- Postural rigidity
- Resting tremor
- Rigidity of movement
- Alterations in hedonic responses
- Alterations in some higher cognitive processes
- Sleep disturbance

---

[16]NO, especially its role in oxidative processes and interaction with reactive oxygen species, has been heavily implicated in Alzheimer's disease. For a good starting point on this I recommend the paper of Balez & Ooi (2016).

So, how is NO implicated in Parkinson's disease (PD)? Well, much of the early evidence from post-mortem studies clearly showed alterations in the expression of nNOS levels in humans who had received a diagnosis of PD during their life. Whether this change was an increase or decrease in nNOS levels, however, seems to be largely dependent on the brain region, with nNOS mRNA levels generally being reduced in areas of the dorsal striatum (specifically the putamen) and increases in nNOS levels generally being found in other territories of the basal ganglia, such as the subthelamic nuclei and in the substantia nigra par compacta. Results relating to the substantia nigra pars compacta are particularly interesting when you consider this contains a large population of dopaminergic neurons. We will come back to this later, but the interaction between NO and dopamine can perhaps be seen as the main way NO and its mechanisms contribute to the pathology of PD. Of course, post-mortem research always comes with the caveat that these changes may be more a consequence of other pathology than a causal factor in the onset of the disease state or disease progression.

So, what about research that may be able to clarify whether or not alterations in NO mechanisms are a causal factor in the pathology of PD. Well, there is some research that shows a dramatic increase in nNOS after the application of MPTP.[17] Interestingly, the research by Chalimoniuk, Langfort, Lukacova, and Marsala (2004) shows that this increase in nNOS is followed by an increase in the activation of sGC. As discussed earlier in the chapter, this results in a myriad of downstream effects. These mechanisms may be the contributing factor to NO's involvement in dopaminergic fibre degeneration in PD. The reason why this is particularly interesting is that many believe that NO's involvement in degenerative disease is generally a result of their interaction with ROS. The research by Chalimoniuk et al. (2004) suggests that this may not be the case and it is in fact an increased activity in the 'normal' mechansims of NO signal transduction. Putting this contention aside for one moment, further support for the fact that NO may play a causal role in the pathology of PD comes from research using nNOS knockout mice. Researchers have found that neural damage is less in these animals in response to MPTP administration. It suggests that NO presence may contribute to a greater negative effect on neuronal health after environmental insults. For more information related to environmental insults, alterations in NO mechanism and degenerative disorders, see the Focus on research box below.

·

---

[17]You may remember from Chapter 5 on dopamine that MPTP is the designer drug discovered to produce Parkinsonian-like symptomology. It is commonly cited as the starting point of the dopamine hypothesis of Parkinson's disease.

Of course, there are other lines of research which also suggest that alteration in NO's mechanisms contribute to the pathology of Parkinson's disease. One good example is research which focuses on gene polymorphisms. These studies have revealed alterations in NO-related genes, such as alterations in the gene for nNOS, which is seen to contribute to PD pathology and certainly seems to produce increased levels of nitrites in those with a PD diagnosis. There is a lot of contradiction in this literature and significant differences in the findings of these studies based upon the ethnicity of the population who participate in the studies. As well as direct gene-related involvement, NO may also be involved via alteration of other genes and gene expression through s-nitrolysation. One example is s-nitrolysation of Parkin.[18] This has been shown to result in increased P53 gene expression. P53 codes for the protein P53 which results in cellular apoptosis. Therefore, what we see with NO is that it may not be its mechanisms that are directly altered *per se*, but because of its affinity with other processes and the gene expression for other proteins, it may play a key role in the neurodegeneration seen in PD, indirectly.

Much of what we have focused on here seems to implicate nNOS as the key NO-related mechanism altered in those with a diagnosis of PD. However, it is important to keep in mind that there are at least three members of the NOS family of enzymes. Indeed, much research suggests that it is not all about nNOS. There is significant evidence for the involvement of other members of the NOS family of enzymes in PD. One particular form that the research literature heavily implicates in the pathology of Parkinson's disease is iNOS. Again, like nNOS, early post-mortem research shows significantly altered levels of iNOS in the brain of those diagnosed with PD. There is also evidence from animal models, using MPTP, that iNOS contributes significantly to the dopaminergic neuron degeneration in this model of PD. Supporting this finding is that iNOS knockout animals are significantly more resistant to dopaminergic neuron degeneration in the MPTP model of PD. It is important to remember that, although known to be expressed in neurons, iNOS is largely responsible for NO synthesis in glia and astrocytes. It is not surprising that alterations in astrocyte function would lead to neurodegeneration. This helps to emphasise the fact that NO mechanisms contributing to disease states are likely to go way beyond their role in neurons and altered neurotransmission.

One thing that I have tried to emphasise throughout this book is that neurotransmitters and their mechanisms rarely operate in isolation. In the context of PD, NO is a good example of this. Much of the research suggests that it is the interaction

---

[18]Parkin is a ligase which aids the survival of neurons, especially dopaminergic neurons. Mutations in its genes have been associated with Parkinson's disease.

between NO's mechanisms and dopamine's mechanisms that result in the behavioural phenotype of PD. I mentioned at the start of this section that alterations in dopamine are well known to play a significant role in the pathology of PD, with dopaminergic neurons being one of the first groups to be lost as the condition progresses. This is a good point to think about the complexity of NO's mechanisms as there are well known interactions between NO and dopamine. It is therefore likely that the symptomology of PD may, at least in part, be due to the loss or dysfunction of the interaction between dopamine and NO mechanisms.

Although it seems clear that altered NO mechanisms appear likely to play a role in PD pathology, the reality is that much of the research literature related to NO and PD focuses on the role NO plays in nitration and the cytotoxic effects of molecules such as peroxynitrite. Much of this literature suggests that these processes can result in significant damage to proteins such as tau and α-synuclein. These proteins have become a hot focus of research on degenerative disorders, especially the literature on PD and Alzheimer's disease. It seems likely, at least to some extent, that the aggregation of these proteins, which is such a common feature of neurodegenerative disorders, is a result of NO products such as the aforementioned peroxynitrite. However, many argue that it may be the other way around, namely that alterations in tau and α-synuclein may result in increased expressions of NO and peroxynitrite. This is yet to be definitively resolved but understanding of the direction of effect here could prove a major breakthrough in our understanding and the treatment of these degenerative disorders. For further reading on this, consult the reference list for this section. However, to aid you, these are good starting points: see Zhang et al. (2018), Kumar et al. (2016), and Paxinou et al. (2001).

Focus on research 🔍

SOD, peroxynitrite and head injuries

Much research has revealed a connection between head injuries and increases in ROS, NO and peroxynitrite. One such study is that of Bayir and colleagues (2007):

Bayır, H., Kagan, V. E., Clark, R. S. B., Janesko-Feldman, K., Rafikov, R., Huang, Z., & Kochanek, P. M. (2007). Neuronal NOS-mediated nitration and inactivation of manganese superoxide dismutase in brain after experimental and human brain injury. *Journal of Neurochemistry, 101*(1), 168–181.

The reason why research on head injuries is of some interest in the case of motor degenerative disorders, such as Parkinson's disease and motor neurone disease (MND) is that

(Continued)

this research suggests that environmental insults, such as head injuries/trauma, may be the root cause of these degenerative diseases. The circumstantial evidence for this comes from the fact that an unusually high percentage of sports people develop motor degenerative disease. For example, MND is also referred to as Lou Garrick disease, as he developed the condition. Lou Garrick was a prominent baseball player for the New York Yankees. It is safe to assume that those who take part in sport, especially contact sports, have a higher incidence of head injury. This could result in increased levels of peroxynitrite and ultimately trigger the train of NO-related events that results in motor degenerative disorders.

A good example that fits with this line of thought is the former Scottish international rugby player Doddie Wier (see www.theguardian.com/sport/2017/jun/20/former-scotland-rugby-player-doddie-weir-diagnosed-with-motor-neurone-disease). Clearly, rugby is a sport where head injuries are hard to avoid during a professional player's career.

Novel topics in nitric oxide research

Nitric oxide and vascular diseases

When we are discussing NO and the vascular system, your thoughts should instantly be drawn to endothelial cells. These cells line the vasculature and express the enzyme eNOS that specifically synthesises NO within these cells. Endothelial cells and eNOS do play a key role in modulating the flow of blood through the vascular system as NO is essential for vasodilation.[19] Recent research suggests that NO derived from eNOS is not the only player in this and NO from neurons (synthesised by nNOS) also plays a vital role. This means that dysfunction in NO-synthesising neurons may also result in dysfunctional vasodilation and, ultimately, neuronal cell death. Recently, the research literature has focused on nNOS and the NO derived from neurons as a key alteration that could play an essential role in a number of vascular diseases, including hypertension, atherosclerosis and vascular dementia, for a good review on the topic see Shabir, Berwick and Franics (2018).

So, what is the evidence that suggests NO derived from nNOS plays a key role in 'normal' vascular homeostasis? Well, there is a rich body of animal research stretching

---

[19]Vasodilation is the relaxing of blood vessels' muscular walls, which results in the widening of the vasculature, allowing a greater volume of blood to pass through.

back to the mid 1990s. Early studies used nNOS-specific antagonists, such as the 7-nitroindasole (7-NI), and found that these antagonists supressed cerebral blood flow (CBF) at baseline as well as reducing a number of other markers of blood flow, such as capillary flow. As is usually the case, the literature suggests that nNOS derived NO does not just act as a vasodilator alone and seems to modulate other mechanisms of vasodilation. For example, a number of studies have found in nNOS knockout animals that the increased medullary blood flow seen in response to angiotensin II is decreased. This suggests that NO derived from nNOS may also play a fundamental role in other mechanisms that maintain vascular homeostasis, as well as being a direct vasodilator itself. Acetycholine is also a well-known vasodilator, with its effects partly being mediated by its impact on NO synthesis. This means that in nNOS knockout animals the application of acetycholine should produce a reduced vasodilatory response. This is indeed what much of the literature using animal models and isolated vessels shows, supporting the key mediatory role that nNOS plays in vasodilation. Of course, a key concern is to ensure that this is also true in humans. Much of the literature supports the conclusion that it is. One influential example study carried out by Seddon and colleagues (2008) used brachial artery infusion of the nNOS inhibitor S-methyl-L-thiocitrulline (SMTC) in healthy male patients. This caused a significant dose-dependent reduction in cerebral blood flow and was partially reversed by administration of the NO synthesising substrate L-arginine.

Next let's discuss how alterations in nNOS and NO derived from this source may contribute to two vascular diseases. First, a classic vascular disease is hypertension.[20] Several studies have identified imbalances in nNOS expression in animal models of hypertension. In fact, the weight of the literature seems to suggest significant increase in nNOS expression in hypertensive animal models, especially in and around arteries. However, there are studies that have shown reduced nNOS levels and reduced functioning of nNOS in the adrenal gland of hypertensive animals. This is the reason why I initially used the word 'imbalance', as it seems that hypertensive animals show different alterations in different regions. More recent studies using hypertensive rats reveals a yet more complex picture for the involvement of nNOS in hypertension. Findings from these studies suggest that it is perhaps the nNOS derived H_2O_2[21] pathway that contribute to the biological symptoms of hypertension, rather than NO alterations caused by changes in nNOS expression.

---

[20]This is a condition characterised by sustained high pressure in the vessels.

[21]H_2O_2 being a biproduct of NO synthesis by nNOS.

Beyond hypertension, there is a good body of research that implicates alterations in NO mechanisms, and alterations in other molecules' mechanisms caused by NO, in degenerative diseases such as the dementias. Early theories related to the cause of the dementias generally conceptualised it as diseases caused by alterations in blood flow. These theories fell out of favour as it became apparent that a number of aberrant and mutated proteins build up in the brains of those with a diagnosis of dementia. However, more recently there has been a move back to the role alterations in the vasculature may play in the dementias. It is logical that if this vascular system is damaged, then reduced blood flow will result in cell death, as it struggles to keep up with the supply of key compounds essential for the healthy metabolic function of the cells. So, what causes the alterations in the vascular system that underpin vascular dementia? Well, as mentioned previously, NO is a vasodilator[22] and, as such, dysfunction in the production of NO would result in a reduced ability for oxygenated blood to be provided to areas with a high metabolic demand. This would ultimately result in cell death (neurons in the brain). As mentioned above, we would perhaps expect this to be related to issues with NO synthesised by eNOS and released by the endothelial vascular cells. However, there seems to be a significant role played by nNOS. For example, convincing research by Choi and colleagues (2018) has found notable pathology in nNOS-expressing interneurons in mouse models of Alzheimer's disease. Several other studies have also found consistent alterations in nNOS expression in those with Alzheimer's disease. One matter of contention is whether this is a cause, consequence or contributory factor in the dementias. Certainly, there is significant evidence that these alterations may be caused by the accumulation of mutated proteins. This would perhaps mean they are more a consequence and contributory factor to the degeneration than they are a cause. Perhaps it is best to think of these changes in nNOS as part of a cycle of deterioration that ultimately results in cell death[23] and progressively increased cognitive impairments that underpin the dementias.

So, what have we learnt from this section? Well, what this section serves to emphasise is that the effects of molecules, commonly considered neurotransmitters and derived from neurons, isn't just to modulate the behaviour of other neurons. As is the case with NO derived from nNOS, these molecules can have profound effects on a range of other

---

[22]A vasodilator is a substance that causes the vascular system to dilate (open), resulting in a greater volume/flow of blood. This is thought to be a key mechanism allowing oxygenated blood and nutrients to be taken to areas of high metabolic demand.

[23]Through reduced vascular control as well as other mechanisms.

cells found in and around the central nervous system. A consequence of alterations in these mechanisms can therefore have profound negative effects on these non-neuronal cells. But further down the chain, this can then return to having profound negative effects on the health of neurons.

Conclusions

Nitric oxide is undoubtedly an interesting molecule and does not conform to many of the expected norms of a neurotransmitter. Its unique qualities make it able to function as a volume transmitter permeating freely into neurons and acting on a range of proteins, lipids and nucleic acids in a manner not unlike the secondary messengers typically formed as a result of a neurotransmitter binding to a metabotropic receptor. It is important to keep in mind that NO's unique qualities mean it plays a plethora of other functional roles beyond neurotransmission, in a range of different tissues and cells, such as vasodilation.

While NO is clearly involved in many cytotoxic processes, it is important to remember that many of these processes do not involve NO directly; rather, they involve secondary molecules produced from the addition of an NO_2 group or as a result of molecules such as peroxynitrite formed as secondary intermediates.

Test Yourself Answers

Test Yourself 9.1

Table 9.2

Nitration	Nitrosylation	Reactive oxygen species
• A non-reversible process that causes inactivation of many lipids, proteins and nucleic acids • Addition of an NO_2 group • Peroxynitrite is a key player in nitration	• A reversible process • NO binds to a metal or a cysteine residue • Binding of NO to soluble guanylyl cyclase • NO binding to amino acids	• O_2^-

Test Yourself 9.2

The correct motor and non-motor symptoms of PD are:

(Continued)

- Brady dyskinesia - motor
- Postural rigidity - motor
- Resting tremor - motor
- Rigidity of movement - motor
- Alterations in hedonic responses - non-motor
- Alterations in some higher cognitive processes - non-motor
- Sleep disturbance - non-motor

Answer: all the above are known motor and non-motor symptoms of PD. Although not always the case, the non-motor symptoms are typically the first to develop followed by the motor symptoms.

References

Key information

Britton, G. (2014). NO-independent modulation of soluble guanylyl cyclase (sGC) activity and function (masters thesis). The University of Texas, Houston, Texas.

Picón-Pagès, P., Garcia-Buendia, J., & Muñoz, F. J. (2019). Functions and dysfunctions of nitric oxide in brain. *Biochimica et Biophysica Acta (BBA) - Molecular Basis of Disease, 1865*(8), 1949–1967. doi:https://doi.org/10.1016/j.bbadis.2018.11.007

Kelm, M. (1999). Nitric oxide metabolism and breakdown. *Biochimica et Biophysica Acta (BBA) - Bioenergetics, 1411*(2), 273-289. doi:https://doi.org/10.1016/S0005-2728(99)00020-1

Vincent, S. R. (2010). Nitric oxide neurons and neurotransmission. *Prog Neurobiol, 90*(2), 246–255. doi:10.1016/j.pneurobio.2009.10.007

Vincent, S. R., & Kimura, H. (1992). Histochemical mapping of nitric oxide synthase in the rat brain. *Neuroscience, 46*(4), 755–784. doi:http://dx.doi.org/10.1016/0306-4522(92)90184-4

Lee, K. E., & Kang, Y. S. (2017). Characteristics of (L)-citrulline transport through blood-brain barrier in the brain capillary endothelial cell line (TR-BBB cells). *J Biomed Sci, 24*(1), 28. doi:10.1186/s12929-017-0336-x

Zand, J., Lanza, F., Garg, H. K., & Bryan, N. S. (2011). All-natural nitrite and nitrate containing dietary supplement promotes nitric oxide production and reduces triglycerides in humans. *Nutrition Research, 31*(4), 262–269. doi:https://doi.org/10.1016/j.nutres.2011.03.008

Jiménez-Jiménez, F. J., Alonso-Navarro, H., Herrero, M., Garcia-Martin, E., & Agúndez, J. (2016). An update on the role of nitric oxide in the neurodegenerative processes of Parkinson's disease. *23*, 2666–2679. doi:10.2174/092986732366616081

Park, C. S., Pardhasaradhi, K., Gianotti, C., Villegas, E., & Krishna, G. (1994). Human retina expresses both constitutive and inducible isoforms of nitric oxide synthase mRNA. *Biochem Biophys Res Commun, 205*(1), 85–91. doi:https://doi.org/10.1006/bbrc.1994.2633

Kapur, S., Bédard, S., Marcotte, B., Côté, C. H., & Marette, A. (1997). Expression of nitric oxide synthase in skeletal muscle: a novel role for nitric oxide as a modulator of insulin action. *Diabetes, 46*(11), 1691–1700. doi:10.2337/diab.46.11.1691

Colasanti, M., Persichini, T., Fabrizi, C., Cavalieri, E., Venturini, G., Ascenzi, P., . . . Suzuki, H. (1998). Expression of a NOS-III-like protein in human astroglial cell culture. *Biochem Biophys Res Commun, 252*(3), 552–555. doi:https://doi.org/10.1006/bbrc.1998.9691

Förstermann, U., Gorsky, L. D., Pollock, J. S., Schmidt, H. H. H. W., Heller, M., & Murad, F. (1990). Regional distribution of EDRF/NO-synthesizing enzyme(s) in rat brain. *Biochem Biophys Res Commun, 168*(2), 727–732. doi:https://doi.org/10.1016/0006-291X(90)92382-A

Buchwalow, I. B., Podzuweit, T., Bocker, W., Samoilova, V. E., Thomas, S., Wellner, M., . . . Lerch, M. M. (2002). Vascular smooth muscle and nitric oxide synthase. *Faseb j, 16*(6), 500–508. doi:10.1096/fj.01-0842com

Esplugues, J. V. (2002). NO as a signalling molecule in the nervous system. *British Journal of Pharmacology, 135*(5), 1079–1095. doi:10.1038/sj.bjp.0704569

Costa, E. D., Rezende, B. A., Cortes, S. F., & Lemos, V. S. (2016). Neuronal nitric oxide synthase in vascular physiology and diseases. *Frontiers in Physiology, 7*(206). doi:10.3389/fphys.2016.00206

Dawson, T. M., Dawson, V. L., & Snyder, S. H. (1992). A novel neuronal messenger molecule in brain: the free radical, nitric oxide. *Ann Neurol, 32*(3), 297–311. doi:10.1002/ana.410320302

Mannick, J. B., & Schonhoff, C. M. (2004). Review NO means no and yes: regulation of cell signaling by protein nitrosylation. *Free Radical Research, 38*(1), 1–7. doi:10.1080/10715760310001629065

Britton, G. (2014). NO-independent modulation of soluble guanlylcyclase (sGC) activity and function. Retrieved from www.researchgate.net/publication/299305604_NO-INDEPENDENT_MODULATION_OF_SOLUBLE_GUANYLYL_CYCLASE_sGC_ACTIVITY_AND_FUNCTION

Derbyshire, E. R., & Marletta, M. A. (2012). Structure and regulation of soluble guanylate cyclase. *Annu Rev Biochem, 81*, 533–559. doi:10.1146/annurev-biochem-050410-100030

Radi, R. (2018). Oxygen radicals, nitric oxide, and peroxynitrite: Redox pathways in molecular medicine. *Proceedings of the National Academy of Sciences, 115*(23), 5839. doi:10.1073/pnas.1804932115

Gómez, R., Caballero, R., Barana, A., Amorós, I., Calvo, E., López, J. A., . . . Delpón, E. (2009). Nitric oxide increases cardiac IK1 by nitrosylation of cysteine 76 of Kir2.1 channels. *Circulation Research, 105*(4), 383–392. doi:10.1161/CIRCRESAHA.109.197558

Choi, Y.-B., Tenneti, L., Le, D. A., Ortiz, J., Bai, G., Chen, H.-S. V., & Lipton, S. A. (2000). Molecular basis of NMDA receptor-coupled ion channel modulation by S-nitrosylation. *Nature Neuroscience, 3*(1), 15–21. doi:10.1038/71090

Nitric oxide and affective disorder

Zhou, Q.-G., Zhu, X.-H., Nemes, A. D., & Zhu, D.-Y. (2018). Neuronal nitric oxide synthase and affective disorders. *IBRO reports, 5*, 116–132. doi:10.1016/j.ibror.2018.11.004

Suzuki, E., Yagi, G., Nakaki, T., Kanba, S., & Asai, M. (2001). Elevated plasma nitrate levels in depressive states. *Journal of Affective Disorders, 63*(1), 221–224. doi:https://doi.org/10.1016/S0165-0327(00)00164-6

Savas, H. A., Gergerlioglu, H. S., Armutcu, F., Herken, H., Yilmaz, H. R., Kocoglu, E., . . . Akyol, O. (2006). Elevated serum nitric oxide and superoxide dismutase in euthymic bipolar patients: Impact of past episodes. *The World Journal of Biological Psychiatry, 7*(1), 51–55. doi:10.1080/15622970510029993

Luiking, Y. C., Engelen, M. P. K. J., & Deutz, N. E. P. (2010). Regulation of nitric oxide production in health and disease. *Current Opinion in Clinical Nutrition and Metabolic Care, 13*(1), 97–104. doi:10.1097/MCO.0b013e328332f99d

Kim, Y.-K., Paik, J.-W., Lee, S.-W., Yoon, D., Han, C., & Lee, B.-H. (2006). Increased plasma nitric oxide level associated with suicide attempt in depressive patients. *Progress in Neuro-Psychopharmacology and Biological Psychiatry, 30*(6), 1091–1096. doi:https://doi.org/10.1016/j.pnpbp.2006.04.008

Jefferys, D., & Funder, J. (1996). Nitric oxide modulates retention of immobility in the forced swimming test in rats. *European Journal of Pharmacology, 295*(2), 131–135. doi:https://doi.org/10.1016/0014-2999(95)00655-9

Olfson, M., Marcus, S. C., & Shaffer, D. (2006). Antidepressant drug therapy and suicide in severely depressed children and adults: A case-control study. *Arch Gen Psychiatry, 63*(8), 865–872. doi:10.1001/archpsyc.63.8.865

Kim, Y.-K., Paik, J.-W., Lee, S.-W., Yoon, D., Han, C., & Lee, B.-H. (2006). Increased plasma nitric oxide level associated with suicide attempt in depressive patients. *Progress in Neuro-Psychopharmacology and Biological Psychiatry*, *30*(6), 1091–1096. doi:https://doi.org/10.1016/j.pnpbp.2006.04.008

Xing, G., Chavko, M., Zhang, L. X., Yang, S., & Post, R. M. (2002). Decreased calcium-dependent constitutive nitric oxide synthase (cNOS) activity in prefrontal cortex in schizophrenia and depression. *Schizophr Res*, *58*(1), 21–30. doi:10.1016/s0920-9964(01)00388-7

Dhir, A., & Kulkarni, S. K. (2011). Nitric oxide and major depression. *Nitric Oxide*, *24*(3), 125–131. doi:10.1016/j.niox.2011.02.002

Ghasemi, M. (2019). Nitric oxide: Antidepressant mechanisms and inflammation. *Adv Pharmacol*, *86*, 121–152. doi:10.1016/bs.apha.2019.04.004

Ghasemi, M., Claunch, J., & Niu, K. (2019). Pathologic role of nitrergic neurotransmission in mood disorders. *Progress in Neurobiology*, *173*, 54–87. doi:https://doi.org/10.1016/j.pneurobio.2018.06.002

Oliveira, R. M., Guimarães, F. S., & Deakin, J. F. (2008). Expression of neuronal nitric oxide synthase in the hippocampal formation in affective disorders. *Braz J Med Biol Res*, *41*(4), 333–341. doi:10.1590/s0100-879x2008000400012

Rao, J. S., Harry, G. J., Rapoport, S. I., & Kim, H. W. (2010). Increased excitotoxicity and neuroinflammatory markers in postmortem frontal cortex from bipolar disorder patients. *Mol Psychiatry*, *15*(4), 384-392. doi:10.1038/mp.2009.47

Cepeda, M. S., Stang, P., & Makadia, R. (2016). Depression is associated with high levels of C-reactive protein and low levels of fractional exhaled nitric oxide: Results from the 2007–2012 National Health and Nutrition Examination Surveys. *J Clin Psychiatry*, *77*(12), 1666–1671. doi:10.4088/JCP.15m10267

Tanda, K., Nishi, A., Matsuo, N., Nakanishi, K., Yamasaki, N., Sugimoto, T., . . . Miyakawa, T. (2009). Abnormal social behavior, hyperactivity, impaired remote spatial memory, and increased D1-mediated dopaminergic signaling in neuronal nitric oxide synthase knockout mice. *Molecular brain*, *2*(1), 19. doi:10.1186/1756-6606-2-19

Zhou, Q. G., Hu, Y., Hua, Y., Hu, M., Luo, C. X., Han, X., . . . Zhu, D. Y. (2007). Neuronal nitric oxide synthase contributes to chronic stress-induced depression by suppressing hippocampal neurogenesis. *J Neurochem*, *103*(5), 1843–1854. doi:10.1111/j.1471-4159.2007.04914.x

Nitric oxide and Parkinson's disease

Jiménez-Jiménez, F. J., Alonso-Navarro, H., Herrero, M., Garcia-Martin, E., & Agúndez, J. (2016). An update on the role of nitric oxide in the neurodegenerative processes of Parkinson's disease. *23*, 2666-2679. doi:10.2174/092986732366616081

Bayır, H., Kagan, V. E., Clark, R. S. B., Janesko-Feldman, K., Rafikov, R., Huang, Z., . . . Kochanek, P. M. (2007). Neuronal NOS-mediated nitration and inactivation of manganese superoxide dismutase in brain after experimental and human brain injury. *Journal of Neurochemistry, 101*(1), 168–181. doi:https://doi.org/10.1111/j.1471-4159.2006.04353.x

Pacher, P., Beckman, J. S., & Liaudet, L. (2007). Nitric oxide and peroxynitrite in health and disease. *Physiological Reviews, 87*(1), 315–424. doi:10.1152/physrev.00029.2006

Picón-Pagès, P., Garcia-Buendia, J., & Muñoz, F. J. (2019). Functions and dysfunctions of nitric oxide in brain. *Biochimica et Biophysica Acta (BBA) - Molecular Basis of Disease, 1865*(8), 1949–1967. doi:https://doi.org/10.1016/j.bbadis.2018.11.007

Hunot, S., Boissière, F., Faucheux, B., Brugg, B., Mouatt-Prigent, A., Agid, Y., & Hirsch, E. C. (1996). Nitric oxide synthase and neuronal vulnerability in Parkinson's disease. *Neuroscience, 72*(2), 355–363. doi:10.1016/0306-4522(95)00578-1

Eve, D. J., Nisbet, A. P., Kingsbury, A. E., Hewson, E. L., Daniel, S. E., Lees, A. J., . . . Foster, O. J. (1998). Basal ganglia neuronal nitric oxide synthase mRNA expression in Parkinson's disease. *Brain Res Mol Brain Res, 63*(1), 62–71. doi:10.1016/s0169-328x(98)00259-9

Chalimoniuk, M., Langfort, J., Lukacova, N., & Marsala, J. (2004). Upregulation of guanylyl cyclase expression and activity in striatum of MPTP-induced parkinsonism in mice. *Biochem Biophys Res Commun, 324*, 118–126. doi:10.1016/j.bbrc.2004.09.028

Schulz, J. B., Matthews, R. T., Muqit, M. M., Browne, S. E., & Beal, M. F. (1995). Inhibition of neuronal nitric oxide synthase by 7-nitroindazole protects against MPTP-induced neurotoxicity in mice. *J Neurochem, 64*(2), 936–939. doi:10.1046/j.1471-4159.1995.64020936.x

Liberatore, G. T., Jackson-Lewis, V., Vukosavic, S., Mandir, A. S., Vila, M., McAuliffe, W. G., . . . Przedborski, S. (1999). Inducible nitric oxide synthase stimulates dopaminergic neurodegeneration in the MPTP model of Parkinson disease. *Nat Med, 5*(12), 1403–1409. doi:10.1038/70978

Gupta, S. P., Kamal, R., Mishra, S. K., Singh, M. K., Shukla, R., & Singh, M. P. (2016). Association of polymorphism of neuronal nitric oxide synthase gene with risk to Parkinson's Disease. *Mol Neurobiol, 53*(5), 3309–3314. doi:10.1007/s12035-015-9274-3

Levecque, C., Elbaz, A., Clavel, J., Richard, F., Vidal, J. S., Amouyel, P., . . . Chartier-Harlin, M. C. (2003). Association between Parkinson's disease and polymorphisms in the nNOS and iNOS genes in a community-based case-control study. *Hum Mol Genet, 12*(1), 79–86. doi:10.1093/hmg/ddg009

Sunico, C. R., Nakamura, T., Rockenstein, E., Mante, M., Adame, A., Chan, S. F., . . . Lipton, S. A. (2013). S-Nitrosylation of parkin as a novel regulator of p53-mediated neuronal cell death in sporadic Parkinson's disease. *Mol Neurodegener, 8*, 29. doi:10.1186/1750-1326-8-29

Stefanis, L. (2012). α-Synuclein in Parkinson's disease. *Cold Spring Harbor Perspectives in Medicine, 2*(2), a009399-a009399. doi:10.1101/cshperspect.a009399

Kumar, A., Leinisch, F., Kadiiska, M. B., Corbett, J., & Mason, R. P. (2016). Formation and implications of alpha-synuclein radical in maneb- and paraquat-induced models of Parkinson's disease. *Mol Neurobiol, 53*(5), 2983–2994. doi:10.1007/s12035-015-9179-1

Zhang, G., Xia, Y., Wan, F., Ma, K., Guo, X., Kou, L., . . . Wang, T. (2018). New perspectives on roles of alpha-synuclein in Parkinson's disease. *Frontiers in Aging Neuroscience, 10*, 370. Retrieved from www.frontiersin.org/article/10.3389/fnagi.2018.00370

Paxinou, E., Chen, Q., Weisse, M., Giasson, B. I., Norris, E. H., Rueter, S. M., . . . Ischiropoulos, H. (2001). Induction of α-synuclein aggregation by intracellular nitrative insult. *The Journal of Neuroscience, 21*(20), 8053–8061. doi:10.1523/jneurosci.21-20-08053.2001

Nitric oxide and vascular diseases

Balez, R., & Ooi, L. (2016). Getting to NO Alzheimer's Disease: Neuroprotection versus neurotoxicity mediated by nitric oxide. *Oxidative medicine and cellular longevity, 2016*, 3806157–3806157. doi:10.1155/2016/3806157

Lee, L., Boorman, L., Glendenning, E., Christmas, C., Sharp, P., Redgrave, P., . . . Howarth, C. (2019). Key Aspects of neurovascular control mediated by specific populations of inhibitory cortical interneurons. *Cerebral Cortex, 30*(4), 2452–2464. doi:10.1093/cercor/bhz251

Shabir, O., Berwick, J., & Francis, S. E. (2018). Neurovascular dysfunction in vascular dementia, Alzheimer's and atherosclerosis. *BMC Neuroscience, 19*(1), 62. doi:10.1186/s12868-018-0465-5

Costa, E. D., Rezende, B. A., Cortes, S. F., & Lemos, V. S. (2016). Neuronal nitric oxide synthase in vascular physiology and diseases. *Frontiers in Physiology, 7*(206). doi:10.3389/fphys.2016.00206

Pacher, P., Beckman, J. S., & Liaudet, L. (2007). Nitric oxide and peroxynitrite in health and disease. *Physiological Reviews*, *87*(1), 315–424. doi:10.1152/physrev.00029.2006

Montécot, C., Borredon, J., Seylaz, J., & Pinard, E. (1997). Nitric oxide of neuronal origin is involved in cerebral blood flow increase during seizures induced by kainate. *J Cereb Blood Flow Metab*, *17*(1), 94–99. doi:10.1097/00004647-199701000-00012

Hudetz, A. G., Shen, H., & Kampine, J. P. (1998). Nitric oxide from neuronal NOS plays critical role in cerebral capillary flow response to hypoxia. *Am J Physiol*, *274*(3), H982–989. doi:10.1152/ajpheart.1998.274.3.H982

Mattson, D. L., & Meister, C. J. (2005). Renal cortical and medullary blood flow responses to L-NAME and ANG II in wild-type, nNOS null mutant, and eNOS null mutant mice. *Am J Physiol Regul Integr Comp Physiol*, *289*(4), R991–997. doi:10.1152/ajpregu.00207.2005

Huang, A., Sun, D., Yan, C., Falck, J. R., & Kaley, G. (2005). Contribution of 20-HETE to augmented myogenic constriction in coronary arteries of endothelial NO synthase knockout mice. *Hypertension*, *46*(3), 607–613. doi:10.1161/01.HYP.0000176745.04393.4d

Seddon, M. D., Chowienczyk, P. J., Brett, S. E., Casadei, B., & Shah, A. M. (2008). Neuronal nitric oxide synthase regulates basal microvascular tone in humans in vivo. *Circulation*, *117*(15), 1991–1996. doi:10.1161/circulationaha.107.744540

Silva, G. C., Silva, J. F., Diniz, T. F., Lemos, V. S., & Cortes, S. F. (2016). Endothelial dysfunction in DOCA-salt-hypertensive mice: role of neuronal nitric oxide synthase-derived hydrogen peroxide. *Clin Sci (Lond)*, *130*(11), 895–906. doi:10.1042/cs20160062

Iadecola, C. (2013). The pathobiology of vascular dementia. *Neuron*, *80*(4), 844–866. doi:10.1016/j.neuron.2013.10.008

Choi, S., Won, J. S., Carroll, S. L., Annamalai, B., Singh, I., & Singh, A. K. (2018). Pathology of nNOS-expressing GABAergic neurons in mouse model of Alzheimer's disease. *Neuroscience*, *384*, 41–53. doi:10.1016/j.neuroscience.2018.05.013

Zhu, X., Smith, M. A., Honda, K., Aliev, G., Moreira, P. I., Nunomura, A., . . . Perry, G. (2007). Vascular oxidative stress in Alzheimer disease. *Journal of the Neurological Sciences*, *257*(1–2), 240–246. doi:10.1016/j.jns.2007.01.039

10

NEURONS OUTSIDE THE CNS AND INTERACTIONS BETWEEN NEUROTRANSMITTER MECHANISMS

Chapter outline

In this chapter we will cover:

- Outline of the enteric nervous system:
    - The peristaltic reflex
    - Neurons and neurotransmitters involved in the peristaltic reflex
    - Other neurotransmitters involved in enteric nervous system function
- Enteric nervous system dysfunction
    - Irritable bowel syndrome (IBS)
    - The microbiome and alteration in neurotransmitters
    - Enteric nervous system dysfunction and 'big brain' dysfunction
- Interactions between neurotransmitter mechanisms
    - Interactions between substance p and other neurotransmitters
    - NOS, SERT and NMDA receptors
    - Cholinergic neurons and dopamine
    - Interactions between GABA and glutamate - E/I balance

Introduction: Neurons outside the central nervous system

In this book we have focused heavily on neurons within the central nervous system (CNS) and predominantly within the brain. There has also been the occasional mention of groups of neurons outside the CNS, such as discussion focused on the dorsal root ganglion when discussing pain research. In this chapter we will look at the largest collection of neurons outside the CNS; these are the neurons of the enteric nervous system. The enteric nervous system (ENS) includes a large variety of neurons, and therefore neurotransmitter-related mechanisms, and is typically referred to as the 'little brain'.

Outline of the enteric nervous system

So where is the enteric nervous system? Put simply, it lines the GI tract (gastrointestinal tract). The GI tract is a collective name for all the structures that process ingested material, moving from the mouth, down through the stomach and intestine to finally be excreted after the necessary nutrients have been removed and absorbed. The GI tract is generally considered to consist of two sections, namely the upper GI tract and the lower GI tract. The upper GI tract is the mouth down to the stomach and the lower GI tract is what we generally refer to as the gut. This starts after the pyloric sphincter of the stomach and ultimately terminates at the anus. Before going further, try to label the diagram in Figure 10.1 to clarify for yourself what structures are found in the GI tract.

──Test Yourself 10.1──

Below is a schematic of the GI tract, which is generally segregated into the upper and lower GI tract. As we are discussing the enteric nervous system in this chapter, we are more interested in the lower GI tract. Have a go at labelling the diagram in Figure 10.1, to ensure you are aware of the components of the GI tract.

Table 10.1

Oesophagus	Stomach	Large intestine
Small intestine	Rectum	Pyloric sphincter of the stomach. This is generally considered the dividing line between the upper GI tract and the lower GI tract

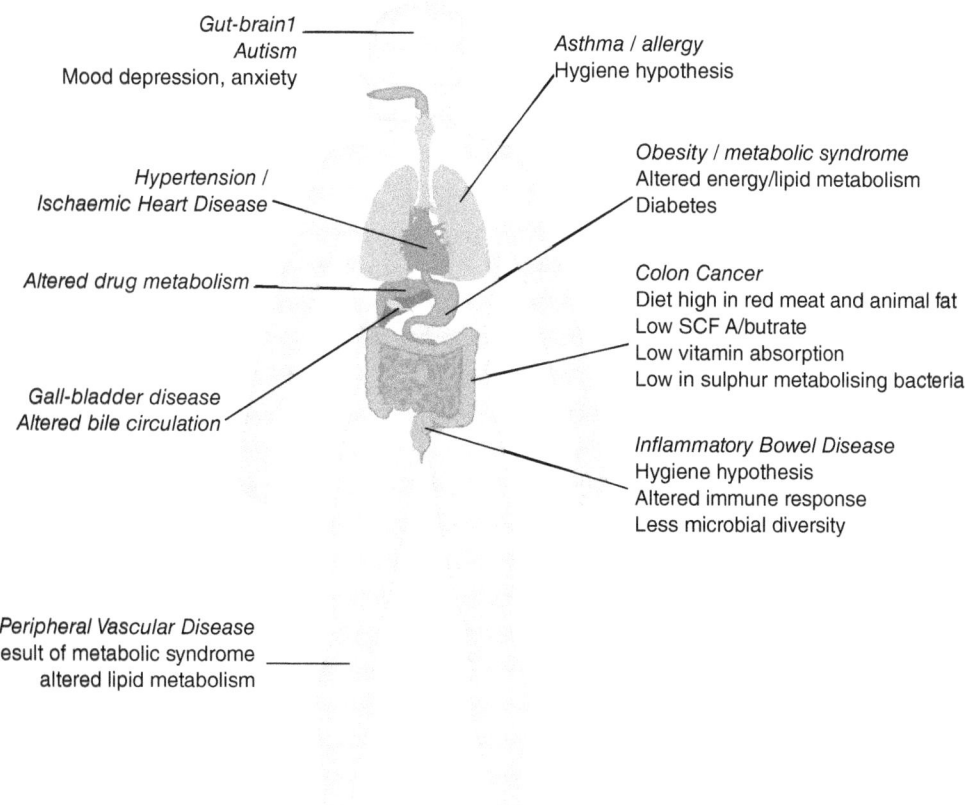

Figure 10.1   Schematic of the GI tract

Source: Boore, J.R.P., Cook, N. and Shepherd, A. (2016) *Essentials of Anatomy and Physiology for Nursing Practice*. London: SAGE Publications Ltd.

Answers to Test Yourself sections are at the end of the chapter.

So, what is the function of the GI tract and how do neurons and neurotransmission control it? Well, essentially the GI tract's key aim is to ingest food, break down this food to release nutrients and eventually excrete any waste left over. We will discuss this process next. But an important point to note is that it is controlled by the collection of neurons that form the enteric nervous system. It is also very important to note that the enteric nervous system is largely autonomous in its function. What do I mean by this? I mean that the 'big brain' up in the CNS has little impact on how the ENS functions. This is to such a degree that many refer to the enteric nervous system as the 'little brain'. This is an important point as for many years it was believed that the ENS was a slave system controlled by the 'big brain', via the vagal nerve that descends from the 'big brain' and innervates the GI tract and ENS at various points. However, this is very much a misnomer as most of the vagal nerve fibres actually ascend from the gut to the brain. It is believed to be about 80:20 in favour of the ascending fibres. Thus, the vast majority of signals being conducted by the vagal nerve are travelling to the brain and not from it. We will discuss the significance of this later in the chapter in the context of 'big brain' dysfunction.

Peristaltic reflex

Peristaltic reflex is the mechanism by which the gut moves content through and along the lower GI tract. It is essentially a series of wave-like muscle contractions, and is controlled by a complex local circuit of neurons within the enteric nervous system. It was originally identified by Bayliss and Starling (1899) and has become known as the 'Law of the Intestine'. It involves the contraction of muscles at one end of the gut with corresponding relaxation of muscle at the other end of the gut. At the same time, longitudinal muscles are triggered by neuronal mechanisms which help to push the contents of the gut forward, ultimately towards expulsion at the anus.

So, how are neurons and neurotransmission involved? Well, the peristaltic reflex is triggered primarily by distention. This is the swelling of the gut that is the result of ingested material travelling from the stomach into the intestine. This distention initially triggers endochromaffin (EC) cells that line the internal surface of the gut wall. These are not neurons but synthesise and release the monoamine neurotransmitter serotonin. It is at this point that the neurons and neurotransmitters of the enteric nervous system become intimately involved in peristalsis. We will return to this in a second.

First, it is important to consider where all these neurons are found? The obvious answer is within the enteric nervous system. The enteric nervous system lines the walls of the lower GI tract. However, it is not one homogeneous layer of tissue and is generally split into around five layers.[1] The first layer is the inner wall of the gut and is called the mucosa. Next is the submucosa, then a band of circular muscle, the myenteric plexus, and finally a band of longitudinal muscle. For further clarity on the structure layers of the enteric nervous system see Figures 10.2.

---

[1]Although only two are really considered the enteric nervous system. These are the submucosa and the myenteric plexus.

Neurons and neurotransmitters involved in peristalsis

Before we discuss the mechanisms, it is worth taking a moment to think about how complex these are in the CNS. You may think that because the moniker of the ENS is the 'little brain' that the number of neurotransmitters, and consequently their mechanisms, may be simpler here. This is not the case. There is a significant level of complexity within the ENS, which is keeping researchers actively occupied in understanding the basic aspects of its physiology.

So, previously we discussed how peristalsis is triggered by distention and how this results in endochromaffin cells (ECs) releasing serotonin. It is at this point that the ENS begins its functional involvement. The release of serotonin from the ECs targets receptors expressed on a group of neurons called intrinsic primary afferent neurons (iPANS). The soma of these neurons is found in the submucosa, although they project deep into the mucosa where they receive this signal. It is generally considered that the serotonin receptors expressed on these iPANs are 5HT3 and 5HT4 receptors.[2] Once bound, this triggers neurotransmitter release from the iPANs. IPANs release a range of neurotransmitters/neuromodulators, but the key player at this stage is acetylcholine. IPANs can directly innervate motor neurons at this point, which innervate the circular and longitudinal muscle of the gut, causing contraction of the muscle. However, the vast majority of iPANs split their dendritic arbours, projecting orally[3] and aborally. The projections innervate ascending and descending interneurons. Orally and aborally, these interneurons ultimately convey the signal to motor neurons, which act as affectors on the circular and longitudinal muscle. It is wise to think of these interneurons as there are often a number connecting the iPANs to the motor neurons and they release a wide variety of different neuromodulators, including acetylcholine, substance P and enkephalin. What this means is that, ultimately, the interneurons can heavily modulate the signal before it arrives at the motor neurons. This can result in different contraction patterns and different types of rhythmic motility within the gut. Their basic function, however, is to activate the motor neurons. Orally, this causes the motor neurons to release acetylcholine and substance P into the muscle. These both have generally excitatory effects on the circular muscle, causing it to contract, narrowing the circumference of the gut. Aborally, these motor neurons release a different selection of neurotransmitters, including the gaseous neurotransmitter nitric oxide and vasointestinal peptide (VIP). This causes the circular muscle aborally to relax.

---

[2] This is interesting as 5HT4 are G-protein coupled, while 5HT3 are ligand-gated ion channels. This means there will be both fast and slow excitatory effects on the iPANs in response to serotonin binding.

[3] Orally means towards the mouth.

As this is all happening iPANs and interneurons orally and aborally are also stimulating motor neurons that innervate the longitudinal muscle. This causes the longitudinal muscle to contract. Contractions in the longitudinal muscle push the material in the gut (that caused the distention) forward. Because the circular muscle at the oral end has contracted, it means that the waste material is pushed in the only direction it can move, which is aborally. Ultimately this happens along the whole length of the lower GI tract and results in excretion of the waste material.

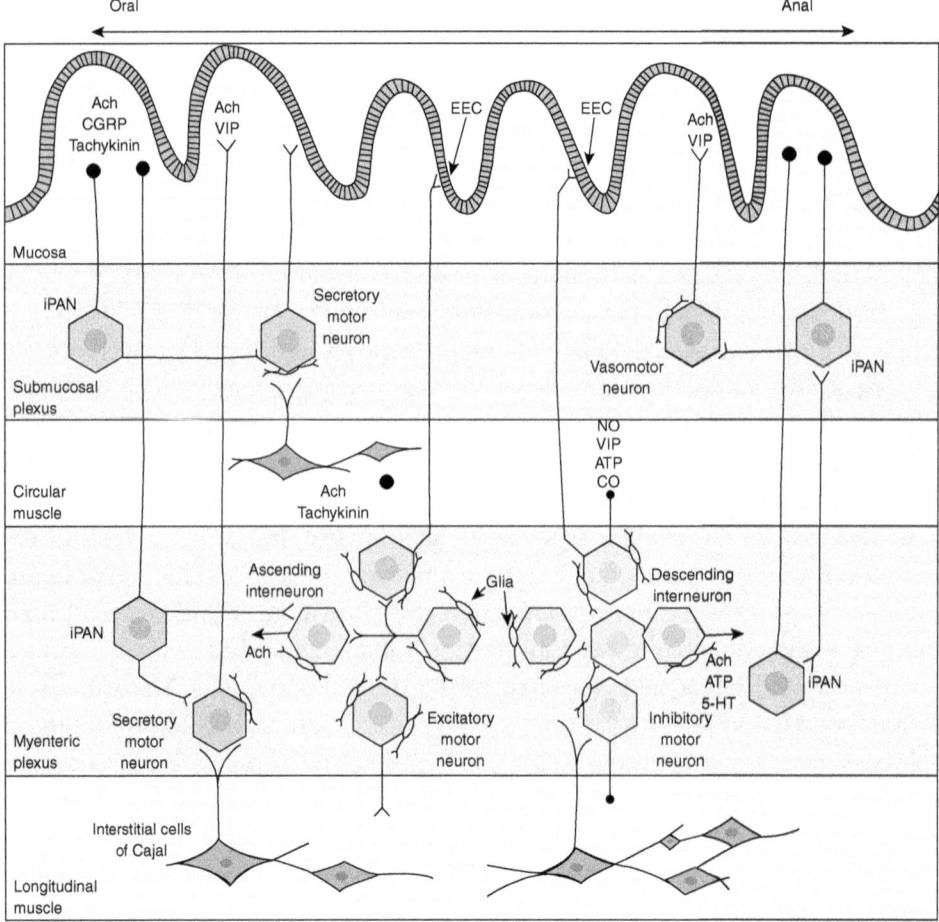

Figure 10.2 A schematic of the layers of the ENS in the intestine

The internal wall of the gut is at the top of the image. As can be seen in the figure, the neuronal makeup of the ENS is complex with a diverse array of neurons playing different roles and expressing a range of different neurotransmitters. IPAN = intrinsic primary afferent neuron; EEC = enteroendocrine cell; Ach = acetylcholine; CGRP = calcitonin gene-related peptide; VIP = vasoactive intestinal peptide; ATP = adenosine triphosphate; NO = nitric oxide; CO = carbon monoxide; 5-HT = serotonin.

Source: Figure and abbreviations are from: Fleming, M. A., Ehsan, L., Moore, S. R., & Levin, D. E. (2020). The enteric nervous system and its emerging role as a therapeutic target. *Gastroenterology Research and Practice, 2020*, 8024171. doi:10.1155/2020/8024171. Copyright © 2020 Mark A. Fleming II et al., licensed under CC BY 4.0.

Other neurotransmitters involved in ENS function

As can be seen in Figure 10.2, there is a wide variety of neuron types, neurotransmitters and consequently receptors actively involved in the function of the ENS. However, the reality is that even Figure 10.2 is somewhat of a simplification, especially when it comes to the neurotransmitters active within the circuits of the ENS. Why not test yourself to identify the neurotransmitters we have named so far?

Test Yourself 10.2

Without looking back at the previous section, try to identify the neurotransmitters mentioned and match them with the types of neurons below. To start you off, one of the big players is **acetylcholine**.

Table 10.2

iPANs:

Interneurons:

Motor neurons:

So, what other neurotransmitters are involved in modulating the circuitry of the ENS? Well, the simple answer is that it is almost as complex as the 'big brain'. Therefore, all the neurotransmitters we have discussed in the preceding chapters of this book are synthesised and released in the ENS. There are also corresponding receptors for them all. What is interesting is that so much is still unknown about classic neurotransmitters in the GI tract, especially their involvement in peristalsis. For example, GABA and its receptors are known to be expressed by both neurons and endocrine cells within the GI tract, although it is still a matter of considerable debate whether GABA and its mechanisms are involved in GI motility, and if so, how. This is also the case for glutamate, with research identifying the expression of glutamate and glutamate receptors (APMA and NMDA) in the GI tract.

Let's now focus on a couple of neuron types to reveal a bit more about the complexity of ENS neurotransmission. Approximately two thirds of neurons in the myenteric plexus synthesise acetylcholine and around half of these also express substance P and neurokinin A. The other third of neurons in the myenteric plexus are known to express VIP and nitric oxide (NO), with some also expressing neuropeptide y. The above include motor neurons, iPANs and interneurons. There are, however, also neurons that synthesise enkephalin and somatostatin. These are believed to be expressed only by interneurons that innervate other interneurons and act as a modulatory system. Orally, enkephalin is co-expressed with acetylcholine and is generally thought to act as a negative feedback mechanism suppressing further activity in those interneurons. Aborally, somatostatin interneurons innervate enkephalin interneurons, but these two populations also separately innervate motor neurons and provide a mechanism to switch between excitatory input to inhibitory input. There is certainly a high level of complexity in the circuitry of the ENS and, importantly, still much key research to do to reveal the basic mechanisms of action.

There are also local neurons within the ENS that innervate the circuitry controlling peristalsis and therefore present other mechanisms of modulating the peristaltic reflex. One such group synthesise and release dopamine. These are only expressed in small numbers and the expression is much lower in the colon than it is in the upper GI tract. These neurons have become of some interest over recent years as studies have presented evidence to suggest they are reduced in patients with Parkinson's disease (PD), implying that there could be some role of these local neurons in PD degeneration. Another group of local neurons releases yet more serotonin. The serotonergic neurons generally target both 5HT4 and 5HT3 receptors. This is interesting as these receptors have somewhat opposing effects on the target neuron, with 5HT4 receptors being G-protein coupled, producing slow excitatory responses, while 5HT3 receptors are ligand-gated ion channels and are generally considered to produce fast excitatory responses. It is important to remember that these serotonergic neurons only account for a very small percentage of the serotonin within the ENS, with the majority[4] being synthesised by ECs. However, recent research evidence suggests they are extremely important for the development and survival of other local neurons within the ENS. One example is the research by Li and colleagues (2011), which suggests that a lack of neuronal serotonin, early in development, results in a loss of dopaminergic and GABAergic local neurons within the ENS. They also suggest that this has profound effects on GI tract motility. It serves to illustrate the common point throughout this book, namely that neurons and neurotransmitters interact and can have profound effects on one another, the behaviour of cells and the organism.

---

[4]Estimates suggesting approximately 95%.

Beyond neurons

Although this book is focused on neurons and neurotransmitters, it is important to acknowledge that there are many other cell types in the gut that play a role in the behaviour of neurons within the ENS and are known to synthesise and release neurotransmitters. The obvious example discussed above are ECs cells that line the mucosa of the lower GI tract, which release serotonin that directly targets receptors on iPANs. However, there are other cells of significance in terms of the function of the ENS and the peristaltic reflex. One well-researched example are Interstitial Cells of Cajal (ICC). These cells are generally located between nerve terminals and the smooth circular muscle of the gut. General belief is that these cells act as essential modulators in the production of muscular contraction and, ultimately, help produce a rhythm to motility within the gut. Many researchers think of these as essentially pacemaker cells.

Enteric nervous system dysfunction

ENS dysfunctions can be devastating and have a major impact on people's quality of life. Despite this, it is a somewhat under-researched area and is perhaps obviously not the most instantly appealing topic in science. Nevertheless, a fair body of research exists that clarifies how alterations in neurotransmitters' mechanisms may be significant causal factors in the dysfunction of the gut/ENS. Much of this research has understandably focused on identifying mechanisms that can be targeted for efficacious pharmaceutical intervention.

Irritable bowel syndrome

Irritable bowel syndrome (IBS) is commonly believed to have two types, which are referred to as IBS-C (IBS-Constipation type) and IBS-D (IBS-Diarrhoea type). Common clinical symptoms for both forms of IBS include alteration in the frequency of stools and the type of stools produced,[5] bloating and abdominal pain.

It seems most likely, due to its primary importance in triggering the peristaltic reflex via release from EC cells, that serotonin and its mechanism would be involved in the dysfunction of the gut, such as IBS. Indeed, there is a reasonable amount of literature showing that serum serotonin levels are altered in IBS, although there is

---

[5] Firm or loose.

difference between IBS-C and IBS-D, with increased levels in IBS-D and reduced levels in IBS-C. More specifically, there seems to be altered serotonin when measured more directly, although there is some contention as to whether this is an increase or a decrease. It is worth keeping in mind that it may not be neuronal mechanisms that cause this alteration in serotonin. Remember that approximately 95% of the body's serotonin is synthesised by EC cells, so perhaps it is an alteration in serotonin synthesised by EC cells that is the causal factor. The weight of the literature seems to suggest that this is not the case, with studies reporting no significant changes in numbers of EC cells and no alteration in the amount of serotonin released from them in those with the clinical symptoms of IBS.

So, is neuronal serotonin involved? A good example study by Li and colleagues (2011) highlights this involvement of neuronal serotonin in IBS. In this study, they found TPH2[6] knockout animals had marked reductions in intestinal transit, but also greatly increased gastric emptying. This very much mirrors the typical clinical symptoms seen in humans with an IBS diagnosis. This study also implicates dopaminergic neuron degeneration as well as serotonin. Therefore, although the root cause may be seen as altered neuronal serotonin levels, the clinical symptoms of IBS may be mediated by alterations in other neurons and their respective neurotransmitters.

Although it is perhaps unlikely that one neurotransmitter alone is responsible for the clinical symptoms of IBS, as evidenced above, there is certainly significant evidence for the involvement of altered neuronal serotonin in IBS. One common serotonin-related mechanism that seems to be altered in IBS is the serotonin reuptake transporter (SERT). Studies, such as that by Coates et al. (2004), have found that SERT levels seem to be reduced in both IBS-D and IBS-C. Research focused on genetic polymorphisms at the population level has provided some support for the suggestion that SERT alterations underpin IBS. For example, an early study by Pata et al. (2002) suggested that there were genotype differences between the groups with an IBS-C diagnosis and those with an IBS-D diagnosis. Specifically, this related to the SERT-LPR (SERT-linked gene polymorphism region), which is known to have a combination of short and long forms, with two short forms being more prominent in those with IBS-C. Although more recent research supports alteration in SERT-LPR in those with IBS, the findings related to long and short forms, relative to IBS-C and IBS-D, are not consistent. So, although there seems to be a definite role for altered SERT in IBS, what role it plays is yet to be conclusively established. Recent research suggests that it likely contributes to inflammation and immune responses in the gut, which in turn result in changes that produce the symptoms of IBS.

---

[6]Tryptophan hydroxylase 2 – this is the neuronal form of the enzyme that helps synthesise serotonin.

The microbiome and alteration in neurotransmitters

An emergent topic in recent years within the research literature is the role of the gut microbiome. The gut microbiome consists of a vast number of different types of bacteria. These bacteria synthesise a large variety of different compounds and molecules, some of which are believed to be beneficial for gut function and others are believed to be bad. Of interest for us is the fact that certain populations are known to synthesise and release a variety of neurotransmitters, including GABA and serotonin. For example, research by Yano et al. (2015) convincingly showed that certain populations of bacteria produce the neurotransmitter serotonin. It is worth stopping and thinking for a moment what this might mean. It could be that higher serotonin levels[7] may result in unnecessary, increased activation of iPANS and therefore the peristaltic response will be triggered without distention and the presence of material within the gut. The increased level of serotonin will also have an impact on the expression of serotonin receptors within the ENS, consequently altering circuit dynamics and resulting in imbalance. It is a reasonable question to ask how does the serotonin from the gut microbiome get into the ENS. Many people think that the starting point in bowel illnesses, such as IBS, may be a genetic propensity towards bowel inflammation (IBD, inflammatory bowel disease). This is hypothesised to cause increased permeability into the ENS and consequently an increased opportunity for products of the microbiome to interact and alter the circuitry of the ENS. However, it is also known that serotonin plays a pivotal role in immune responses and inflammation, so it may be the likely trigger for inflammation which makes the mucosa more permeable to the products of the microbiome.

As ever, it is also worth remembering that it is unlikely to all be about altered serotonin and its mechanisms. Another neurotransmitter we have previously discussed that is impacted by the microbiome is nitric oxide (NO). Research, such as that by Gangula et al. (2015), has shown that the presence of certain bacteria in the microbiome that are a consequence of periodontitis[8] can result in reduced levels of nNOS expression in the ENS. Nitric oxide is one of the key neurotransmitters at the neuro-muscular junction aborally, resulting in relaxation of the muscle at this point. Therefore, reduced nNOS levels will likely result in a reduced ability for the muscle to relax and for waste material to move smoothly through the lower GI tract.

---

[7]As a result of high levels of serotonin-synthesising bacteria in the gut.
[8]Infection in the gums/mouth.

So, where does all this research on the gut microbiome lead? Well, the simple answer is to potentially efficacious treatments for gut disease, such as IBS, by altering the microbiome. Fundamentally, this should mean with targeted antibiotics that can get to the gut, we can potentially tackle diseases of the gut. Currently, the majority of treatments for IBS target the serotonergic system of the ENS and, although efficacious to some degree, these can be seen as perhaps tackling the consequence rather than the cause.

Enteric nervous system dysfunction and 'big brain' dysfunction

For many years there has been a well-known relationship between 'big brain' dysfunction and GI tract dysfunction, with many people diagnosed with a number of developmental, degenerative or psychiatric disorders[9] displaying altered gut function. The general consensus was that this was largely a consequence of alterations in the brain. However, interesting research has emerged that suggests that gut dysfunction may precede 'big brain' dysfunction and therefore may be the 'cause' of 'big brain' dysfunction, rather than the consequence.

Research in many fields, such as research on autism spectrum disorder (ASD), Alzheimer's disease (AD), schizophrenia and depression, to name a few, has suggested that alterations in the function of the ENS and the population of bacteria that constitute the microbiome within the gut seem to be a causal factor in the development of these conditions. Examples of research include that by Sharon et al. (2019). This study involved injecting faeces from humans who had a diagnosis of ASD into mice, which were then bred and their offspring were measured for ASD-like traits, such as increased repetitive behaviours and altered social interactions. Another study focused on Parkinson's disease (Kim et al., 2019) and involved injections of mutated alpha-synuclein[10] into the lower GI tract of mice. After injections, these alpha-synuclein molecules were found in the brains of the mice and these animals developed both motor and non-motor symptomology of Parkinson's disease. Although very much at the start of research in this area, there is some good-quality research that draws very solid conclusions. To help develop your further reading in this area, I strongly recommend the following primary research papers:

---

[9]Including autism spectrum disorder (ASD), Parkinson's disease and schizophrenia.

[10]This is a protein that is known to be in Lewy bodies. Lewy bodies are a biological hallmark, found in the brains of those with a Parkinson's disease diagnosis.

1. Guida, F., Turco, F., Iannotta, M., De Gregorio, D., Palumbo, I., Sarnelli, G., & Maione, S. (2018). Antibiotic-induced microbiota perturbation causes gut endocannabinoidome changes, hippocampal neuroglial reorganization and depression in mice. *Brain, Behavior, and Immunity, 67*, 230–245. doi:10.1016/j.bbi.2017.09.001
2. Sharon, G., Cruz, N. J., Kang, D.-W., Gandal, M. J., Wang, B., Kim, Y.-M., & Mazmanian, S. K. (2019). Human gut microbiota from autism spectrum disorder promote behavioral symptoms in mice. *Cell, 177*(6), 1600–1618.e1617. doi:https://doi.org/10.1016/j.cell.2019.05.004
3. Kim, S., Kwon, S. H., Kam, T. I., Panicker, N., Karuppagounder, S. S., Lee, S., & Ko, H. S. (2019). Transneuronal propagation of pathologic alpha-Synuclein from the gut to the brain models Parkinson's disease. *Neuron, 103*(4), 627–641.e627. doi:10.1016/j.neuron.2019.05.035

For more research reviews on this topic, which should help give you a wider appreciation of the research that has currently been conducted into this topic, see the reference list.

Focus on research methods

Searching for relevant literature

One of the most important skills to develop in the final years of an undergraduate degree and during post-graduate study is effective literature searches. This will help you to understand the current state of the literature, develop an appreciation for the complexity of the literature, allow you to identify gaps in the literature that you could focus your research on, and identify the key players in terms of research groups. A great place to start searching for primary research and up-to-date literature reviews is Pubmed.gov: https://pubmed.ncbi.nlm.nih.gov/

Try putting the following search terms into Pubmed, followed by either ASD, schizophrenia, depression or Parkinson's disease:

Search terms: Microbiome, Gut, Brain.

It is important to remember to start a search with a small number of specific targeted words. You can select good search terms from reading books such as this.

Interactions between neurotransmitter mechanisms

In the first part of this final chapter, I have emphasised the fact that neurotransmitters and their mechanisms play an essential role in neuronal behaviour and the organism's behaviour as a whole, beyond the central nervous system. I have done this by exploring the roles of neurotransmitters in the enteric nervous system of the gut and considered how alterations in these mechanisms cause dysfunction of the gut, but also may contribute to 'big brain' dysfunction. The main reason for doing this is because we have concentrated on the brain in this book and it is important to remember that there are neurons throughout our body, which means that alterations in neuronal mechanisms can have profound effects on the organism outside the brain and CNS. Hopefully, the previous focus on the ENS has helped you appreciate this.

Next, I wish to spend a little time emphasising the importance of interactions between neurons, neurotransmitters and their mechanisms. As we have discussed in almost every section of this book, it is important to always remember that neurotransmitters and their respective mechanisms do not function in isolation. Indeed, the complexity of their interaction is the key thing that keeps many neuroscientists occupied as it not only often makes research difficult to conduct, but it also makes interpretation of the findings rich with caveats. In this section we are going to look at a few examples of interactions between different neurotransmitters and/or their mechanisms. The aim is really to solidify the message that things are complex when considering the effects of neurotransmitters on both the behaviour of the neuron and the behaviour of the whole organism.

Before we do this, it might be worth taking a moment to think about the content of this book so far by attempting the following Test Yourself section.

Test Yourself 10.3

Complete the following box with all the details you can remember. I have started you off by giving you the first letter(s) of each neurotransmitter we have covered.

Table 10.3

Neurotransmitters covered	Synthesising precursor and enzymes	Receptors
G		
GI		
A		
D		
S		
S P		
E O		
N O		

Interaction between substance P and other neurotransmitters

Tachykinin receptors are known to be expressed in high concentration in regions with large populations of serotonergic neurons, such as the raphe nuclei, and in regions with high populations of dopaminergic neurons, such as the ventral tegmental area (VTA). This is no coincidence and a number of studies have shown that substance P and its respective tachykinin receptors, especially NK1R and NK3R, mediate the activity of both dopaminergic and serotonergic neurons, resulting in modulation of the release of both monoaminergic neurotransmitters.

This is specifically known to be the case with regard to projections innervating the striatum, with NK1R agonist injected into the VTA producing increases in dopamine release in the striatum, and with immunohistochemical studies showing that these projections express a high level of NK1 receptors. This is particularly important as the striatum, especially its ventral eminence,[11] is well known to be involved in affective responses, reward-related behaviour and is often referred to as the main area hijacked by drugs of abuse. This, therefore, could be one contender for a mechanism subverted by drugs of abuse. However, it becomes more complicated as substance P seems to have a distinct effect on dopamine release within specific regions of the striatum. These regions, while not visible under the light microscope, become visible when you stain the striatum for molecules such as substance P or μ-opioid receptors (MORs) (see Figure 10.3 for an example immunohistochemical image). These regions are referred to as the striosomes and matrix.[12] Brimblecome and Cragg (2015, 2017) revealed that substance P agonists, especially NK1Rs, increase the release of dopamine within the striosomes but have no seeming effect on the release of dopamine within the matrix. This has key potentially functional implications as the striosomes are believed by some to govern different aspects of behaviour to the matrix. It is yet to be fully elucidated, but adds an interesting extra layer of complexity for the interaction of substance P, its receptors and dopamine in the modulation of behaviour, such as drug addiction.

---

[11]The nucleus accumbens.

[12]Striosomes stain heavily for MORs and substance P and appear like islands scattered within the matrix compartment, which does not stain heavily for MORs or substance P.

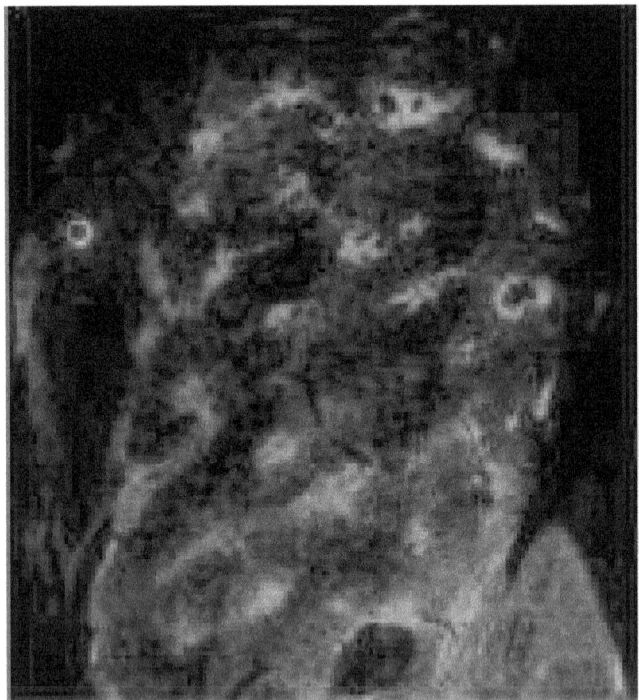

Figure 10.3   Image of the striatum labelled for MORs

The bright red areas are those areas that express high concentrations of MORs and are referred to as striosomes. The darker areas surrounding are referred to as the matrix.

Source: Reprinted from: Banghart, Matthew R., Neufeld, Shay Q., Wong, Nicole C., & Sabatini, Bernardo L. (2015). Enkephalin disinhibits mu opioid receptor-rich striatal patches via delta opioid receptor. *Neuron, 88*(6), 1227–1239. Copyright (2015), with permission from Elsevier.

Focus on research 🔍

Stephanie Cragg

Stephanie Cragg is a leading professor in the investigation of dopamine transmission and has published extensively on the topic, especially centred on its role in and modulation within the basal ganglia circuitry. She has been Professor of Neuroscience in the Department of Physiology, Anatomy and Genetics since 2014 and is highly influential.

I strongly recommend checking out her Orchid profile and her group's website at the University of Oxford. Both reveal the depth and breadth of her work:

https://orcid.org/0000-0001-9677-2256
www.dpag.ox.ac.uk/research/cragg-group

As mentioned above, tachykinin receptors are also expressed in high concentrations in regions other than the VTA and on neurons other than dopaminergic neurons. One specific region is the raphe nuclei.[13] It is not just that they are expressed here; they have also been shown to modulate serotonin release here, with studies using pharmacological antagonism of NK1 receptors and genetic deletion studies both finding increased serotonergic activity. It gets somewhat more complicated here, however, as a number of studies suggest that the modulation of serotoninergic neurons by NK1Rs in the raphe nuclei is indirect and a result of their modulation of local glutamatergic innervation. Therefore, the situation may be one where local NK1Rs, when activated, decrease activity in glutamatergic neurons, which results in reduced excitatory innervation by these neurons onto serotonergic populations. Serotonin, glutamate and substance P signalling have all been implicated in addictive processes. Thus, the full picture may lie in an interaction between these neurotransmitters as well as the previously outlined involvement of dopamine.

Cholinergic neurons and dopamine

For many years people have believed that dopamine release into the striatum had a primary importance in modulating reward-related behaviour. Recently, however, several studies have shown that dopamine's impact on reward-related behaviour is at least partially modulated by the interaction between dopaminergic projections into the striatum and striatal cholinergic interneurons. Despite only making up approximately 1% of the neuron populations found in the striatum, cholinergic interneurons (CINs) are typically found in close proximity to dopaminergic terminals. A number of researchers proposed the idea that dopaminergic and cholinergic neurons in the striatum acted as opposing mechanisms, switching striatal output neurons. This was based on the observations that at the postsynaptic site, dopamine and acetylcholine application has opposing effects on these outputs. However, more recently, research suggests that their joint activity may be more cooperative and facilitatory. One proposed mechanism of interaction centres on the known expression of nicotinic and muscarinic acetylcholine receptors on dopaminergic terminals within the striatum. Agonisms of both these receptors on the dopaminergic terminals have revealed that they modulate the firing pattern of dopaminergic neurons and alter the release of

---

[13]A major source of serotonergic neurons in the brain.

dopamine from these terminals, essentially acting as a fine-tuning mechanism. What is particularly interesting about this mechanism of dopamine modulation is that they are not homogeneous across the striatum. For example, nicotinic α6 receptors seem to play a much greater role in modulating dopamine in the dorsal striatum than they do in the ventral striatum. This suggests a potentially segregated role in specific types of behaviour.

It is not only within the striatum that dopaminergic and cholinergic neurons interact, and it is not only within the striatum that their interaction is known to modulate reward-related behaviour. Research evidence also points towards their interaction within the VTA. As previously mentioned, the VTA contains a large population of dopaminergic neurons that project into multiple territories. It has been a long-standing assertion that the activity of these neurons modulates reward-related behaviour. Recent evidence suggests that it is in fact cholinergic neurons that modulate these VTA dopaminergic neurons and, at the least, contribute to their involvement in reward-related behaviour. Several studies using optogenetic methods have shown that stimulating cholinergic neurons in the pedunculopontine (PPN), and laterodorsal tegmental nuclei (LDT) that innervate the VTA, results in alteration in the firing patterns of VTA dopaminergic neurons and modulates reward-related behaviour in animals. However, although it is evident that acetylcholine release from these neurons modulates dopaminergic neuron firing, many researchers have argued that this may be due to acetylcholine's modulation of a second neurotransmitter, which in turn then modulates dopaminergic neurons. However, recent evidence suggests that acetylcholine release into the VTA has direct effects on dopaminergic neurons. Research by Durand-de Cuttoli et al. (2018) suggests that this is specifically modulated by β2 containing nicotinic receptors expressed on the dopaminergic neurons. But what effect does this have on behaviour? Well, as discussed above, much research implicates this in reward-related behaviour. There is also a significant amount of research that suggests that this fine-tuning of dopaminergic neurons by nicotinic receptors is a key mechanism underpinning certain aspects of drug addiction, especially addiction to nicotine. This makes logical sense as drugs of abuse hijack the 'normal' reward circuits. The role of acetylcholine and nicotinic receptors in drug abuse presents novel and exciting mechanisms that can be targeted to modulate dopaminergic responses to drugs of abuse and potentially modulate the negative aspects of substance abuse.

Interaction between NOS and serotonin mechanisms

As mentioned above in the section focused on interactions between substance P and other neurotransmitters, serotonin is largely released by serotonergic neurons located in the raphe nuclei. If you cast your mind back, you will recall that we specifically dealt with serotonin in Chapter 6. In this chapter we extensively discussed its mechanisms and touched upon the role of SERTs (serotonin reuptake transporters) in bringing serotonin back into the presynaptic neuron from the extracellular space. SERTs have been of considerable research interest, especially to the pharmacology community, as one of the key drug treatments we have for affective disorders acts upon them. These are called selective serotonin reuptake inhibitors (SSRIs). However, as previously mentioned, these are not efficacious for all, and consequently alterative mechanisms to module serotonin have been explored. One interesting fact is that nNOS, the enzyme that synthesises nitric oxide (NO), can be found bound to the C terminus of SERTs, which are on their intracellular domain. What is most interesting about this is that the binding of nNOS to SERT's carboxyl terminus results in a reduced trafficking of the reuptake transporter to the cell membrane, as can be seen in Figure 10.4. So, what does this mean? Well, ultimately it should mean reduced reuptake of serotonin and increased serotonin in the extracellular space. However, it has also been shown that serotonin can drive the synthesis of NO in these neurons. So, although nNOS reduces the trafficking of SERTs to the membrane, the SERTs that get there aid the transport of serotonin back across the membrane and, in so doing, trigger the synthesis of NO. These interactions get even more complicated as the synthesis of NO triggers protein kinase G, which also modulates the action of SERTs. This interaction provides a lot of potentially novel mechanisms that could contribute towards the development of more efficacious antidepressants. Indeed, a number of researchers have explored this over the last ten years and have used a range of different compounds that act on NOSs. There is general agreement that NOS targeting has some efficacy in reducing the symptoms of depression. However, the issue is developing a specific compound, which specifically targets the correct type of NOS, as many believe the most efficacious NOS-targeting compounds might actually target other types of NOS, such as iNOS.

An interesting point about serotonergic neurons is that not all of them express nNOS. In fact, in areas such as the striatum, serotonergic neurons do not seem to express NO/nNOS. However, the serotonergic neurons that express nNOS have been found to be more sensitive to the neurotoxicity of amphetamine. These neurons make up about a third of the raphe nuclei serotonergic neurons and are generally seen to project cortically. This segregation could have interesting functional implications.

Finally, it is not just that nNOS is bound to serotonin transport. In a role reversal, serotonin receptors are found to be expressed on nNOS-expressing GABAergic interneurons. These are typically referred to as low threshold spike interneurons (LTSIs). Serotonin receptors are believed to inhibit these neurons, with this inhibition being mediated specifically by $5HT_{2C}$ receptors. Of course, it is not only these nNOS

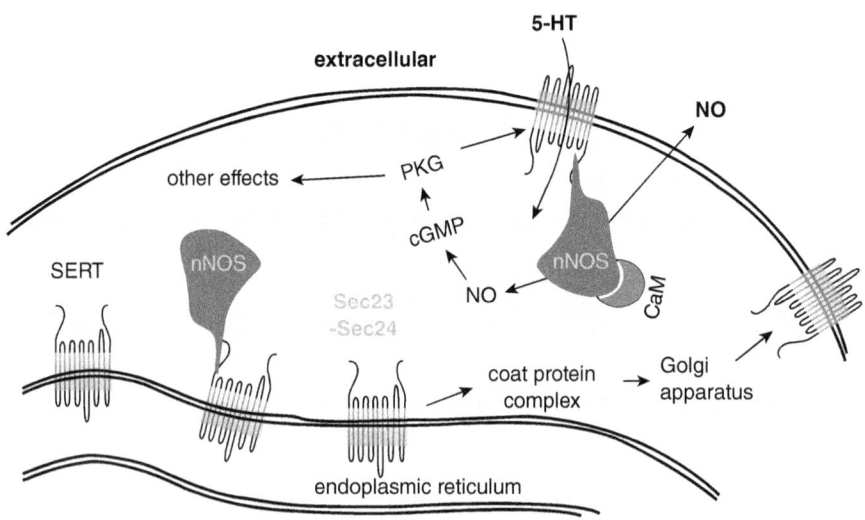

Figure 10.4 The interaction between nNOS and serotonin transporters

nNOS can be found bound to the intracellular domain of serotonin receptors, but, as illustrated in this figure, it is also known to be bound to the serotonin transporter (SERT). Activity at the reuptake channel can activate NO synthesis by nNOS and consequently NO can modulate the expression of serotonin reuptake transporters in the phospholipid membrane.

Source: Figure reprinted with permission from: Garthwaite, J. (2007). Neuronal nitric oxide synthase and the serotonin transporter get harmonious. *Proceedings of the National Academy of Sciences, 104*(19), 7739. doi:10.1073/pnas.0702508104. Copyright (2007) National Academy of Sciences, USA.

neurons that express serotonin receptors. For example, serotonin receptors are also expressed on PV+ (fast spiking) interneurons where they have been shown to have excitatory effects. It is worth keeping this two-way interaction in mind as it reveals much about the complexity of interaction between neurotransmitters.

NOS and glutamate (NMDA receptors) – spinal growth via interaction with NMDA receptors

You may recall from Chapter 9 on nitric oxide, that there are several subtypes of the enzyme that synthesises NO within the neuron (nNOS). Two of these sub-types are found free in the intracellular cytoplasm, while two are found bound to proteins (such as SERT, discussed above) and intracellular aspects of the neuron's phospholipid membrane. One specific membrane-bound protein nNOS is bound to are NMDA receptors, as can be seen in Figure 10.5. nNOS are bound to NMDA receptors by a PDS-95 (postsynaptic density protein[14]). If you cast your mind back, you may remember that these NMDA receptors are ionotropic glutamate receptors. The interaction between nNOS and NMDA receptors is believed to be an essential mechanism that enables NMDA receptors to modulate cellular long-term potentiation. As can be seen in Figure 10.5, one way this interaction may facilitate long-term potentiation is through NO indirectly triggering spine growth once synthesised as a response to glutamate binding to the NMDA receptor (Picón-Pagès, Garcia-Buendia, & Muñoz, 2019). This increase in spine growth increases the connectivity and interaction between the NMDA-expressing neuron and those neurons it is targeted by and targets.

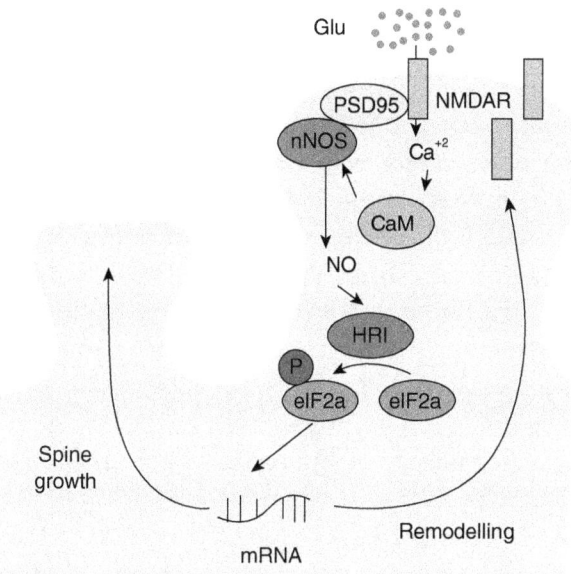

Figure 10.5   NO effect on spine protein translation

Glutamatergic signalling triggers NO release. NO activates the enzyme HRI that phosphorylates the eIF2-alpha. P-eIF2-alpha induces the translation of these mRNAs bearing several AUG in their 5'-UTR, which are proteins needed for the proper function of the spines and allowing their growth.

Source: Picón-Pagès, P., Garcia-Buendia, J., & Muñoz, F. J. (2019). Functions and dysfunctions of nitric oxide in brain. *Biochimica et Biophysica Acta (BBA) – Molecular Basis of Disease, 1865*(8), 1949–1967. doi:https://doi.org/10.1016/j.b badis.2018.11.007. Licensed under CC BY 4.0.

---

[14]This is what is referred to as a scaffolding protein.

As mentioned above, in the section on the interaction between nNOS and SERT, the interaction between these molecules is a potential hot topic for investigation of novel antidepressant therapeutics. Interestingly, the interaction between nNOS and NMDA receptors as a potential target for antidepressants has also been considered. One particularly interesting study by Doucet et al. (2013) found that the application of molecular inhibitors that break the connection between nNOS and the PSD-95 domain, which couples it to the NMDA receptor, significantly reduced depressive-like behaviour in animal models, such as reduced immobility times in the tail suspension test. This is a really good example of how our understanding related to the complexity of the interaction between neurotransmitters and their mechanisms has, and will, enrich our ability to treat and remediate.

Interaction between GABA and glutamate - E/I balance

An area of significant research growth over recent years has centred on the role that alterations in excitatory and inhibitory communication in the brain plays in atypical behavioural states. There is a wide variety of atypical states where these alterations are emerging as contributory factors and, in some cases, causal factors. This includes degenerative disorders, such as Alzheimer's disease, developmental conditions such as autism spectrum disorder (ASD), and psychiatric conditions, such as major depressive disorder (MDD). The basic pathology seems to be a shift towards increased excitation and reduced inhibition.[15] Many believe this shift is a result of reduced GABAergic inhibitory control, which results in hyperexcitability of glutamatergic projections.

There are many mechanisms of interaction between GABAergic and glutamatergic neurons that could be dysfunctional, resulting in altered E/I balance. One could be due to reduced expression and/or functionality of GABAergic interneurons, which act as local brakes on glutamatergic projection neurons. There is a significant amount of animal research that suggests this is the case in ASD. Several studies have found that there are reduced numbers of specific GABAergic interneuron populations in animal models of ASD (Gao & Penzes, 2015). This is typically seen in a group of GABAergic interneurons referred to as parvalbumin positive interneuron (PV+). Research using PV knockout animals also supports their role in the symptomology of ASD, as these PV knockout

---

[15]This is typically referred to as altered E/I balance.

animals display behaviour, such as reduced social interaction and increased repetitive behaviours, which is indicative of the symptomology of a human diagnosis. Alterations in these GABAergic interneurons, and subsequent increased excitability of glutamatergic projections, have also been found in other atypical states, such as Alzheimer's disease (AD), where research has found a reduced number of GABAergic interneurons, such as the previous mentioned PV+ and another group referred to as somatostatin positive interneurons. There are also distinct changes in what are referred to as gamma band oscillations[16] in AD patients. These gamma band oscillations are believed to be a result of the interaction between glutamatergic and GABAergic neurons. Therefore, any alterations reveal functional changes in interaction between glutamatergic neurons and GABAergic interneurons. What is fascinating about this is that these changes are observed a long time before many of the behavioural and cognitive symptoms of AD as well as before many of the biological markers of AD, such as the build-up of amyloid beta. This is interesting because the logical conclusion it suggests is that these changes pre-date many of the biological alterations that are currently believed to be the main causal factors in AD. The fact they predate them suggests that they are more likely to be the primary causal factors.

As well as alterations in GABAergic interneurons and their subsequent interaction with glutamatergic neurons, alterations in E/I balance could also be a result of reduced GABA receptors expressed on glutamatergic neurons. This would result in a reduced ability of GABA interneurons to inhibit glutamatergic neurons, resulting in hyperexcitability of brain circuits. A number of studies have indeed found altered levels of GABA receptors in animal models of both ASD and AD. However, many think that this may be due to changes in the morphology of glutamatergic neurons rather than GABA receptor synthesis *per se*. Several studies have shown that the number spines[17] expressed on glutamatergic projection neurons are found to be altered in atypical behavioural states, such as ASD. This ultimately affects the modulation levels of innervating GABAergic interneurons. Yet other studies have found that rather than reduced spines, the morphology of the spines changes in atypical behavioural states, resulting in an altered surface area and, ultimately, an altered level of GABAergic (inhibitory) modulation from interneurons.

---

[16]These are patterns of neural activity in the 30–80Hz range. This essentially means neuronal activity that shows 'spikes' at 30–80 per second.

[17]These are the physical projections on the neuron that are targeted by other neurons. In theory, the more spines expressed the more innervation/modulation by other neurons.

The interaction between GABA and glutamate is essential to maintain a balanced state within the brain. A number of key mechanisms, such as the number of GABAergic interneurons, GABA receptor expression levels on glutamatergic neurons and glutamatergic spine numbers and morphology may contribute to a range of atypical behavioural states. This provides a range of exciting new target mechanisms to balance excitatory and inhibitory communication, potentially reducing the symptomology of a myriad of atypical behavioural states. However, it may not be simply about the interaction between GABA and glutamate. As mentioned above, many of these GABAergic interneurons also express other neurotransmitters that are typically co-released with GABA, such as the neuropeptide nitric oxide and somatostatin. Changes in these other neurotransmitters' mechanisms of action may also contribute to altered E/I balance in atypical behavioural states. Yet, still several other mechanisms may also contribute that are not directly related to GABA or glutamate, but may alter their release. One good example of this is altered expression of calcium-binding proteins. Studies have found that these are reduced in PV+ interneurons. All we can say, as ever, is that the interactions are complex and myriad. What we can perhaps be certain of is that excitation and inhibition need to be kept controlled within tight parameters within the typical brain.

Conclusion

The purpose of this final section, of the final chapter, was to try to instil the important point that neurotransmitters and their mechanisms do not operate in isolation. Hopefully, the focus on this small number of interactions between neurotransmitters serves well to illustrate this point. This should always be kept at the back of your mind when doing further reading. Ultimately, this should also help you to cast a critical eye on research, especially that which presents conclusions on the function of specific neurotransmitters to modulate the behaviour of the cell, and of the whole organism in isolation. Remember, neuroscience is inherently reductionist in its methods, and single studies should always be considered in the context of the wider research literature. Fundamentally complex behaviour is typically underpinned by complex machinery. This is partly due to the nature of evolution and the fact that the neuronal circuits of our brains were built upon as we evolved as a product of need, rather than design. Each layer adds an extra amount of complexity both to our behaviour and to the interwoven physiological architecture.

Test Yourself Answers

Test Yourself 10.1

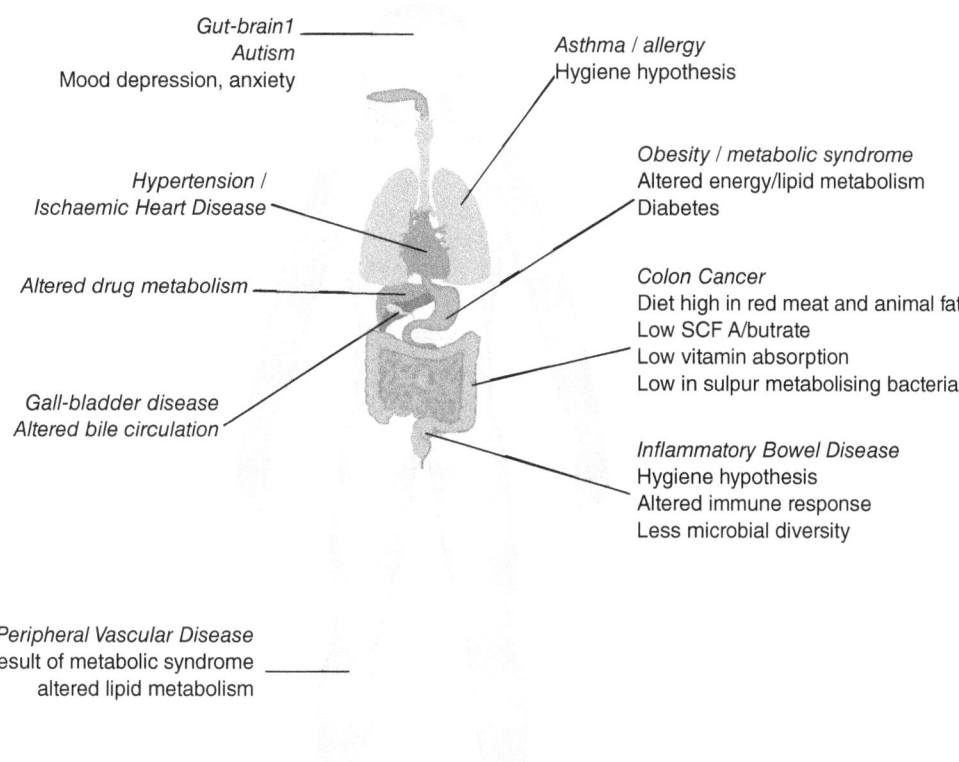

Gut-brain1 _____
Autism
Mood depression, anxiety

Asthma / allergy
Hygiene hypothesis

Hypertension /
Ischaemic Heart Disease

Obesity / metabolic syndrome
Altered energy/lipid metabolism
Diabetes

Altered drug metabolism

Colon Cancer
Diet high in red meat and animal fat
Low SCF A/butrate
Low vitamin absorption
Low in sulpur metabolising bacteria

Gall-bladder disease
Altered bile circulation

Inflammatory Bowel Disease
Hygiene hypothesis
Altered immune response
Less microbial diversity

Peripheral Vascular Disease
Result of metabolic syndrome _____
altered lipid metabolism

Figure 10.6 Labelled schematic of the GI tract

Source: Boore, J. R. P., Cook, N., & Shepherd, A. (2016). *Essentials of anatomy and physiology for nursing practice.* London: Sage.

Test Yourself 10.2

Table 10.4

iPANs:
5HT4 receptors
Acetylcholine
Substance P

(Continued)

Table 10.4   (Continued)

Interneurons:

Acetylcholine

Substance P

Enkephalin

Motor neurons:

Acetylcholine and substance P (oral)

Nitric oxide and vasointestinal peptide (aboral)

Test Yourself 10.3

Table 10.5

Neurotransmitters covered	Synthesising precursor and enzymes	Receptors
GABA	GAD65 and GAD67 Glutamate	$GABA_a$, $GABA_b$ and $GABA_c$
Glutamate	Glutaminase Glutamine	NMDA, AMPA, kainite and a range of metabotropic receptors
Acetylcholine	ChaT Acetyl-CoA and choline	Nicotinic and muscarinic receptors
Dopamine	TH and DDC Tyrosine and L-DOPA	D1-like and D2-like
Serotonin	TH (tryptophan hydroxylase) and AAADC Tryptophan	5HT1 - 7 receptors, many of which have further subtypes
Substance P	Neuropeptide so synthesised because of transcription and translation	NK1 (key for SP) NK2 and NK3 receptors
Endogenous opioids	Neuropeptide so synthesised because of transcription and translation	MORs, DORs and KORs
Nitric oxide	NOS - nNOS, eNOS, iNOS	sGC

Above is not a definitive list of everything we have covered, but it includes many of the points from each chapter. If you have put something as an answer that is not present in the table above, why not look back to the chapter and check if you are correct.

References

The enteric nervous system

Key information

Bayliss, W. M., & Starling, E. H. (1899). The movements and innervation of the small intestine. *The Journal of Physiology, 24*(2), 99–143. doi:10.1113/jphysiol.1899.sp000752

Benarroch, E. E. (2007). Enteric nervous system. *Neurology, 69*(20), 1953. doi:10.1212/01.wnl.0000281999.56102.b5

Grider, J. R. (2003). Neurotransmitters mediating the intestinal peristaltic reflex in the mouse. *Journal of Pharmacology and Experimental Therapeutics, 307*(2), 460. doi:10.1124/jpet.103.053512

Gershon, M. D. (1999). The enteric nervous system: a second brain. *Hosp Pract (1995), 34*(7), 31–32, 35–38, 41–32 passim. doi:10.3810/hp.1999.07.153

Fleming, M. A., Ehsan, L., Moore, S. R., & Levin, D. E. (2020). The enteric nervous system and its emerging role as a therapeutic target. *Gastroenterology Research and Practice, 2020*, 8024171. doi:10.1155/2020/8024171

Auteri, M., Zizzo, M. G., & Serio, R. (2015). GABA and GABA receptors in the gastrointestinal tract: from motility to inflammation. *Pharmacological Research, 93*, 11–21. doi:https://doi.org/10.1016/j.phrs.2014.12.001

Swaminathan, M., Hill-Yardin, E. L., Bornstein, J. C., & Foong, J. P. P. (2019). Endogenous glutamate excites myenteric calbindin neurons by activating group I metabotropic glutamate receptors in the mouse colon. *Frontiers in Neuroscience, 13*, 426. Retrieved from www.frontiersin.org/article/10.3389/fnins.2019.00426

Seifi, M., & Swinny, J. D. (2016). Immunolocalization of AMPA receptor subunits within the enteric nervous system of the mouse colon and the effect of their activation on spontaneous colonic contractions. *Neurogastroenterol Motil, 28*(5), 705–720. doi:10.1111/nmo.12768

Furness, J. B., Callaghan, B. P., Rivera, L. R., & Cho, H. J. (2014). The enteric nervous system and gastrointestinal innervation: integrated local and central control. *Adv Exp Med Biol, 817*, 39–71. doi:10.1007/978-1-4939-0897-4_3

Neuhuber, W., & Wörl, J. (2018). Monoamines in the enteric nervous system. *Histochem Cell Biol, 150*(6), 703–709. doi:10.1007/s00418-018-1723-4

Bulbring, E., & Lin, R. C. (1958). The effect of intraluminal application of 5-hydroxytryptamine and 5-hydroxytryptophan on peristalsis; the local production of 5-HT and its release in relation to intraluminal pressure and propulsive activity. *The Journal of Physiology, 140*(3), 381–407. Retrieved from pubmed.ncbi.nlm.nih.gov/13514713 www.ncbi.nlm.nih.gov/pmc/articles/PMC1358765/

Bertrand, P. P., & Bertrand, R. L. (2010). Serotonin release and uptake in the gastrointestinal tract. *Autonomic Neuroscience, 153*(1), 47–57. doi:https://doi.org/10.1016/j.autneu.2009.08.002

Enteric nervous system dysfunction

Gangula, P., Ravella, K., Chukkapalli, S., Rivera, M., Srinivasan, S., Hale, A., Kesavalu, L. (2015). Polybacterial Periodontal Pathogens Alter Vascular and Gut BH4/nNOS/NRF2-Phase II Enzyme Expression. *PLoS ONE, 10*(6), e0129885. doi:10.1371/journal.pone.0129885

Mawe, G. M., Coates, M. D., & Moses, P. L. (2006). Review article: intestinal serotonin signalling in irritable bowel syndrome. *Alimentary Pharmacology & Therapeutics, 23*(8), 1067–1076. doi:10.1111/j.1365-2036.2006.02858.x

Coates, M. D., Mahoney, C. R., Linden, D. R., Sampson, J. E., Chen, J., Blaszyk, H., . . . Moses, P. L. (2004). Molecular defects in mucosal serotonin content and decreased serotonin reuptake transporter in ulcerative colitis and irritable bowel syndrome. *Gastroenterology, 126*(7), 1657–1664. doi:10.1053/j.gastro.2004.03.013

Grundy, D. (2008). 5-HT system in the gut: roles in the regulation of visceral sensitivity and motor functions. *Eur Rev Med Pharmacol Sci, 12* Suppl 1, 63–67.

Beyak, M. J. (2010). Visceral afferents — Determinants and modulation of excitability. *Autonomic Neuroscience, 153*(1), 69–78. doi:https://doi.org/10.1016/j.autneu.2009.07.019

Crowell, M. D. (2004). Role of serotonin in the pathophysiology of the irritable bowel syndrome. *British Journal of Pharmacology, 141*(8), 1285–1293. doi:10.1038/sj.bjp.0705762

Hammerle, C. W., & Surawicz, C. M. (2008). Updates on treatment of irritable bowel syndrome. *World Journal of Gastroenterology, 14*(17), 2639–2649. doi:10.3748/wjg.14.2639

Li, Z., Chalazonitis, A., Huang, Y. Y., Mann, J. J., Margolis, K. G., Yang, Q. M., . . . Gershon, M. D. (2011). Essential roles of enteric neuronal serotonin in gastrointestinal motility and the development/survival of enteric dopaminergic neurons. *J Neurosci, 31*(24), 8998–9009. doi:10.1523/jneurosci.6684-10.2011

Terry, N., & Margolis, K. G. (2017). Serotonergic mechanisms regulating the GI tract: Experimental evidence and therapeutic relevance. *Handb Exp Pharmacol, 239*, 319–342. doi:10.1007/164_2016_103

Pata, C., Erdal, M. E., Derici, E., Yazar, A., Kanik, A., & Ulu, O. (2002). Serotonin transporter gene polymorphism in irritable bowel syndrome. *Am J Gastroenterol, 97*(7), 1780–1784. doi:10.1111/j.1572-0241.2002.05841.x

Wang, Y. M., Chang, Y., Chang, Y. Y., Cheng, J., Li, J., Wang, T., . . . Wang, B. M. (2012). Serotonin transporter gene promoter region polymorphisms and serotonin transporter expression in the colonic mucosa of irritable bowel syndrome patients. *Neurogastroenterol Motil, 24*(6), 560–565, e254–565. doi:10.1111/j.1365-2982.2012.01902.x

Jin, D. C., Cao, H. L., Xu, M. Q., Wang, S. N., Wang, Y. M., Yan, F., & Wang, B. M. (2016). Regulation of the serotonin transporter in the pathogenesis of irritable bowel syndrome. *World Journal of Gastroenterology, 22*(36), 8137–8148. doi:10.3748/wjg.v22.i36.8137

Yano, J. M., Yu, K., Donaldson, G. P., Shastri, G. G., Ann, P., Ma, L., . . . Hsiao, E. Y. (2015). Indigenous bacteria from the gut microbiota regulate host serotonin biosynthesis. *Cell, 161*(2), 264–276. doi:10.1016/j.cell.2015.02.047

Grubelic Ravić, K., Paić, F., Vucelić, B., Brinar, M., Čuković-ćavka, S., Božina, N., . . . Nikuševa Martić, T. (2018). Association of polymorphic variants in serotonin re-uptake transporter gene with Crohn's disease: a retrospective case-control study. *Croatian Medical Journal, 59*(5), 232–243. doi:10.3325/cmj.2018.59.232

Enteric nervous system dysfunction and big brain dysfunction

Anderson, G., Noorian, A. R., Taylor, G., Anitha, M., Bernhard, D., Srinivasan, S., & Greene, J. G. (2007). Loss of enteric dopaminergic neurons and associated changes in colon motility in an MPTP mouse model of Parkinson's disease. *Experimental Neurology, 207*(1), 4-12. doi:10.1016/j.expneurol.2007.05.010

Liang, S., Wu, X., Hu, X., Wang, T., & Jin, F. (2018). Recognizing depression from the microbiota-gut-brain axis. *International journal of molecular sciences, 19*(6), 1592. doi:10.3390/ijms19061592

Tremlett, H., Bauer, K. C., Appel-Cresswell, S., Finlay, B. B., & Waubant, E. (2017). The gut microbiome in human neurological disease: A review. *Ann Neurol, 81*(3), 369–382. doi:10.1002/ana.24901

Sampson, T. R., Debelius, J. W., Thron, T., Janssen, S., Shastri, G. G., Ilhan, Z. E., . . . Mazmanian, S. K. (2016). Gut microbiota regulate motor deficits and neuroinflammation in a model of Parkinson's Disease. *Cell, 167*(6), 1469–1480. e1412. doi:10.1016/j.cell.2016.11.018 10.1016/j.cell.2016.11.018.

Pulikkan, J., Mazumder, A., & Grace, T. (2019). Role of the gut microbiome in autism spectrum disorders. *Adv Exp Med Biol, 1118*, 253–269. doi:10.1007/978-3-030-05542-4_13

Liddle, R. A. (2018). Parkinson's disease from the gut. *Brain Res, 1693*(Pt B), 201–206. doi:10.1016/j.brainres.2018.01.010

Cheung, S. G., Goldenthal, A. R., Uhlemann, A.-C., Mann, J. J., Miller, J. M., & Sublette, M. E. (2019). Systematic review of gut microbiota and major depression. *Frontiers in Psychiatry, 10*, 34–34. doi:10.3389/fpsyt.2019.00034

Mangiola, F., Ianiro, G., Franceschi, F., Fagiuoli, S., Gasbarrini, G., & Gasbarrini, A. (2016). Gut microbiota in autism and mood disorders. *World Journal of Gastroenterology, 22*(1), 361–368. doi:10.3748/wjg.v22.i1.361

Zheng, P., Zeng, B., Liu, M., Chen, J., Pan, J., Han, Y., . . . Xie, P. (2019). The gut microbiome from patients with schizophrenia modulates the glutamate-glutamine-GABA cycle and schizophrenia-relevant behaviors in mice. *Sci Adv, 5*(2), eaau8317. doi:10.1126/sciadv.aau8317

Interactions between neurotransmitters

Interaction between substance p and other neurotransmitters

Brimblecombe, K. R., & Cragg, S. J. (2015). Substance P weights striatal dopamine transmission differently within the striosome-matrix axis. *J Neurosci, 35*(24), 9017–9023. doi:10.1523/jneurosci.0870-15.2015

Brimblecombe, K. R., & Cragg, S. J. (2017). The striosome and matrix compartments of the striatum: A path through the labyrinth from neurochemistry toward function. *ACS Chem Neurosci, 8*(2), 235–242. doi:10.1021/acschemneuro.6b00333

Schank, J. R. (2020). Neurokinin receptors in drug and alcohol addiction. *Brain Research, 1734*, 146729–146729. doi:10.1016/j.brainres.2020.146729

Schank, J. R. (2014). The neurokinin-1 receptor in addictive processes. *J Pharmacol Exp Ther, 351*(1), 2-8. doi:10.1124/jpet.113.210799

Kirby, L. G., Zeeb, F. D., & Winstanley, C. A. (2011). Contributions of serotonin in addiction vulnerability. *Neuropharmacology, 61*(3), 421–432. doi:10.1016/j.neuropharm.2011.03.022

Blomeley, C., & Bracci, E. (2008). Substance P depolarizes striatal projection neurons and facilitates their glutamatergic inputs. *J Physiol, 586*(8), 2143–2155. doi:10.1113/jphysiol.2007.148965

Nikolaus, S., Huston, J. P., & Hasenöhrl, R. U. (1999). Reinforcing effects of neurokinin substance P in the ventral pallidum: mediation by the tachykinin NK1 receptor. *European Journal of Pharmacology, 370*(2), 93–99. doi:https://doi.org/10.1016/S0014-2999(99)00105-3

Sandweiss, A. J., McIntosh, M. I., Moutal, A., Davidson-Knapp, R., Hu, J., Giri, A. K., . . . Vanderah, T. W. (2018). Genetic and pharmacological antagonism of NK1 receptor prevents opiate abuse potential. *Mol Psychiatry, 23*(8), 1745–1755. doi:10.1038/mp.2017.102

Cholinergic neurons and dopamine

Durand-de Cuttoli, R., Mondoloni, S., Marti, F., Lemoine, D., Nguyen, C., Naudé, J., . . . Mourot, A. (2018). Manipulating midbrain dopamine neurons and reward-related behaviors with light-controllable nicotinic acetylcholine receptors. *eLife, 7,* e37487. doi:10.7554/eLife.37487

Threlfell, S., Clements, M. A., Khodai, T., Pienaar, I. S., Exley, R., Wess, J., & Cragg, S. J. (2010). Striatal muscarinic receptors promote activity dependence of dopamine transmission via distinct receptor subtypes on cholinergic interneurons in ventral versus dorsal striatum. *The Journal of Neuroscience, 30*(9), 3398–3408.

Threlfell, S., Lalic, T., Platt, N. J., Jennings, K. A., Deisseroth, K., & Cragg, S. J. (2012). Striatal dopamine release is triggered by synchronized activity in cholinergic interneurons. *Neuron, 75*(1), 58–64. doi:10.1016/j.neuron.2012.04.038

Picciotto, M. R., Zoli, M., Rimondini, R., Léna, C., Marubio, L. M., Pich, E. M., . . . Changeux, J.-P. (1998). Acetylcholine receptors containing the β2 subunit are involved in the reinforcing properties of nicotine. *Nature, 391*(6663), 173–177. doi:10.1038/34413

Tapper, A. R., McKinney, S. L., Nashmi, R., Schwarz, J., Deshpande, P., Labarca, C., . . . Lester, H. A. (2004). Nicotine activation of alpha4* receptors: sufficient for reward, tolerance, and sensitization. *Science, 306*(5698), 1029–1032. doi:10.1126/science.1099420

Prado, V. F., Janickova, H., Al-Onaizi, M. A., & Prado, M. A. M. (2017). Cholinergic circuits in cognitive flexibility. *Neuroscience, 345,* 130–141. doi:https://doi.org/10.1016/j.neuroscience.2016.09.013

Dorst, M. C., Tokarska, A., Zhou, M., Lee, K., Stagkourakis, S., Broberger, C., . . . Silberberg, G. (2020). Polysynaptic inhibition between striatal cholinergic interneurons shapes their network activity patterns in a dopamine-dependent manner. *Nature Communications, 11*(1), 5113. doi:10.1038/s41467-020-18882-y

Dautan, D., Souza, A. S., Huerta-Ocampo, I., Valencia, M., Assous, M., Witten, I. B., . . . Mena-Segovia, J. (2016). Segregated cholinergic transmission modulates dopamine neurons integrated in distinct functional circuits. *Nature Neuroscience, 19*(8), 1025–1033. doi:10.1038/nn.4335

Interactions between nNOS and serotonergic mechanisms

Garthwaite, J. (2007). Neuronal nitric oxide synthase and the serotonin transporter get harmonious. *Proceedings of the National Academy of Sciences, 104*(19), 7739. doi:10.1073/pnas.0702508104

Chanrion, B., Mannoury la Cour, C., Bertaso, F., Lerner-Natoli, M., Freissmuth, M., Millan, M. J., . . . Marin, P. (2007). Physical interaction between the serotonin transporter and neuronal nitric oxide synthase underlies reciprocal modulation of their activity. *Proceedings of the National Academy of Sciences, 104*(19), 8119. doi:10.1073/pnas.0610964104

De Silva, D. J., French, S. J., Cheung, N. Y., Swinson, A. K., Bendotti, C., & Rattray, M. (2005). Rat brain serotonin neurones that express neuronal nitric oxide synthase have increased sensitivity to the substituted amphetamine serotonin toxins 3,4-methylenedioxymethamphetamine and p-chloroamphetamine. *Neuroscience, 134*(4), 1363–1375. doi:https://doi.org/10.1016/j.neuroscience.2005.05.016

Lu, Y., Simpson, K. L., Weaver, K. J., & Lin, R. C. S. (2010). Coexpression of serotonin and nitric oxide in the raphe complex: Cortical versus subcortical circuit. *The Anatomical Record, 293*(11), 1954–1965. doi:https://doi.org/10.1002/ar.21222

Cains, S., Blomeley, C. P., & Bracci, E. (2012). Serotonin inhibits low-threshold spike interneurons in the striatum. *J. Physiol.-London, 590*(10), 2241–2252. doi:10.1113/jphysiol.2011.219469

Blomeley, C. P., & Bracci, E. (2009). Serotonin excites fast-spiking interneurons in the striatum. *The European Journal of Neuroscience, 29*(8), 1604–1614. doi:10.1111/j.1460-9568.2009.06725.x

Joca, S. R. L., Sartim, A. G., Roncalho, A. L., Diniz, C. F. A., & Wegener, G. (2019). Nitric oxide signalling and antidepressant action revisited. *Cell and Tissue Research, 377*(1), 45–58. doi:10.1007/s00441-018-02987-4

Interactions between nNOS and glutamate

Doucet, M. V., Levine, H., Dev, K. K., & Harkin, A. (2013). Small-molecule inhibitors at the PSD-95/nNOS interface have antidepressant-like properties in mice. *Neuropsychopharmacology, 38*(8), 1575–1584. doi:10.1038/npp.2013.57

Picón-Pagès, P., Garcia-Buendia, J., & Muñoz, F. J. (2019). Functions and dysfunctions of nitric oxide in brain. *Biochimica et Biophysica Acta (BBA) - Molecular Basis of Disease, 1865*(8), 1949–1967. doi:https://doi.org/10.1016/j.bbadis.2018.11.007

Zhou, Q.-G., Zhu, X.-H., Nemes, A. D., & Zhu, D.-Y. (2018). Neuronal nitric oxide synthase and affective disorders. *IBRO reports, 5*, 116–132. doi:10.1016/j.ibror.2018.11.004

Girouard, H., Wang, G., Gallo, E. F., Anrather, J., Zhou, P., Pickel, V. M., & Iadecola, C. (2009). NMDA Receptor activation increases free radical production through nitric oxide and NOX2. *The Journal of Neuroscience, 29*(8), 2545. doi:10.1523/JNEUROSCI.0133-09.2009

Wang, W., Zhou, T., Jia, R., Zhang, H., Zhang, Y., Wang, C., . . . Xue, W. (2019). NMDA receptors and L-arginine/nitric oxide/cyclic guanosine monophosphate pathway contribute to the antidepressant-like effect of Yueju pill in mice. *Bioscience Reports, 39*(9). doi:10.1042/BSR20190524

Haj-Mirzaian, A., Amiri, S., Amini-Khoei, H., Haj-Mirzaian, A., Hashemiaghdam, A., Ramezanzadeh, K., . . . Dehpour, A. R. (2018). Involvement of NO/NMDA-R pathway in the behavioral despair induced by amphetamine withdrawal. *Brain Res Bull, 139*, 81–90. doi:10.1016/j.brainresbull.2018.02.001

Interactions between GABA and glutamate: altered e/I balance

Rubenstein, J. L., & Merzenich, M. M. (2003). Model of autism: increased ratio of excitation/inhibition in key neural systems. *Genes Brain Behav, 2*(5), 255–267.

Waller, R., Mandeya, M., Viney, E., Simpson, J. E., & Wharton, S. B. (2020). Histological characterization of interneurons in Alzheimer's disease reveals a loss of somatostatin interneurons in the temporal cortex. *Neuropathology, 40*(4), 336–346. doi:10.1111/neup.12649

Wöhr, M., Orduz, D., Gregory, P., Moreno, H., Khan, U., Vörckel, K. J., . . . Schwaller, B. (2015). Lack of parvalbumin in mice leads to behavioral deficits relevant to all human autism core symptoms and related neural morphofunctional abnormalities. *Translational Psychiatry, 5*, e525. doi:10.1038/tp.2015.19 www.nature.com/articles/tp201519#supplementary-information

Filice, F., Vörckel, K. J., Sungur, A., Wöhr, M., & Schwaller, B. (2016). Reduction in parvalbumin expression not loss of the parvalbumin-expressing GABA interneuron subpopulation in genetic parvalbumin and shank mouse models of autism. *Mol Brain, 9*, 10. doi:10.1186/s13041-016-0192-8

Rapanelli, M., Frick, L. R., Xu, M., Groman, S. M., Jindachomthong, K., Tamamaki, N., . . . Pittenger, C. (2017). Targeted interneuron depletion in the dorsal striatum produces autism-like behavioral abnormalities in male but not female mice. *Biol Psychiatry, 82*(3), 194–203. doi:10.1016/j.biopsych.2017.01.020

Zhang, W., Zhang, L., Liang, B., Schroeder, D., Zhang, Z.-W., Cox, G. A., . . . Lin, D.-T. (2016). Hyperactive somatostatin interneurons contribute to excitotoxicity in neurodegenerative disorders. *Nature Neuroscience, 19*(4), 557–559. doi:10.1038/nn.4257

Goutagny, R., Gu, N., Cavanagh, C., Jackson, J., Chabot, J. G., Quirion, R., . . . Williams, S. (2013). Alterations in hippocampal network oscillations and theta-gamma coupling arise before Aβ overproduction in a mouse model of Alzheimer's disease. *Eur J Neurosci, 37*(12), 1896–1902. doi:10.1111/ejn.12233

Bi, D., Wen, L., Wu, Z., & Shen, Y. (2020). GABAergic dysfunction in excitatory and inhibitory (E/I) imbalance drives the pathogenesis of Alzheimer's disease. *Alzheimer's & Dementia, 16*(9), 1312–1329. doi:https://doi.org/10.1002/alz.12088

Ferguson, B. R., & Gao, W.-J. (2018). PV interneurons: Critical regulators of E/I balance for prefrontal cortex-dependent behavior and psychiatric disorders. *Front Neural Circuits, 12*, 37–37. doi:10.3389/fncir.2018.00037

Lauterborn, J. C., Scaduto, P., Cox, C. D., Schulmann, A., Lynch, G., Gall, C. M., . . . Limon, A. (2021). Increased excitatory to inhibitory synaptic ratio in parietal cortex samples from individuals with Alzheimer's disease. *Nature Communications, 12*(1), 2603. doi:10.1038/s41467-021-22742-8

Yang, E.-J., Ahn, S., Lee, K., Mahmood, U., & Kim, H.-S. (2016). Early behavioral abnormalities and perinatal alterations of PTEN/AKT pathway in valproic acid autism model mice. *PLOS ONE, 11*, e0153298. doi:10.1371/journal.pone.0153298

Hutsler, J. J., & Zhang, H. (2010). Increased dendritic spine densities on cortical projection neurons in autism spectrum disorders. *Brain Res, 1309*, 83–94. doi:10.1016/j.brainres.2009.09.120

INDEX

Page numbers in *italics* refer to figures.